SYSTEM SIMULATION
PROGRAMMING STYLES AND LANGUAGES

INTERNATIONAL COMPUTER SCIENCE SERIES

Consulting editors **A D McGettrick** University of Strathclyde
J van Leeuwen University of Utrecht

OTHER TITLES IN THE SERIES

Programming in Ada (2nd Edn.) *J G P Barnes*

Computer Science Applied to Business Systems *M J R Shave and K N Bhaskar*

Software Engineering (2nd Edn.) *I Sommerville*

A Structured Approach to FORTRAN 77 Programming *T M R Ellis*

The Cambridge Distributed Computing System *R M Needham and A J Herbert*

An Introduction to Numerical Methods with Pascal *L V Atkinson and P J Harley*

The UNIX System *S R Bourne*

Handbook of Algorithms and Data Structures *G H Gonnet*

Office Automation: Concepts, Technologies and Issues *R A Hirschheim*

Microcomputers in Engineering and Science *J F Craine and G R Martin*

UNIX for Super-Users *E Foxley*

Software Specification Techniques *N Gehani and A D McGettrick* (eds.)

The UNIX System V Environment *S R Bourne*

Data Communications for Programmers *M Purser*

Prolog Programming for Artificial Intelligence *I Bratko*

Modula-2: Discipline & Design *A Sale*

Introduction to Expert Systems *P Jackson*

Local Area Network Design *A Hopper, S Temple and R C Williamson*

PROLOG *F Giannesini, H Kanoui, R Pasero and M van Caneghem*

Programming Language Translation: A Practical Approach *P D Terry*

SYSTEM SIMULATION
PROGRAMMING STYLES AND LANGUAGES

Wolfgang Kreutzer

University of Canterbury
New Zealand

ADDISON-WESLEY
PUBLISHING
COMPANY

Sydney · Wokingham, England · Reading, Massachusetts
Menlo Park, California · Don Mills, Ontario
Amsterdam · Singapore · Tokyo · Madrid
Bogota · Santiago · San Juan

Cover graphic by kind permission of Apollo Computer Inc.
Typeset by Setrite of Hong Kong and Quorum Technical Services Ltd.
Printed in Great Britain by The Bath Press.

British Library Cataloguing in Publication Data
Kreutzer, Wolfgang
System simulation: programming styles and languages.
1. Computer simulation – Programming
I. Title
001.4′24 QA76.9.C65

ISBN 0-201-12914-0

Library of Congress Cataloging in Publication Data
Kreutzer, Wolfgang.
System simulation.

Bibliography: p.
Includes index.
1. Digital computer simulation. 2. Electronic digital computers – Programming. 3. Programming languages (Electronic computers) I. Title.
QA76.9.C65K74 1986 003 85-28579
ISBN 0-201-12914-0

ABCDEF 89876

Preface

This is a book about simulation programming. It shows the application of different modelling styles to a suite of simple examples, using a variety of specification and programming tools. Monte Carlo methods, continuous system simulation, discrete-event simulation and combined simulation are discussed and exemplified.

There are many textbooks on simulation. Most of these, however, concentrate on its statistical aspects. Their coverage of programming is often unsatisfactory. Since system simulation is a collection of a wide variety of techniques, employed in many different contexts, this is not altogether surprising. Simulation encompasses too many distinct paradigms and methods to treat all of them at sufficient depth. Therefore often a single paradigm and a single programming tool are emphasized. This approach has the advantage of being able to describe most aspects of simulation methodology (system identification, model design, model implementation, model validation and model experimentation) within a reasonable number of pages. Discussion of alternative tools and techniques is often restricted to brief overviews. As a result, the importance of programming methodology is de-emphasized compared to other aspects of the simulation technique. Choosing a particular programming system is rarely justified beyond reasons of familiarity and availability. This is unfortunate because appropriate programming tools are vital for a successful simulation project.

Methodology is certainly important, but every effort should also be made to ensure that the tools used to apply this methodology support rather than get in the way of its conceptual framework. Since simulation should normally be a method of last resort, models tend to be large and complicated pieces of software. Managing complexity in a disciplined manner is therefore at the heart of simulation programming.

Many widely used programming systems, while appropriate at their time, have outlived their usefulness today. Simulation programming should incorporate both theoretical results and practical tools emerging from current computer science research. Investigations about strongly typed languages and their effects on program reliability, or the productivity enhancements achievable through friendly environments and exploratory programming styles, are of particular relevance.

The decision as to which world view, modelling style and language to use for a particular modelling task is never an easy one. Availability, costs, familiarity, personal fluency and time constraints will undoubtedly be among the major influences, in addition to less tangible personal preferences and dislikes. This is as it should be, as long as there is a general awareness of the whole spectrum of relevant tools and techniques, their characteristics, strengths and weaknesses. This book lays a foundation.

It is targeted at students, practitioners, teachers, researchers and hobbyists looking for a comprehensive introduction to the programming aspects of simulation. It assumes that the reader has some familiarity with programming in an ALGOL-like language. PASCAL and SIMULA are the main programming tools, although a wide spectrum of other languages, such as SIMPAS, SIMSCRIPT, DYNAMO, GPSS, DEMOS, DISCO, ADA, SMALLTALK and PROLOG, are briefly discussed and demonstrated. An object-oriented programming style is advocated throughout – as far as possible within the constraints of PASCAL. Graphics and light-hearted examples are used wherever possible. Toolboxes of PASCAL-based utility programs are provided, as are solutions to some of the exercises for each of the four modelling styles. A final chapter summarizes the most popular simulation programming tools and discusses likely future developments with particular reference to knowledge-based systems and modelling environments on personal work stations.

Chapter 1 justifies the use of modelling and simulation as a powerful problem-solving technique. Two different styles of simulation are identified, with a set of basic tools and techniques applicable to both. The first style is its classical application as a vehicle for numerical predictions; the second has traditionally been used by artificial intelligence as a powerful means for exploring and understanding symbolic models of complex systems.

The rest of this chapter briefly discusses the relationship between cognitive tools and complexity, as well as justifying the use of PASCAL, SIMULA and object-oriented programming.

Chapter 2 introduces four paradigmatic modelling styles. Small PASCAL programs show the associated programming techniques.

Data collection devices and distribution sampling are discussed for Monte Carlo methods. Continuous system simulations are exemplified by system dynamics and differential equation models, using levels, state variables, equations and clock objects. Simple time slicing is used to animate them.

The treatment of discrete-event simulation centres on queueing networks. It deals with alternative representations for entities, relationships, clocks, agendas, events, activities and processes. *Next event* sequencing and open and closed queueing networks are demonstrated.

A brief section on combined models focuses on the distinction between time and state events.

Chapter 3 surveys a wide range of software systems. Their historical roots and different lines of development are identified. This chapter focuses on specialized tools with a reasonable degree of flexibility. The use of general-purpose languages, such as FORTRAN, for simulation programming, although still common practice, is argued against. SIMPAS and SIMSCRIPT serve as examples for relatively low-level event-oriented preprocessors and compilers. Section 3.4 shows how application-oriented scenarios, offering automatic report generation, good debugging aids and model descriptions in terms of 'natural' concepts and operations, can be easily built. As SIMULA is advocated as a base for such language extensions, a tutorial discussion of SIMULA's key concepts and the building of application-specific contexts in an object-oriented style is included. The *stable-marriage* problem is chosen as an example.

Monte Carlo scenarios (state-transition graphs and class MCcontext), system dynamics scenarios (system dynamics diagrams, DYNAMO and class CScontext), queueing network scenarios (activity cycle diagrams, SIMULA, GPSS and DEMOS) and combined scenarios (class DISCO) are briefly demonstrated.

Queueing networks are studied from both machine- and material-oriented viewpoints. GPSS' transaction flow model is shown as lacking in both flexibility and execution efficiency. DEMOS shows how a number of more highly structured schemes can be used for the description of complex process interaction patterns (producer–consumer, capacity-constrained resources, condition queues, master–slave relationships).

The final two sections deal with the advantages and disadvantages of program generators (CAPS) and likely new base languages for future simulation tools (ADA, SMALLTALK, PROLOG).

Chapter 4 summarizes the programming tools and techniques discussed in the book and surveys likely developments in simulation programming over the next decade.

Chapter 5 contains a number of simulation problems for each of the four programming styles. These may be attempted using either the PASCAL toolboxes in Appendix 1 or any appropriate simulation programming system. Appendix 4 offers some solutions, together with sample data sets and results.

There are a number of appendices. These contain program shells and utilities (object declarations and method definitions) for statistical instrumentation, Monte Carlo methods, system dynamics, discrete event, queueing network and combined simulations. There is a list of SIMULA implementations and addresses of organizations by which these were developed, an annotated list of textbooks on simulation methodology, and solutions to some of the simulation problems in Chapter 5.

This book may be used in a number of contexts. It provides a reference for the simulation specialist, summarizing essential information on a broad spectrum of tools. It also offers a wide-ranging discussion of simulation programming techniques and their application in an object-oriented style.

The use of PASCAL in Chapter 2 makes it accessible to a large readership. It can therefore serve as a primary text in a course on simulation programming, following or running concurrently with a more general introduction to simulation methodology. It can also be used as a supplementary textbook in a classical simulation course – for example, in computer and information science, management science, the engineering or social sciences. In this context, it may complement such texts as Mitrani (1982), Bratley *et al.* (1983), Fishman (1978) or Payne (1982). The discussion of different programming systems also makes it a good reference for courses on principles and comparison of programming languages.

Finally, the book should be particularly well suited for self-study by hobbyists or professionals, using the PASCAL toolboxes and exercises.

Acknowledgements

The author would like to thank the University of Canterbury for offering a pleasant and stimulating working environment, and the German Research Community (DFG) for assistance during part of the research leading to this book. Paul Ashton and John Marychurch contributed solutions to exercises.

New Zealand
September 1985

W. Kreutzer

Contents

Chapter 1
Introduction

1.1 System simulation: The Art and Science

The history of mankind is a history of model building. Rivett (1972)

The craft of modelling is central to the way in which we understand the world around us. It has a long history in science and everyday life. A human's model of the world depends on knowledge and reasoning strategies encoded both in the brain's hardware and in symbol structures acquired during a lifetime of experience; it guides the way we perceive, classify and react to our environment. The act of model building explores alternative ways to represent this knowledge, while analysis and simulation are used in its application. Disciplines such as system theory, mathematics, philosophy, psychology and linguistics focus on different aspects of modelling. Rivett (1972), Klir (1972), Davis and Hersch (1981), Sloman (1978), Savage (1978), Piatelli–Palmarini (1980), Olsson (1980) and Pylyshyn (1984) offer a selection of interesting perspectives for readers who would like to explore the roots of the underlying concepts.

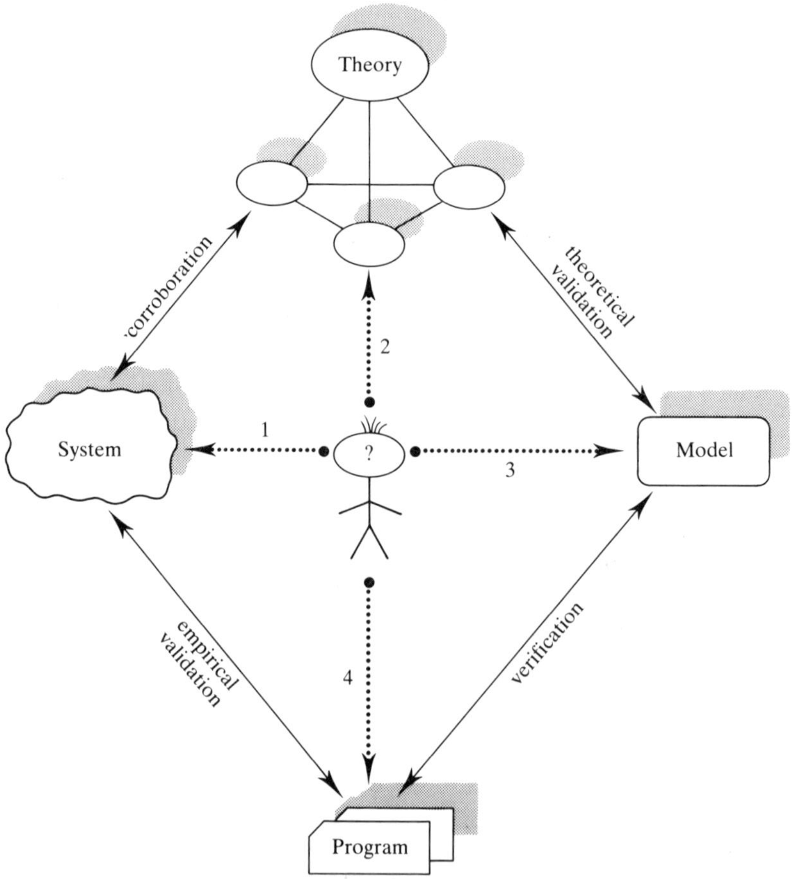

Figure 1.1 Systems, theories, models and programs

Figure 1.1 gives a much simplified account of modelling as a mental activity. Four distinct phases are identified. The first of these is often referred to as **system identification**. It establishes relationships between an observer and some part of reality. A **system** is defined as a collection of objects, their relationships and behaviour relevant to a set of purposes, characterizing some relevant part of reality. This definition stresses the fact that modelling ought to be a purpose-driven activity. There is never a best or even 'correct' model. Its usefulness always depends on objectives and context of its application.

The second phase is that of **system representation**, where symbolic images of objects, relationships and behaviour patterns are bound into structures as part of some larger mental framework of beliefs, background assumptions and theories of the problem solver.

The completely disinterested observer is only a convenient fiction. As humans we cannot avoid being theory driven. Our theories can be descriptive, used for explanation and prediction, or they can be normative, used for prescribing what ought to be done. Different kinds of theories yield different kinds of models.

The third phase deals with **model design**. A **model** is an 'appropriate' representation of structures and processes of a miniworld, instantiating some aspects of a theory. Humans continually explore models to reflect on situations, be it consciously (Gedanken experiments) or subconsciously. Models can quickly grow to form very complicated structures. Control and constraint of this complexity lie at the heart of any modelling method.

As a general rule models should be kept as simple as possible. There is an old philosophical proverb called 'Occam's razor':

> It is vain to multiply entities without need.

Koopman (1979) has reformulated this as a basic principle of cognitive economy in model building:

> Complications in models are not to be multiplied beyond the necessity of practical application and insight.

Model complexity can range from a one-to-one image of reality to a few symbols on the back of an envelope. One should always strive for clarity through simplicity; although simplicity is by itself an elusive concept. The complexity of a model can best be described as a relation between what is to be represented, the medium of representation and an interpreting object. There seems to be a natural tendency to strive for a minimum set of concepts, on one hand, and a high degree of visualization on the other. The first helps with reasoning about the essence of complex systems, while the second helps with the perception of a system as a whole.

Any useful model has to simplify and idealize. Abstraction and aggregation are important techniques in keeping models of complex phenomena intellectually manageable. Great care must be taken to preserve only those characteristics of a system that are essential. This again depends on a model's purposes, since these characteristics and their meaning are always relative to some spatio-temporal framework of concepts.

The fourth phase of the modelling process assumes that a computer is used for model interpretation. It may be referred to as the **model coding** stage. This stage seeks a formal representation of symbol structures and their transformations into data structures and computational procedures in some programming language. Models must be analyzed to be useful; that is, to further the understanding of some facet of a system or to predict the effects of alternative courses of action. Such analyses may be deductive, based on a theory about certain classes of structures, or they may be inductive, based on observations of a model in operation. Both strategies assume that the model's validity can be established in some way. On a rational basis validity can be proved through correspondence with the theory a model is designed to instantiate; that is, by showing how its behaviour can be logically derived from purely formal properties of its representation. A model's validity may also be empirically corroborated, via a set of correspondences between its own and the observed behaviour of the structure or process it represents. The ultimate test, however, is always a pragmatic one: Does it serve its purpose? This may involve an improved understanding of some system, better decision making or prediction of future events.

In all of these cases there are quite different criteria, according to which success will ultimately be judged.

Verification seeks correspondences between a model and its programmed representation. This can again be established purely empirically or on some rational basis. Empirical corroboration is achieved through extensive testing, while a proof of correctness according to formal specifications is the rational alternative. The current state of the art in program proving, however, makes this a rather futile approach for any but the simplest of models. Something less than certainty has to be accepted, just as the ultimate truth of theories and specifications must always remain in some doubt. These too can only be increasingly corroborated but never conclusively established. Corroboration of models may be attempted through unsuccessful falsification or in some other way, depending on one's personal philosophical outlook. Brown (1977) gives an interesting and instructive account of falsificationism and competing philosophies of science, whose basic principles should be familiar to any practising scientist.

Models can be categorized in a number of ways, the broadest scheme classifying them as iconic, analogue or symbolic. A globe is a good example of an **iconic model** of the earth's surface. **Analogue models** represent systems through structures or processes whose external characteristics are different from what is represented, but can be associated through some abstract notion of similarity. In this sense, for instance, a hydraulic model can be viewed as an analogue representation of an economy. Flows of goods, money, services and information are mirrored in a different medium.

This book concentrates on **symbolic models**. Symbolic representations are sometimes called **Fregean representations**, after Gottlob Frege, a

German mathematician who invented the *Begriffsschrift*, a formal symbol system for abstract representation of concepts. Symbols are used to denote objects, relationships, actions and processes. A mental association is established between these symbols and the concepts they represent, although there is usually no direct visual resemblance. Newtonian physics, for example, offers a mathematical model of the solar system. A set of simple symbols and equations describe the dynamic behaviour of the sun and its planets at a purely abstract level.

Symbolic models can be either descriptive or analytical. **Analytical models** offer general solutions for whole families of systems and purposes. They are basically deductive, answering questions on how to achieve some desired outcome. To accomplish this, analytical models must be based on some strong theory. If adequate theories of this kind exist for the class of systems under study and are able to answer the kind of questions envisaged, then analytical models can provide solutions very efficiently. Their main disadvantage is that their design requires a great deal of insight into the studied system and often a considerable degree of mathematical maturity. This difficulty can be somewhat avoided by using ready-made models, although a fair amount of understanding about their validity and limitations is still advisable.

Unfortunately most analytical models involve making extensive simplifying assumptions in order to be mathematically tractable. Most of these preserve the fiction of independence of events, leading to some degree of linear behaviour. Non-linear processes are generally very difficult to analyze. The assumptions involved cannot often be justified for adequate representation of a complex system to solve the problem in hand. Analysts also typically tend to employ those approaches they feel most competent to deal with, often yielding to a strong temptation to formulate problems in terms of these, although they involve invalid simplification. Such misapplication of analytical models should of course be detected during model validation. But validation, although crucial for any degree of confidence in a model, is difficult. It is a relative notion at the best of times. Often it is not done properly, or not at all. A model is never valid as such unless it is a perfect image of a system. Validity can be established only for a stated use to which a model should be put, but it is often tempting to extend the use of existing models outside their range of applicability.

Descriptive models, on the other hand, are mainly speculative. They offer only symbolic representations of some problem space, without giving any guidance on how to search it. Use of descriptive models is therefore an inductive, experimental technique for exploring 'possible worlds' through computational processes. The term **simulation** is used for this kind of model animation, summarizing a wide range of methods. Any exploration of states and patterns of behaviour under a chosen experimental frame is a simulation experiment in this sense.

An **experimental frame** defines a set of circumstances – observed variables and their levels, initial state, input schedules, termination conditions, specifications for data collection and compression – under which a system or model will be observed. This is of course related to purposes and resources, computational or otherwise. Using descriptive models involves searching for solutions through a finite number of experiments to which a model can be subjected. Running a simulation under a given experimental frame corresponds to one such experiment. If the model contains stochastic components, its analysis poses statistical problems similar to the ones encountered in other empirical sciences.

Simulation experiments can be computationally expensive because they may require a great number of model executions. However, the spectacular drop in hardware costs over the last few years has made this much less of a problem than it used to be. The great attraction of simulation lies in its generality. Mathematically strong theories about a system are not needed to describe it in a sufficiently formal way. Simulation models can be relatively easy and fun to build, given a good methodology and notation. They do not require any extensive mathematical background and can in principle cope with any degree of complexity and interdependence between model components. It is important to be aware, however, that this flexibility can be a two-edged sword. It is all too easy to succumb to the temptation to include excessive detail; particularly if the system under study is not well understood. This temptation should of course be resisted, a recommendation which is easier made than adhered to.

The wide range of possible applications and modelling techniques makes it very difficult to give general guidelines on which class of models to use. This question has to be settled in a particular situation and for a specific purpose. The only general rule follows from Occam's razor: a good model is the simplest one which can still be justified. Simple back-of-an-envelope and analytical models are fine, as long as questions are simple and the models can be properly validated. There are, however, a great number of systems and circumstances under which simulation is the method of last resort. Often no other techniques exist to build and explore models of complex, highly interconnected, irregular systems. Simulation may serve as a vehicle to gain sufficient insight to construct simpler analytical models, focusing on those aspects of a system that prove to be relevant. Experimentation with a real system may not be possible; it may be non-existent, unobtainable, too dangerous or too costly. Exploring design alternatives for bridges or nuclear power plants are obvious examples. In addition, simulation modelling may well be the only possible method to analyze processes with very short or very long time frames, like studying the behaviour of subatomic particles or the evolution of galaxies.

Simulation modelling can serve two considerably different purposes. Its more traditional application is in decision making. Here the emphasis is on generating numeric information, which is used, for instance, for comparisons, tuning and prediction of a system's performance. Exploring the effects of routing policies and priority rules for a set of tasks processed by a network of server stations is a classical example for this style of simulation studies. Such models are in principle still analytical in outlook. Simulation is used to solve a model of some well-defined problem, but only because no suitable analytical techniques exist or because none are applicable for other reasons. The usefulness of the model lies in the accuracy of its predictions. This is the kind of simulation that is part of the methodological repertoire of operations research.

A different kind of simulation strives not so much for numeric solutions, but for improved understanding of complex systems. It supports a more exploratory, speculative style of modelling. A quote from Hamming (1962) summarizes the essence of this modelling style:

> The purpose of modelling is insight not numbers.

In this context relatively little is known about a system; a well-defined problem may not even exist. Simulation is used at a prototheoretical stage, as a vehicle for thought experiments. The purpose of a model lies in the act of its construction and exploration, and in the resultant, improved intuition about the system's behaviour, essential aspects and sensitivities. The model may never be used to generate any numerical results at all. Sloman (1978) calls this style of conceptual analysis and model exploration **experimental epistemology**. This kind of simulation may well be the only way to check the consistency and understand all consequences of complex theories. Artificial intelligence miniworlds [see Winston (1984), Rich (1983)] are simulations in this sense. Many people believe that the potential of simulation for this style of computational exploration of symbol structures will in future become much more important than its traditional application as an aid to decision making.

Computer-aided simulation is a programming technique for model animation. Figure 1.2 shows the life cycle of a simulation project. A number of phases are identified, as well as a number of cycles through which they are related.

One important aspect of computer-aided simulation revolves around the activities of model and program design, implementation, verification and validation. At the centre of this is the art of simulation programming, which is the topic of this book. Some prototypical modelling styles and tasks can be identified, leading to a number of programming styles and utilities common to all or large families of simulation models.

A different facet of the simulation method is rooted in the fact that it is an experimental technique, comprising aspects like model calibration and data collection as well as experimental designs and output analysis. Most models use stochastic processes to represent subsystems whose detailed operation is thought to be irrelevant to their purpose. Simulations of this kind are sampling experiments in a statistical sense. The theory of stochastic processes and specialized statistical techniques must be used to define and analyze them. This is often considered the scientific aspect of simulation methodology and is correspondingly emphasized by most textbooks. Statistical methods can answer questions on how to start and stop simulation runs,

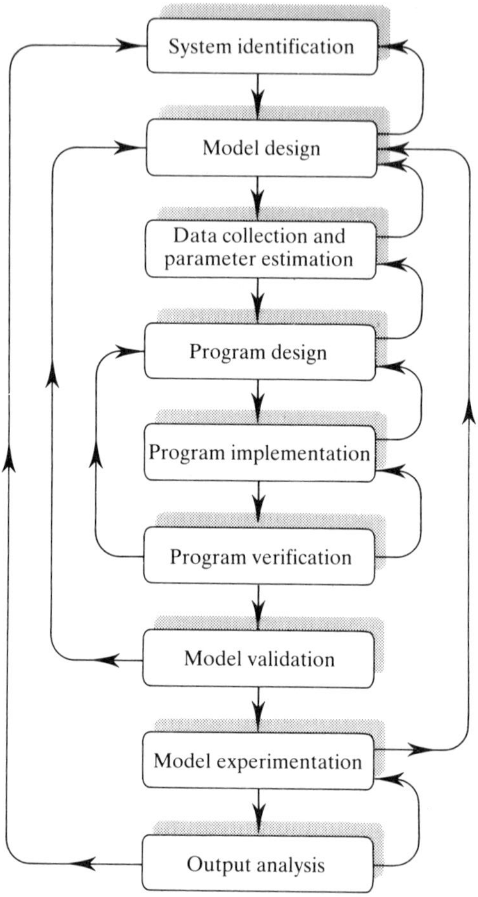

Figure 1.2 Nine aspects of simulation methodology

and how to analyze their results. Kleinjen (1974, 1975, 1979, 1982) gives comprehensive summaries of these techniques.

This aspect of simulation is of course very important, particularly if it is used as a decision-making tool. While there is some merit in saying that 'Simulations carried out without adequate statistical support are not merely useless but dangerous' [Bratley *et al.* (1983); p. 6] this fails to tell the whole story. A simulation consists of a combination of a programmed representation of some model and suitable experimental designs for its exploration. A sound understanding of characteristics, advantages and disadvantages of different modelling and programming methods and tools is just as important as statistical method, although this is often not sufficiently acknowledged.

Simulation programs are complex pieces of software, which must be built in a well-disciplined, well-structured and reliable fashion. Presenting a small collection of sample models coded in FORTRAN or GPSS treats this aspect too lightly. This book tries to balance the score: it does not attempt to cover the statistical background for proper use of simulation as a tool for sampling experiments. A number of other texts [Mitrani (1982), Fishman (1978), Law and Kelton (1982)] cover this area well. It rather aims to complement these textbooks, discussing a number of modelling styles and programming techniques that were developed to cope with the complexities of system simulation. Managing complexity is at the heart of simulation programming and has guided the development of its tools since the early 1960s. Graphical symbol systems and pseudo-codes also have a role to play. This repertoire of specialized tools makes it possible to bridge, in a reliable fashion, large conceptual distances between a modeller, his or her mental model of a system and a computer executing its computational image.

1.2 Notation as a tool of thought

> The limits of my language are the limits of my world. Wittgenstein (1961), par. 5.6.

There are intimate relationships between thought and language and language and methodology. This idea is well expressed in Wittgenstein's quotation. As demonstrated convincingly by Amarel (1968), Hayes and Simon (1976), Korf (1980), Huber (1983) and others, the encoding of problems into different representations greatly influences the effort needed to solve them.

The importance of appropriate conceptual and notational tools that guide and structure a modelling activity during all its phases can therefore hardly be overstated. Well-designed system description and programming languages are not simply notations used to communicate with computer systems. They are also cognitive devices that structure perception, representation and reasoning about classes of problems.

Green (1980), Coombs and Alty (1981), Thomas and Carroll (1981), Pylyshyn (1984) and Hofstadter (1985) elaborate on some of the foundations for this viewpoint. Simplicity, expressiveness, flexibility and efficiency of modelling methods are at least partly conflicting goals between which suitable tradeoffs have to be made. In the long run nothing is gained from restricting a tool to too low a level. Improved machine efficiencies can be achieved by binding languages closer to machine structures. This can, however, result in disproportionate increases in development, debugging and maintenance effort.

The final success of any tool depends critically on user acceptance and satisfaction. If a programming notation fails to provide effective and natural modelling styles, separate methods, often ill-structured and inconsistent, have to be manually mapped into low-level abstractions. This may require highly error-prone transformations across large conceptual distances. To avoid unacceptable misdesigns, some analysis is needed of human information processing. Cognitive tools must not be designed without previous thought about effective environments for mental activities – something that computer scientists have done for too long already.

Interpretation, aggregation and abstraction occur at all levels, from low-level sensory experience to high-level mental reflection. Even at its most basic level human intuition is always bound into some spatio-temporal framework. States and their changes are perceived. This frame forms the foundation for all mental activities: recursively perceiving, classifying, grouping and relating complex phenomena expanded in space (entity constellations) or time (event sequences).

Information-processing abilities are constrained by the fact that only a very limited amount of structural and especially functional complexity can be dealt with directly. This may be because short-term memory is scarce, although it does not apparently depend on the complexity of the concept a stored symbol denotes. This memory limitation handicaps any attempts to simultaneously cope with large and highly interconnected sets of phenomena. To comprehend complicated structures and processes we must use chunking and hierarchies; strategies that enable us to limit our span of attention to small, conceptually closed segments of some complex system. This view has now been widely accepted within the computing community [Denning (1974)]. Its background has been elaborated by Simon (1969), Gaines (1976), Anderson (1980) and others.

Effective symbol systems must fit an individual's overall conceptual framework, offering familiar ideas and mnemonics. Multi-level grouping, where clusters of related concepts are aggregated and encoded by single symbols at a higher semantic level, can accomplish this task. The importance of such encapsulation strategies has often not been fully recognized by existing simulation methods and languages. Modelling and simulation in general still lacks satisfactory theoretical and methodological foundations. Some progress has been made and is reported by Zeigler (1976, 1984), Zeigler *et al.* (1979), Doyle (1976), Kindler (1979) and others, but it has not yet been incorporated in language designs.

Much can be learned from representation and structuring of domain-specific knowledge, a field that is explored by the artificial-intelligence community. Brachman and Smith (1980) and Barr *et al.* (1981) provide a broad perspective on current research.

Notational devices can be classified in a variety of ways. They can be formal or informal, one dimensional or graphical. Informal tools like pseudo-codes and graphical symbol systems are especially useful during system identification and model design, whereas formal tools must be used for final representation and coding.

A number of **pseudo-codes** have been proposed to aid in design and documentation of simulation models. DELTA [Andersen and Kyng (1978)], [Overstreet and Nance (1983)] and SMSDL [Frankowski and Franta (1980)] are interesting examples. Simple arrow diagrams [Roberts *et al.* (1983)], data structure diagrams [Bachman (1969)], life-cycle diagrams [Hutchinson (1975a)] and Petri nets [Peterson (1977)] have also found wide application.

Advantages of **graphical representations** are that structural connectivity and symmetry are emphasized. Things that are conceptually related appear to be close. They also provide a rich syntax to visually define such concepts as links, flows, directions,. . . . For most people this seems to be easier to grasp than procedural encodings, which are more abstract. The reason seems to lie in the inherent parallelism of the human visual system and the corresponding importance of visualization as a problem-solving tool.

The main disadvantage of graphical notation is that one must use a hierarchy of levels for more complex structures (to fit them on a single page). Some of the more specialized graphical notations will be demonstrated when looking at scenario languages.

In stressing the priority of methodology over notation it has sometimes been argued that one should program into a language, not in it. This sweeps the problem somewhat under the carpet. There is still a need for a web of concepts to comprehend, express and use any method; and there is surely a need for a method for guidance. Since language is possibly the most important tool of thought, a method can never be truly independent of some formal or informal representation. Such representations must be reflected in some kind of notation, which should carry rather than hinder the expression of relevant ideas.

A good notation should be a catalyst. There is a sore need for more research into relationships and trade-offs between mind, concepts, notation and representation. Unfortunately neither mathematics nor psychology, the two disciplines naturally turned to for advice, provides much guidance. Polya (1957, 1962) and Wickelgren (1974) are notable exceptions. Although this topic is difficult and seems to straddle traditional boundaries between disciplines, it should be at the core of cognitive psychology and computer science. Wittgenstein (1953, 1956), Sloman (1978), Hofstadter (1979), Olsson (1980) and Pylyshyn (1984) offer many stimulating ideas and insights; as does the literature on knowledge representation for artificial intelligence programs.

The psychology of programming is now beginning to establish itself as a respectable subject. Shneiderman (1980), Smith and Green (1980), and Sheil (1981, 1983) are first steps in this direction.

For many purposes classical mathematical concepts do not seem rich enough, whereas natural language is too unstructured and ambiguous [Hill (1983)]. Computer-programming languages provide formal concepts and notations to describe classes of objects and processes at fairly high levels. Translator programs can automate the process of mapping from higher to lower levels, a method that is far safer and less error prone than manual transformations by the model builder. Appropriate symbol systems (notations, programming languages, etc.) can make certain things seem easy and natural. Some suggest, or help with the visualization of solutions, whereas others seem to get in the way of what one is trying to express.

The author does not believe in the merit of some universal language, be it for programming or any other task. Even natural language is not equally well suited to everything. This is well demonstrated by specialized jargons for different areas of human discourse (scientists, engineers, lawyers, programmers (!), etc.). Languages are cognitive tools that should be appropriate for the tasks to which they are put. Multi-purpose tools can work quite satisfactorily for a wide range of tasks, but some tasks will always require the higher power and precision of more specialized ones.

For this reason a range of tools can and should be used to design and implement simulation models, as well as for any other complex programming task. Graphical symbol systems, pseudo-codes and specialized programming languages all have their place in a simulator's toolbox, alongside a rich repertoire of methodological knowledge about their safe application. Ideally such a toolbox should be constructed around a small set of simple, well-behaved building blocks and bound into a frame that is founded on some uniform and natural metaphor.

Comparing languages, including programming languages, for goodness or badness is quite a difficult undertaking. Beauty is always in the eyes of the beholder. Quality is a relationship between a tool, a user and an application, with complicated interactions between them. To paraphrase Winograd (1983): To ask whether LISP is a better language than PASCAL is somewhat like asking whether a jellyfish is a better animal than an elephant. I like both, but each in its proper place. Both have different strengths and weaknesses and one is possibly better suited to certain tasks than the other; although even that will be difficult enough to establish empirically. What can be done is to identify different programming paradigms and styles [Floyd (1979)], define and exemplify their characteristics, and discuss their strengths and weaknesses.

A number of textbooks have been written to compare programming languages, offering interesting insights into trade-offs between basic principles involved in language design, use and translation. Abelson *et al.* (1985), Barron (1977), Tennent (1981) and MacLennan (1983) are some of the author's favourites.

Any recommendation on what style and tool to use is necessarily made on a somewhat subjective basis. This is how it should be, as long as the range of alternatives, and some measure of their potential and limitations, is known. It is what this book is trying to provide for the field of simulation programming.

Some personal bias will always remain. The author believes that object-oriented programming, as originated in SIMULA and more recently popularized by SMALLTALK and other so-called 'actor' languages, has the best potential for describing complex systems in a well-structured and natural way. Its advantages are not limited to simulation modelling. Rather they extend to programming as a cognitive activity in general and may well become more widely appreciated in future. After all, it is quite easy to metaphorically view all programming activity as computer animations of miniworlds. Since it seems that both model building and conceptual closure are central to the way we cope with the complexities of the world around us, it seems only reasonable to demand that computer systems should be made to support such problem-solving styles – embedded in safe, flexible and user-friendly programming environments. Well-designed interfaces for man–machine communication should offer complete contexts for interactive model design and execution. Graphical displays of model structure, state, behaviour and results should be possible in such a framework, which should also unify a wide range of visually pleasing symbol systems in a window-based, interaction-driven modelling environment. SMALLTALK [Goldberg and Robson (1983), Goldberg (1984), Krasner (1983)] and JADE [Unger *et al.* (1984)] are representative for such attempts to add new dimensions to the modelling process.

1.3 Building simulation toolboxes: An object-oriented approach

Chapters 2 and 3 discuss and exemplify a number of different programming styles and tools for system simulation. Chapter 2 demonstrates the characteristics of Monte Carlo methods, quasi-continuous system simulation, discrete-event simulation of queueing networks, and combined discrete–continuous simulation. This is accomplished by studying miniworlds: simple scenarios used to show each technique's basic approach. All of these are taken from the imaginary realm of DreamWorld. Although the exact location of DreamWorld is unknown, it has given birth to many myths and legends about places where witchcraft, unicorns and magic are facts of everyday life. Manguel and Gualdalupi (1980) give an entertaining account of a great number of these.

To discuss programming techniques we need a programming tool. PASCAL was chosen because it is the language taught by most computer-science departments around the world. This is a compromise, justified mainly by PASCAL's popularity. It is by no means an ideal language for simulation programming. In particular it lacks concepts for data encapsulation and concurrency, such as those offered by SIMULA and newer languages like MODULA, ADA or SMALLTALK.

This limitation is unfortunate because the object-oriented programming style these features can support is particularly appropriate for the design and implementation of simulation programs. Chapter 3 tries to demonstrate this. Other relevant disadvantages of PASCAL are its lack of standard facilities for separate compilation, which makes the development of portable libraries somewhat difficult, and the fact that records cannot hold references to executable objects (functions and procedures).

On the other hand, good PASCAL compilers are widely (and cheaply!) available on both large mainframes and personal computers. PASCAL also offers flexible, safe control and data structures, encouraging good coding styles. User-definable types and flexible rules for naming objects are also among its assets. In spite of its shortcomings it is still an attractive tool for making an introductory treatment of simulation programming accessible to a large number of students.

Even though PASCAL is used to discuss the basic techniques covered in Chapter 2, the use of other, special-purpose languages is explored in Chapter 3. SIMULA is the primary tool here. In spite of its age it is still the most expressive and elegant language for system simulation. SIMULA offers most of PASCAL's features, with a few exceptions (such as no user-defined types). Above all it can be used in an object-oriented programming style, which permits easy and conceptually clean development of special-purpose application packages. This is demonstrated by systems like DEMOS and DISCO.

Appendix 1 contains the code for six PASCAL-based toolboxes, written to support statistical instrumentation, Monte Carlo methods, quasi-continuous simulation, discrete-event simulation (with event, activity and process orientation), queueing network scenarios and combined discrete–continuous simulation. Concepts and utility procedures provided by these toolboxes are used throughout Chapter 2. They are primarily intended for teaching purposes. There are only simple debugging aids and efficiency considerations have a low priority.

Descriptive, hopefully meaningful identifiers are used to make the code largely self-documenting. Every effort is made to emulate an object-oriented style, although PASCAL sometimes makes this difficult to achieve.

In object-oriented frameworks operations are clustered around objects. The idea of objects is similar to the idea of abstract data types, although the two notions have different roots. Objects are modules representing a closure of information about some conceptual item's state and all actions that are allowed to access (read or change) this information. In the context of simulation programming, objects are often referred to as entities – symbolic images of things in the real world.

Objects receive and react to messages. This is the way in which computations take place. Objects have safe, well-defined interfaces, implemented through a set of methods whose execution can be triggered through parameterized messages. Local data structures may be manipulated only by internal methods, not from outside an object. In this way objects can only be asked to perform operations that are part of their instruction set (the so-called message protocol). They are at liberty to accept or decline a task.

Method implementations are private to the objects they are encapsulated in. For example, *Sample* might be used as the name of a method to obtain a number randomly drawn from some distribution object. Its implementation for objects of type *normal distribution* may look significantly different from that for objects of type *uniform distribution*.

In PASCAL classes of objects will be represented by record types, and methods by functions and procedures that manipulate objects of a given class. For instance, objects of class *aHistogramType* will be used for data collection. Objects of this type respond to procedures like *InitHistogram* (*aHistogram*, *noOfClasses*, *lowerBound*, *upperBound*), *UpdateHistogram* (*aHistogram*, *observedValue*) and *ShowHistogram* (*aHistogram*). In PASCAL, unfortunately, an object type and its methods cannot be clustered into a single textual module. Data-structure declarations and function or procedure declarations are different entities, defined in different parts of a program. We can declare classes of objects and their attributes as record types, instances of which can be created at compile or run time (for pointer-based structures). For all of our models this will normally be done by a procedure called *ReadAndInitialize*.

Methods must be placed in the program's global environment. They cannot be made local to objects. It follows that messages cannot be passed directly to objects; instead, the name of an object has to be passed as a parameter to a procedure that manipulates it.

To use the PASCAL-based toolboxes the necessary data and procedure declarations must somehow be included in the program. If no method for separate compilation exists, this may be done by straight insertion, using a text editor. Sometimes it is also necessary for the programmer to declare certain standard constants and data types. An object of type monitor, called *TheMonitor*, will be assumed as a globally accessible structure in all toolboxes. *TheClock*, a global clock type object, is also always required, although it is never used by the Monte Carlo toolbox. These assumptions are made purely for convenience. If they were explicitly imported or exported via parameter lists these entities would have to be passed to almost all of the methods, since they are so frequently referred to.

Use of any of the predefined entities always follows the same pattern. First, variables of the relevant type must be declared. These variables must be initialized (*procedures Init...*). They can then be operated on and, finally, their statistics (if any) are reported (*procedures Report...*).

The outer shells of the simulation programs are also stereotyped as much as possible:

1. initialize model structure and read data;
2. perform model interpretation (main simulation loop);
3. report results of the simulation.

There is always a *ReadAndInitialize* procedure, which instantiates and initializes all entities and their data structures. Note that model parameters are always defined by compile-time constants. This is done solely for the sake of clarity. It should not be taken as a recommendation for actual practice, where the flexibility to experiment with different parameter settings without recompilation is important, and model parameters should be read from a file at program execution time. All the simulations also have a main loop which cycles over a series of simulation experiments. It is guarded either by a specified sample size, a given length of a simulation run (clock time), or both. A *WriteResults* procedure is used to generate reports on a model's behaviour.

Program shells for different modelling styles are summarized in Appendix 1.

Chapter 2
Life on Dream World: Four Paradigms of System Simulation

"I THINK YOU SHOULD BE MORE EXPLICIT HERE IN STEP TWO."

2.1 Monte Carlo methods: Sampling possible worlds

The title of this chapter contains the term 'paradigm'. It was originally coined by Kuhn (1962) and is somewhat hard to define, since it denotes a whole range of related ideas. We will use the term according to the fifth and tenth of the meanings discussed by Masterman (1970). From this viewpoint a paradigm encompasses a whole coherent scientific tradition together with some accepted examples of scientific practice; its laws, theories, applications and tools. Floyd in his 1979 Turing lecture on 'The paradigms of programming' used the term to refer to programming styles, associated with prototypical data types, operations and control structures. At a more elementary level it can also be used to identify different models of computation and the schools or cultures built around them. Procedural programming, functional programming, logic programming and object-oriented programming are some of these. Kupka (1983) describes a number of others.

Modelling styles are paradigms in this sense. Monte Carlo methods, continuous, discrete-event and combined simulation all have their own methodologies, tools, traditions, user communities and prototypical applications. This section describes and exemplifies each of these world views.

Monte Carlo methods are traditionally used to estimate probabilities of a model's states through sample-driven experimentation. This is a static, distribution sampling kind of simulation, which makes no causal assumptions about a model; only statistical correlations between input and output values are explored. Objects are represented by variables and relationships by frequencies or probability distributions. Results are often summarized in frequency tables. Monte Carlo methods can be employed either to study complex phenomena that are insufficiently well understood to be amenable to more detailed cause/effect analyses, or when solutions to non-probabilistic models are too complicated or computationally too expensive. The first kind of model often arises in the context of studying social or economic phenomena (such as economic forecasting), whereas the second kind is well exemplified by numerical methods for solving multiple integrals.

The name 'Monte Carlo methods' is derived from the Mediterranean city of Monte Carlo, home of the world's most famous casino. John von Neumann, who 'coined' the name, clearly saw a similarity between roulette, the prototypical game of chance, and the way sampling experiments can be performed by computer. Since random processes are involved in the generation of both winning states of a roulette wheel and solution states of a Monte Carlo simulation, this seems to be quite an appropriate metaphor.

Hammersley and Handscomb (1964) offer one of the earliest and still one of the most comprehensive discussions of Monte Carlo methods. As in other areas of computing, there is no uniform terminology. Some people restrict the term Monte Carlo to solutions of strictly deterministic models through sampling experiments. Others apply it to any program using random numbers. We will use it to label all static, distribution sampling types of simulation.

Our first example revolves around the manufacture of lovepotions. Any association between technique and subject matter is, of course, purely accidental.

On DreamWorld the production of potions is a highly evolved art, requiring a multitude of jealously guarded secret ingredients and incantations. This scenario introduces Manuel, proud owner of the only lovesickness treatment centre on the island of Gont. Manuel sells potions whose quality has recently been rumoured to fluctuate. To help him dispel such rumours, we will conduct a Monte Carlo experiment and establish an effectiveness profile to guide his customers' expectations.

Through thorough theoretical study and dedicated field work we have discovered that Eye of Newt seems to be a critical ingredient of all love-potion recipes. It seems that if it is available, Manuel's potions are always excellent. Unfortunately it is quite often in short supply, so that Manuel has to use some unknown substitute, which leads to potions of somewhat uneven quality. Extensive trials have shown that Eye of Newt is used in approximately 30% of all potions. If other ingredients are substituted, potion quality is found to be excellent, good, marginal or poor, with equal probability. Fortunately it is already well known that the expected effectiveness of these potions, although it very much depends on the personality of the recipient, can generally be described by a normal distribution over a scale from 0 to 100 cupids. The parameters (mean and sigma) associated with excellent, good, marginal and poor potions were measured at: (90, 10), (80, 20), (50, 30) and (30, 40).

Program 2.1 models this scenario as a PASCAL program, using a toolbox of sampling procedures (see Appendix 1). Table 2.1 shows a summary of all procedures and functions in this toolbox. According to what was previously stated about object-oriented programming, these are grouped around a number of data structures they manipulate. Relevant object classes include a state sampling monitor and a number of probability distributions.

The monitor is a sample-driven interpreter that uses a function called *AnotherSampleP* to advance its state across a specified number of samples. This is a 'generator' object, producing a new value each time it is invoked. The fact that PASCAL modules cannot preserve state information across invocations means that generators cannot be implemented through value-returning functions. Function *AnotherSampleP* therefore causes a side effect on the monitor's *samplesTaken* slot to increment the number of samples each time the loop is entered. It also returns **true** as long as this count has not yet reached a value specified as the sample size.

The structure of a sample-driven interpreter is simple. It is initialized to some sample size (*ReadAndInitialize*) and guards a typical sampling loop:

while *AnotherSampleP* **do**
(1) sample from distributions;
(2) record results;

Since simulation is an experimental technique, data collection devices are

```
program PotionAnalysis (INPUT, OUTPUT);

  const
    MaxHistoClasses  = 50;
    MaxPlotPts       = 50; {needed to compile the whole stats library}
    QualityCategories =  4;

  type
    aQualityType = (excellent, good, marginal, poor);
    anIndexType  = 1..QualityCategories;

    {insert type declarations from the Monte Carlo toolbox}
    {insert type declarations from the statistics     toolbox}

  var
    TheClock           : aStaticClockType;
    TheMonitor         : aStateSamplingMonitorType;
    IsNewtAvailableGen : aDrawGenType;
    OneToFourGen       : aRandIntGenType;
    NormalDists        : array [aQualityType] of aNormalGenType;
    QualitySampled     : aQualityType;
    Effectiveness      : real;
    Results            : aHistogramType;

    {insert function & procedure declarations from the Monte Carlo toolbox}
    {insert function & procedure declarations from the statistics     toolbox}

  procedure ReadAndInitialize;
  {instantiates the model's entities}
    var
      qualityIndex : anIndexType;
    begin
      InitStaticClock            (TheClock);
      InitStateSamplingMonitor (TheMonitor          , 1000);
      InitDrawGen                (IsNewtAvailableGen , 12345,  0.3);
      InitRandIntGen             (OneToFourGen       ,67891, 1,  4);
      for qualityindex := 1 to QualityCategories do
        case qualityIndex of
          1: InitNormalGen (NormalDists [excellent], 23457, 90, 10);
          2: InitNormalGen (NormalDists [good]     , 89123, 80, 20);
          3: InitNormalGen (NormalDists [marginal], 45679, 50, 30);
          4: InitNormalGen (NormalDists [poor]     , 34567, 30, 40)
        end; {case}
      InitHistogram (Results, 0, 100, 10);
    end; {ReadAndInitialize procedure}

  procedure WriteReport;
  {reports the results of a simulation experiment}
    begin
      WRITELN ('Report on Random Number Generation');
```

Program 2.1 A Monte Carlo model of lovepotion effectiveness

```
      WRITELN (' =====================');  WRITELN;
      WRITELN ('Newt availability was ');          WRITELN;
      ReportOnDrawGen (IsNewtAvailableGen);   WRITELN;
      WRITELN ('Potion quality was    ');          WRITELN;
      ReportOnRandIntGen (OneToFourGen);      WRITELN;
      WRITELN ('Cupid samples for the 4 quality categories were: '); WRITELN;
      ReportOnNormalGen (NormalDists [excellent ]);
      ReportOnNormalGen (NormalDists [good      ]);
      ReportOnNormalGen (NormalDists [marginal]);
      ReportOnNormalGen (NormalDists [poor      ]);
      WRITELN; WRITELN;
      WRITELN ('Manuel''s Lovepotions: a Quality Profile (in cupids).');
      WRITELN ('********************'); WRITELN;
      ShowHistogram (Results);
   end; {WriteReport procedure}

begin {***** MAIN program *****}

   ReadAndInitialize;

   {Simulation loop: take as many samples as specified}

   while AnotherSampleP (TheMonitor) do
      begin
         if Draw (IsNewtAvailableGen) then
            QualitySampled := excellent
         else
            begin
               case RandInt (OneToFourGen) of
                  1: QualitySampled := excellent;
                  2: QualitySampled := good;
                  3: QualitySampled := marginal;
                  4: QualitySampled := poor
               end; {case}
            end; {if}

         Effectiveness := Normal (NormalDists [QualitySampled]);

         if Effectiveness < 0 then
            Effectiveness := 0;
         if Effectiveness > 100 then
            Effectiveness := 100;
         UpdateHistogram (Results, Effectiveness);
      end; {of sampling loop}

   WriteReport;

end.
```

Table 2.1 The Monte Carlo library

Object	: *aStaticClockType*;
Methods	: none {dummy declaration};

Object	: *aStateSamplingMonitorType*;
Methods	: *InitStateSampling Monitor* (*aStateSamplingMonitor*, *samples*); *AnotherSampleP* (*aStateSamplingMonitor*): *BOOLEAN*; *ReportOnSamplesTaken* (*aStateSamplingMonitor*);

{all distribution objects share the following internal method:
GenerateZeroToOneNumbers (*seed*): *REAL*;}

Object	: *aDrawGenType*;
Methods	: **private** *MeanOfDrawGenQ* (*aDrawGen*): *REAL*; **public** *InitDrawGen* (*aDrawGen*, *seedling*, *percentTrue*); *Draw* (*aDrawGen*): *BOOLEAN*; *ResetDrawStats* (*aDrawGen*); *ReportOnDrawGen* (*aDrawGen*);

Object	: *aRandIntGenType*;
Methods	: **private** *MeanOfRandIntGenQ* (*aRandIntGen*): *REAL*; **public** *InitRandIntGen* (*aRandIntGen*, *seed*, *low*, *high*); *RandInt* (*aRandIntGen*): *INTEGER*; *ResetRandIntStats* (*aRandIntGen*); *ReportOnRandIntGen* (*aRandIntGen*);

needed in any kind of simulation. Graphical representations of values for some key parameters are often particularly useful. Frequency diagrams (also called bar charts or histograms) are especially appropriate to summarize the results of sampling experiments. Data collectors for counting occurrences of states (*Count*), observing mean and deviation of some time series (*Tally*), and constructing frequency diagrams for some variable's values (*Histogram*) are available from a data collection toolbox and are summarized in Table 2.2. There are predefined procedures and functions to initialize these objects,

Table 2.1 (cont.)

Object	: *aUniformGenType*;
Methods	: **private** *MeanOfUniformGenQ* (*aUniformGen*): *REAL*; **public** *InitUniformGen* (*aUniformGen*, *seed*, *low*, *high*); *Uniform* (*aUniformGen*): *REAL*; *ResetUniformStats* (*aUniformGen*); *ReportOnUniformGen* (*aUniformGen*);

Object	: *aNegExpGenType*;
Methods	: **private** *MeanOfNegExpGenQ* (*aNegExpGen*): *REAL*; **public** *InitNegExpGen* (*aNegExpGen*, *seed*, *meanRate*); *NegExp* (*aNegExpGen*): *REAL*; *ResetNegExpStats* (*aNegExpGen*); *ReportOnNegExpGen* (*aNegExpGen*);

Object	: *aNormalGenType*;
Methods	: **private** *MeanOfNormalGenQ* (*aNormalGen*): *REAL*; **public** *InitNormalGen* (*aNormalGen*, *seed*, *mean*, *sigma*); *Normal* (*aNormalGen*): *REAL*; *ResetNormalStats* (*aNormalGen*); *ReportOnNormalGen* (*aNormalGen*);

sample the distributions, update the data collectors and produce reports on their behaviour.

This data collection toolbox provides some very basic procedures for collecting frequency data and for the plotting of histograms. More sophisticated tools may offer automatic scaling [see for example Dyck *et al*. (1982)] and interfaces to plotters or graphics terminals. Procedures for fitting theoretical distributions to empirical data would be valuable additions. Law and Kelton (1982) survey and discuss theoretical properties for a number of

Table 2.2 The statistics library

Object	: *aCounterType*;
Methods	: **private** *CounterValueQ* (*aCounter*): *INTEGER*; **public** *InitCounter* (*aCounter*); *UpdateCounter* (*aCounter*); *ResetCounterStats* (*aCounter*); *ReportOnCounter* (*aCounter*);

Object	: *aTallyType*;
Methods	: **private** *TalliedMeanQ* (*aTally*): *REAL*; *TalliedSigmaQ* (*aTally*): *REAL*; *TalliedMinQ* (*aTally*): *REAL*; *TalliedMaxQ* (*aTally*): *REAL*; **public** *InitTally* (*aTally*); *UpdateTally* (*aTally*, *obs*); *ResetTallyStats* (*aTally*); *ReportOnTally* (*aTally*);

well-known distributions that are particularly relevant for system simulation.

Note that for this model the number of samples is set to 1000, and that different random-number seeds (12345, . . .) and parameters (i.e. 30% probability that Eye of Newt is unavailable) are specified for distribution objects.

Seeds are required for the process of **random-number generation**, which is at the methodological heart of Monte Carlo methods. There is a large body of literature on advantages and disadvantages of different ways to generate random numbers on digital computers. Jansson (1966), Fishman (1978) and Bratley *et al.* (1983) are good introductions. Even though the bibliography by Nance and Overstreet (1972) is now somewhat dated, it is still useful, since basic techniques for random-number generation have not changed much since the early 1970s.

The oldest strategies for random-number generation used manual methods (coin flipping, die rolling, card shuffling, slips of paper from a hat,

Table 2.2 (cont.)

Object	: *aHistogramType*;
Methods	: **public**
	InitHistogram (*aHistogram*, *low*, *high*, *noOfClasses*);
	UpdateHistogram (*aHistogram*, *obs*);
	ResetHistogramStats (*aHistogram*);
	ShowHistogram (*aHistogram*);

Object	: *aMeasureType*;
Methods	: **private**
	MeasuredMeanQ (*aMeasure*): *REAL*;
	MeasuredSigmaQ (*aMeasure*): *REAL*;
	MeasuredMaxQ (*aMeasure*): *REAL*;
	public
	InitMeasure (*aMeasure*);
	UpdateMeasure (*aMeasure*, *aStateChange*);
	ResetMeasureStats (*aMeasure*);
	ReportOnMeasure (*aMeasure*);

Object	: *aChronographType*;
Methods	: **public**
	InitGraph (*aChronograph*, *maxXobs*, *maxYobs*, *maxXcoord*, *maxYcoord*);
	StorePoint (*aChronograph*, *xObs*, *yObs*, *printSymbol*);
	ResetGraphStats (*aChronograph*);
	ShowGraph (*aChronograph*);

etc.) or analogue devices (spinning discs, pulse generators, emissions from radioactive sources, etc.).

These methods were not very practical for electronic computers, where large numbers of random numbers must be efficiently generated. Stored tables of random numbers, produced by one of the above methods, could and have been used instead. The 'One million random digits' published by the RAND Corporation (1955) is the most popular of these. The main disadvantage of such tables is the storage space they require and the I/O traffic they generate.

Because of these difficulties algorithmic methods to generate random sequences were proposed as early as 1951. Due to the limits of computer technology at that time, memory and processor efficiencies were essential; apart from the obvious requirement of randomness.

John von Neumann (1951) suggested the **mid-square method**. Starting from a given seed, a number was squared and the n middle digits were taken to yield the next 'pseudo-random number'. This process was then repeated any desired number of times. It soon became evident that for a practically useful generator this procedure was much too slow and too prone to cycle or degenerate to zeros.

Nowadays, so-called **congruential generators** are almost universally used. These were first proposed by Lehmer in 1951. Starting from some given seed value (rn_0), congruential methods grow random numbers in a recursive fashion. Their basic recurrence relation is:

$$rn_{i+1} = (rn_i \cdot a + c)(\text{modulo } M), \qquad i = 1,\ldots,n.$$

rn_0, a, c and M should be positive integers.

Numbers produced by these generators are uniformly distributed between 0 and $(M - 1)$. If c is set to zero, then this generator is called a **multiplicative congruential generator**, which is the most commonly used type. A multiplicative congruential generator generates reasonably random sequences of numbers, given some appropriate choice of parameters (seed: rn_0; multiplier: a; modulus: M). Other generators in this family are the combined ($c \neq 0$) and additive congruential generators.

Multiplicative congruential generators have the great advantage of being fast and compact. Only one multiplication and one modulo division must be performed. Modulo division can often be 'simulated' by using fixed-point integer multiplication yielding $2 * M$ bits, discarding the M high-order bits and keeping the M low-order ones. These become the next random number. Only the last of the generated numbers needs to be kept in memory. Of course, such generators produce an exact replica of a given sequence when started with the same parameter setting. Although this seems to run against the grain of our intuitive notion of randomness (which may be ill-conceived anyway), it is actually of great advantage for simulation experiments. In the context of classical experimental designs simulation runs must be replicated, changing only one factor with everything else (including random numbers!) kept exactly the same.

Randomness is a philosophically very challenging but also a very elusive concept. Most people would maintain that, for a process to be random, its future behaviour must be impossible to predict from its past. There should be no recurring patterns whatsoever. This is obviously impossible to achieve by algorithmic methods. For computer simulations we therefore use a very pragmatic method to establish the randomness of a number sequence. A generator is said to produce **pseudo-random numbers** of acceptable quality if they pass a set of statistical tests for uniformity and independence. These tests are designed to show the absence of any obvious patterns. Of course there is a virtually unlimited variety of patterns that could be tested for, and since the notion of randomness is based on the idea of the absence of *any* pattern, the randomness of a number sequence can never be conclusively established. In fact there is no known generator whose numbers pass all of even the most common statistical tests. We can only accept a generator as being good enough if it passes most of the more straightforward ones.

The design of a good random-number generator is quite a difficult art, requiring a sound foundation in number theory. It should certainly not be approached in a random fashion. The behaviour of linear congruential random-number generators has been extensively studied and theoretically analyzed. Marsaglia (1972), Hull and Dobell (1964), and MacLaren and Marsaglia (1965) summarize some important results. Chapter 3 of Knuth (1969) is still the most complete and authoritative source on random-number generation and testing.

Almost all computer manufacturers provide random drawing procedures as part of their system libraries. These normally use some variant of a congruential generator. Some of them are, unfortunately, not particularly good, but they can often be tuned to produce acceptable sequences by adjusting their parameter settings (seed, a, c, M). Knuth (1969; p. 15) gives some recommendations on their choice. In general, a should be $8^t \pm 3$, where t may be any positive integer. It should also be close to $(2^{M/2})$. If M is 2^{35}, $a = (2^{17} + 3)$ is reported to be a good selection [Rubinstein (1981; p. 23); Hull and Dobell (1964); MacLaren and Marsaglia (1965)]. The seed (rn_0) should be relatively prime to M, not too small and odd; M should be large and, for convenience, a power of two. It is often set to 2^k, where k is the computer's word length. The last choice has two advantages. It increases a generator's efficiency because modulo division can now be simulated by taking the overflow of the multiplication as the next random number. It also preserves the potential for a long period, since a sequence will repeat itself as soon as it hits the same number again. Birtwistle *et al.* (1984) describe a nice method to automatically generate seed values in a well-spread manner.

If a generator must be written in PASCAL instead of assembly language, the overflow method mentioned above cannot be used. In PASCAL, overflow is an error condition that leads to unpredictable results. Sedgewick (1983; p. 37) shows a nice trick to avoid overflow by breaking the multiplication into a sequence of operations. A PASCAL function for its implementation is also presented. This is the default (0, 1) generator included in the Monte Carlo toolbox. It may be replaced by a custom-built method for a specific machine, taken for example from some library. Library generators typically produce uniformly distributed numbers over a range from 0 to $M - 1$. Dividing by M normalizes these to an interval 0 to 1. The normalized random numbers can then be plugged into functions that transform them to samples from a variety of distributions with given parameters or frequency profiles. The toolbox contains functions to sample from binary (*Draw* true/false), integer uniform (*RandInt*), continuous uniform (*Uniform*), negative exponential (*NegExp*) and normal (*Normal*) distributions. Note that the negative exponential generator takes a rate (arrival or service) as its parameter, and that the normal generator yields only an approximation. This approximation is, however, accurate enough for most purposes. The methods for sampling these distributions were adapted from Fishman

```
Report on random-number generation
********************************

Newt availability was:

1000 samples were taken by this binary distribution. The sample mean was: 0.30

Potion quality was:

702 samples were taken by this uniform integer distribution. The sample mean was: 2.48

Cupid samples for the four quality categories were:

488 samples were taken from this normal distribution. The sample mean was: 89.75
162 samples were taken from this normal distribution. The sample mean was: 81.05
173 samples were taken from this normal distribution. The sample mean was: 48.94
177 samples were taken from this normal distribution. The sample mean was: 28.10

Manuel's lovepotions: a Quality Profile (in cupids)
*******************

Until 1.00, 1000 observations were made.

This tally's characteristics were as follows:

Mean      :  70.63
Sigma     :  29.69
Min. obs. :   0.00
Max. obs. : 100.00

Overflow  was:  0
Underflow was:  0

  0.000
         ************
         ************    0.07%
         ************
 10.000
         *****
         *****    0.03%
         *****
 20.000
         *****
         *****    0.03%
         *****
 30.000
         *******
         *******    0.04%
         *******
 40.000
         ********
         ********    0.05%
         ********
 50.000
         ******
         ******    0.04%
         ******
 60.000
         ************
         ************    0.07%
         ************
 70.000
         *******************
         *******************    0.12%
         *******************
 80.000
         ***************************************
         ***************************************    0.24%
         ***************************************
 90.000
         ************************************************
         ************************************************    0.30%
         ************************************************
100.000
```

Figure 2.1 A lovepotion analysis

(1978), where descriptions for a number of other generators can also be found. Procedures and functions to sample from empirical distributions defined by frequency tables would be useful additions to the toolbox. Their implementation is left as an exercise.

The accuracy of distribution hypotheses and the quality of the associated random variates is crucial to the validity of any model containing stochastic components. Random-number generators known to be reasonably good should therefore be used whenever possible. Even then it is a good idea to run at least a few elementary tests of the generated number sequences. A simple chi-square test can ensure that no unexpected patterns occur, even if the user has confidence in the generator. A Monte Carlo toolbox can and should be augmented by utilities for such simple statistical tests (chi-square, frequency, serial, poker, gap, runs, etc.) to establish uniformity and independence properties of number sequences [see Sedgewick (1983)]. Solomon (1983) is a good source for some such programs and the associated statistical theory needed to apply them. They are written in FORTRAN but can be easily converted to PASCAL. During debugging of complex models it is also advisable to use a degenerate generator (one that produces deterministic values), so that one can concentrate on verifying the flow of the model logic without any complications produced by stochastic behaviour.

A quote from Sedgewick (1983; p. 45) nicely summarizes the most important principles of random-number generators and their use:

> The moral is: do your best to use a good generator based on the mathematical analysis and the experience of others; just to be sure, examine the numbers to make sure that they 'look' random; if anything goes wrong, blame the random number generator!

Using utilities from the toolbox, the model (Program 2.1) becomes quite straightforward. After declaring the necessary entities (a monitor, some distributions, a histogram, some auxiliary variables), these are instantiated (procedure *ReadAndInitialize*), 1000 samples from the three kinds of distributions are taken within a loop controlled by the model monitor, and the results of the experiment are printed as a histogram (procedure *WriteReport*). Figure 2.1 shows the program's output.

Exercises

1. Write a random-number generator based on the mid-square method. You may want to compare its number sequences with our congruential generator. A visual inspection of number pairs plotted in two-dimensional space should be sufficiently telling.

2. The additive congruential generator is defined by the following formula:

$$X_j = X_{j-1} + X_{j-n} \text{ (modulo } m)$$

where n is some appropriately chosen lag [see, for example, Maisel and

Gnugnoli (1972)]. This generator has the potential for long periods, but it must store the last *n* numbers. Write a procedure to test this generator with different values for *n*, comparing its quality and performance with that of the multiplicative congruential one.

3. A large variety of methods has been proposed to test the quality of random numbers [see, for example, Knuth (1969)]. Implement a number of simple tests (chi-square, frequencies, runs, gaps, etc.) and use them to compare a number of generators.

4. Write the necessary procedures for sampling from empirical distributions represented by pairs of *x* and *y* values. You should consider both step-valued and linearly interpolating generators. Maisel and Gnugnoli (1972) give some guidelines.

5. Patch the toolbox routine for histogram plotting to allow automatic scaling of the *x* and *y* axis, given that a specific number of frequency classes has been explicitly requested. You will have to store all data points and their frequencies before you can determine class boundaries.

2.2 System simulation: Exploring possible worlds

The remaining three of the four paradigms introduce a so-called sequencing variable, whose values are used to bind state transitions into processes. The next step is to explore state-space traversals generated by their dynamic execution. Usually the sequencing dimension is model time, represented by some monotonously increasing **clock**. Simulation in this sense follows a model's time paths, summarizing its behaviour through a number of well-chosen measures. Such models can be deterministic or stochastic, depending on whether sampling from probabilistic submodels is required.

In the simplest case the clock's value is advanced by time slicing. We call this clock mechanism **synchronous** because time is incremented by some fixed-length interval; the degree of resolution of the model. The whole model is evaluated again and again, leading to a new state after each time slice. Appropriate choice of the length of this time slice ($\triangle t$) is important because it determines both a model's accuracy and its computational efficiency. Typical examples of simple dynamic models of the deterministic and stochastic variety are the 'Game of Life' and 'Drunken Cockroach' problems in Chapter 5.

This modelling strategy can be applied to space slicing as well. Most often time- and space-slicing aspects of a model will be interwoven, unless one of the dimensions is deliberately abstracted from. Modelling heat flow in a building is a good example of a so-called compartment model, where the house may be divided into a grid of areas, and state changes must be computed in both the time and space dimensions.

Simple time-slicing models can be extended in two directions. The time increment can be made very small, tending towards a 'quasi-continuous' view of the world. This is the basic approach of **continuous-system simulation**. Or the clock can be changed from synchronous to asynchronous operation, in which time jumps from one cluster of relevant state transitions to the next. Only changes in state are important – any intervening time periods are of no interest. This is the basic world view of **discrete-event simulation**.

Both paradigms have their typical applications and dedicated user communities. Continuous or quasi-continuous simulation is popular among engineers, physicists, biologists, hydrologists, economists and social scientists. Either they study relatively simple systems at fairly high levels of detail, at which changes are smooth and naturally described by differential equations, or they aggregate and abstract highly complex systems into low-resolution models, where trends in overall system behaviour can again be approximated by constant rates of change. It is in the range of models between those two extremes where a discrete-event simulation's potential for capturing details and complex interactions is required. This world view is therefore popular with operations researchers and computer scientists.

Good examples of detailed continuous models are the behaviour of an automobile suspension or an autopilot. Highly aggregated continuous models are well exemplified by predator–prey models of an ecology (see Chapter 5) or the flow of goods, services and money in an economy.

Queueing models – i.e. the flow of messages through a communication line connected to a digital computer system under different loads – are typical examples of discrete-event simulations.

Communication barriers between the user communities of these modelling styles have existed for quite some time. Only recently have we seen the emergence of a new paradigm, **combined discrete–continuous simulation**. This world view tries to incorporate both the capability for modelling processes at smooth rates of change and for capturing irregularities in state-transition patterns. Many problems can be naturally expressed within this framework. A nice example is a model of a (continuously) burning house, used to explore various fire-fighting strategies characterized by discrete events (see Chapter 5).

Sections 2.2.1 to 2.2.3 will discuss these paradigms and the programming styles associated with them.

2.2.1 States and processes: Quasi-continuous simulation

Continuous-modelling techniques stem from a long tradition of building analogue devices for system modelling and simulation. Closer to modern times analogue computers have often been used for system simulation.

The operation of these hard-wired machines can be easily emulated on digital, stored-program computers, with some loss of speed and a considerable gain in programming convenience and computational accuracy. The first special-purpose languages for continuous-system simulation were

FORTRAN-based simulators of a typical analogue computer, providing operations such as summation, multiplication, function generation and integration of flows.

Continuous models are usefully employed if system behaviour depends more on aggregate process flows rather than on individual states and their changes. The predominant activities of the model cause smooth changes in its variables. This view results in potentially infinite state spaces, but these can be approximated to any desired degree of resolution. Accuracy is largely dependent on the choice of time increment.

Relationships between objects are reflected in rates of change. Typically, rates of change are represented by differential equations, which could, at least in principle, be solved analytically. The effort involved in this may, however, be so great that computational techniques become preferable. It is often implicitly assumed that a system can be adequately represented as being both stable and linear. If non-linearities and/or stochastic components are introduced, simulation is the method of last resort.

A **differential equation** describes the rate of change in some quantity as a function of some other quantity. In linear differential equations the dependent variable and its derivatives are multiplied by constant coefficients and then added together. No powers higher than 1 may occur and multiplies must not themselves depend on y. If the dependent variable or any of its derivatives appear in any other form an equation is said to be non-linear. When more than one independent variable occurs in a differential equation it is called a partial differential equation.

Partial differential equations can be used to represent models as continuous hyperspaces. Karplus (1977) calls this style **distributed-parameter systems**. There are no discrete state variables: the model is defined entirely in terms of multi-dimensional state fluctuations. An example of a problem that seems to fit naturally into such a framework [Gordon (1978a; p. 59)] is a model of heat flow in a three-dimensional body. Here there are four independent variables (the three space dimensions and time) and one dependent variable (temperature).

Karplus (1977) classes models with discrete state variables whose behaviour can be described by a set of difference or ordinary differential equations as **lumped-parameter systems**. This is the most popular frame for quasi-continuous simulation in the social and engineering sciences.

Very often a system's behaviour can be captured through a so-called feedback model, postulating a coupling relationship between its inputs and outputs. Feedback can be either positive, reinforcing the system's behaviour, or negative, resulting in a dampening of the system's response. Negative-feedback systems exhibit a kind of self-regulating behaviour as shown by a room thermostat. Feedback models are often used to explain engineering systems, as well as biological, economic or social phenomena. There is a modelling school, founded by J. W. Forrester, which studies feedback models to explain dominant trends in system behaviour. It has spawned a whole philosophy of looking at systems and organizations as webs of control loops. Individual events are highly aggregated, so that they can be treated as continuous flows. This aggregation allows emphasis of trends in the system's overall response. This paradigm is often referred to as **industrial** or **system dynamics** (see Section 3.4.3).

In its simplest form, rates of change can be represented by difference instead of differential equations. Using a digital computer for simulation then reverts to time-indexed evaluation of all equations in the model, calculating a new state as a result of changes during the last time interval. New rates may also be computed, and are assumed to remain constant during the next interval. By using sufficiently small increments even quasi-continuous systems with non-linear characteristics can be adequately represented. One should, however, stress that great care must be taken in dealing with rates that are changing over time. The results of such a simulation may crucially depend on the sequence in which equations were evaluated.

The next scenario exemplifies a simple quasi-continuous simulation. We will study the seasonal rise and fall of lovesickness, one of DreamWorld's most common and fatal diseases [adapted from Rubinstein (1975; pp. 457–62)].

The island of Gont, which we visited previously, is subject to periodical outbreaks of lovesickness among its inhabitants, which may for instance be triggered by the arrival of a lonely and lovesick sailor. Polls and thorough statistical analyses have established that during previous outbreaks Gont's population could be classified into three categories. There are those who were romantically inclined (and thus receptive to outbursts of lovesickness), those who were lovesick already, and those who were perfectly rational and immune to temptation. Lovesickness is highly contagious. Once established it spreads at a rapid rate, say 0.1% of the product of the romantic and lovesick populations per day. This happens through contacts between receptive and already infected persons, and it must be borne in mind that the people of Gont are very social. A bout of lovesickness usually lasts for eight days, after which it vanishes completely, leaving a headache, a stack of bills for Manuel's services, and an immunity that is assured for about two years. Note that a DreamWorld year has 350 days.

Even though lovesickness keeps Manuel in business, it has unpleasant consequences for the island's economy. Currently little is known about the exact nature of the disease. All statisticians always became too involved themselves to gather this crucial data. Therefore it has been decided to perform a simulation study of the disease's progress with a view of using it to time the proper administration of preventive medication.

This scenario is a good example of a typical feedback model. Figure 2.2 shows the disease's time path for a stable population of 1000 citizens, all of who were initially romantically inclined. Note that the relationships between romantic, lovesick and sane populations lead to cyclic behaviour, with the

Lovecycles on DreamWorld
****

+ stands for the romantic population's level.
* stands for the lovesick population's level.
o stands for the sane population's level.

120 plot points were collected.

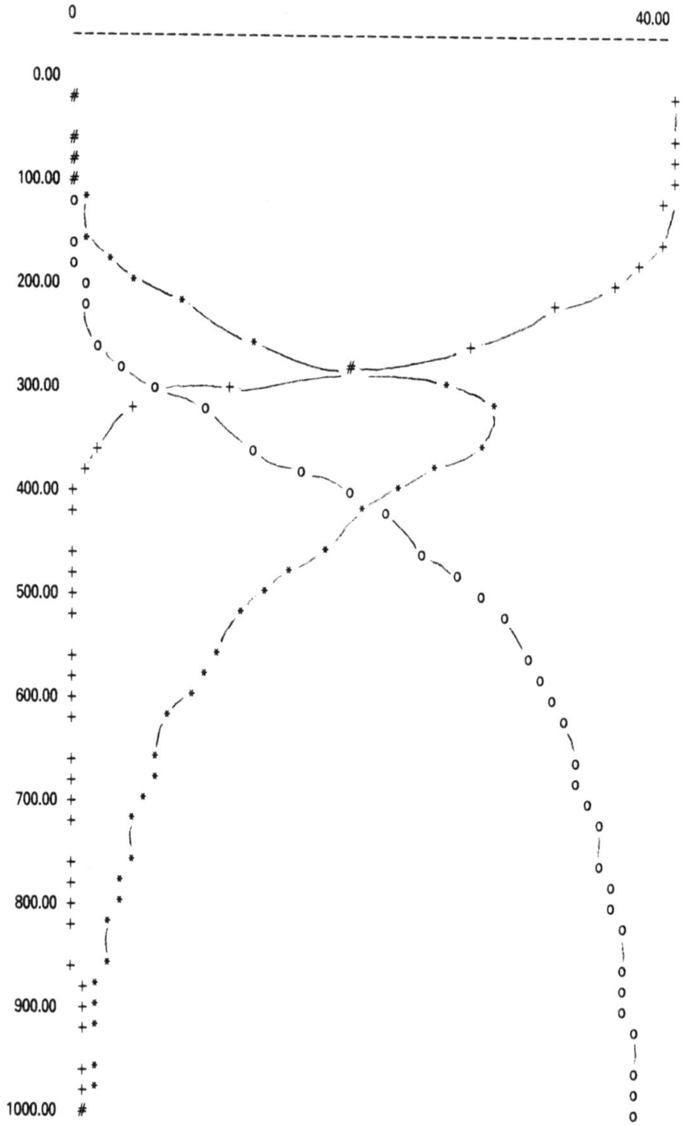

Figure 2.2 A chronograph of Gont's lovecycles

disease dying out after 40 days, and leaving the population immune for another two years.

To implement this scenario in PASCAL, Program 2.2 uses predefined data types and utilities from a quasi-continuous simulation toolbox (Appendix 1). It provides a synchronous clock entity, a simple time-slicing monitor for controlling a series of model evaluations and a data type for representing quasi-continuously changing variables. The clock is called synchronous because at each tick it changes its value by a constant increment, defining its resolution.

Since the model is a dynamic one the time paths of some variables' values are of interest. A graphical data type called a *chronological graph* is part of the toolbox. It can be used to plot states against time. There is also a *measure* object, which gathers time integrals of a variable's values and report its time-averaged mean and deviation.

Using the methods defined for these classes of objects in Table 2.3, the processing cycle of a simple time-slicing monitor is easily written. It is controlled by the clock's value and resolution:

while *AnotherTickP* **do**

(1) step or integrate variables;
(2) record results;

All the model's state variables will be integrated for each time step, and the simulation will be terminated as soon as a predetermined time is reached.

The process of **integration** is at the heart of continuous-system simulation. Continuous simulation packages or languages therefore typically provide a number of fixed and variable step-width methods for numerical integration. Rates must then of course be representable as integrable functions. In the example program we restrict ourselves to a simple procedure, which at each invocation updates a state variable's level by applying some function (*StepState*). This simple scheme uses the global clock's resolution to determine the step width, implementing an integration of difference equations. Practically useful continuous simulation toolboxes should support more sophisticated integration methods. Atkinson and Harley (1983) and Davis and Rabinowitz (1984) give guidelines and programs for such library procedures. They also discuss the technical background necessary for understanding the potential and limitations of their use.

Chronograph objects can be used to plot states against time, as shown in Figure 2.2. They are initialized by an *InitGraph* procedure, data is collected by invoking procedure *StorePoint*, and a call on *ShowGraph* produces a plot on a line printer. This is an unsophisticated and much simplified graph-plotting utility. Dyck *et al.* (1982) describe a more sophisticated version.

The structure of the model in Program 2.2 is again quite simple. A clock, the monitor, the chronograph and the three populations are declared and instantiated by procedure *ReadAndInitialize*. The resolution of the clock is set to one, the length of the simulation to 40 days. The chronograph's plotting area is defined as a grid of 50 * 50 points, where labels for the x and y axes range from 0 to 40 days and 0 to 1000 people respectively.

Three equations define the rates of change for three kinds of state transitions (romantic to lovesick; lovesick to sane; sane to romantic).

The main program implements the simulation's control loop. It is driven by *TheClock* under control of *TheMonitor*. *AnotherTickP* is again a generator function, similar to *AnotherSampleP* in the previous scenario. Through a side effect it changes the clock's hand by one resolution unit and returns **true** as long as the simulation's duration (recorded by a predefined attribute of the global monitor) has not yet been reached. During this cycle

```
program LoveCycles (INPUT, OUTPUT);

  const
    MaxPlotPts      =   50;
    MaxHistoClasses =   50;
    RomanceSymbol   =  '+';
    LovesickSymbol  =  '*';
    SaneSymbol      =  'o';
    ProbOfInfect    = 0.001; {0.1% per potential contact between romantic & }
                             {lovesick population      }
    InLoveTime      =    8; {lovesickness lasts 8 days }
    SanityTime      =  700; {immune for 2 Dreamworld years }

  type
    PosIntType = 0..MAXINT;

    {insert type declarations from the continuous-simulation toolbox}
    {insert type declarations from the statistics            toolbox}

  var
    TheClock                                : aSynchClockType;
    TheMonitor                              : aTimeSlicingMonitorType;
    RomanticPop, LovesickPop, SanePop       : aStepVarType;
    InLoveRate  , OutLoveRate, OutSaneRate  : PosIntType;
    Chronograph                             : aChronographType;

    {insert function & procedure declarations from the continuous-simulation toolbox}
    {insert function & procedure declarations from the statistics            toolbox}

  procedure ReadAndInitialize;
  {instantiates the model's entities     }
  {The BASIC TIME UNIT is 1 day !}
    begin
      InitSynchClock          (TheClock   ,   1);
      InitTimeSlicingMonitor (TheMonitor,  40);
      InitStepVar        (RomanticPop, 1000);
      InitStepVar        (LovesickPop ,    1);
      InitStepVar        (SanePop      ,    0);
      InitGraph  (Chronograph, 40, 1000, 50, 50);
    end; {of ReadAndInitialize procedure}
```

Program 2.2 *A quasi-continuous model of Gont's lovecycles*

the state of each of the three step variables representing the three sectors of the population is recorded and updated. After termination the changes in their levels are plotted against time (Figure 2.2).

Exercises

1. Write procedures for integrating a model described by a set of ordinary differential equations.

```
procedure WriteReport;
{reports the results of a simulation run}
  begin
    WRITELN('***** LoveCycles on DreamWorld *****');
    WRITELN;
    WRITELN (RomanceSymbol, ' stands for the romantic population''s level.');
    WRITELN (LovesickSymbol, ' stands for the lovesick  population''s level.');
    WRITELN (SaneSymbol     , ' stands for the sane      population''s level.');
    WRITELN;
    ShowGraph (Chronograph);
  end; {of WriteReport procedure}

begin {***** MAIN program *****}

  ReadAndInitialize;

  {Simulation loop: cycle through the specified number of time epochs}

  while AnotherTickP (TheClock) do
    begin
      InLoveRate  := ROUND (ProbOfInfect * RomanticPop.level *
                                    LovesickPop.level);
      OutLoveRate:= LovesickPop.level div InloveTime;
      OutSaneRate := SanePop     .level div SanityTime;

      StorePoint (Chronograph, TheClock.NOW, RomanticPop.level ,
                  RomanceSymbol);
      StepState  (RomanticPop.level, OutSaneRate - InLoveRate    );
      StorePoint (Chronograph, TheClock.NOW, LovesickPop.level ,
                  LovesickSymbol);
      StepState  (LovesickPop.level, InLoveRate - OutLoveRate)   ;
      StorePoint (Chronograph, TheClock.NOW, SanePop.level       ,
                  SaneSymbol);
      StepState  (SanePop.level, OutLoveRate - OutSaneRate        );
    end; {of cycle}

  WriteReport;

end.
```

Table 2.3 The continuous simulation library

Object	: *aSynchClockType*;
Methods :	**public** *InitSynchClock* (*aSynchClockType*, *resolution*); *AnotherTickP* (*aSynchClockType*): *BOOLEAN*; *HowLongSimulatedQ* (*aSynchClockType*): *REAL*;

Object	: *aTimeSlicingMonitorType*;
Methods :	**public** *InitTimeSlicingMonitor* (*aTimeSlicingMonitorType*, *simLength*);

Object	: *aStepVarType*;
Methods :	**public** *InitStepVar* (*aStepVarType*, *anIntegerQtty*); *StepState* (*aStepVarType*, *anIntegerQtty*); *ResetStepVar* (*aStepVarType*);

Object	: *aContVarType*;
Methods :	**public** *InitContVar* (*aContVarType*, *aRealQtty*); *UpdateState* (*aContVarType*, *aRealQtty*); *ResetContVar* (*aContVarType*);

2. Patch the toolbox procedure for plotting chronological graphs, so that it automatically performs scaling of data points on to a specified plotting area.

3. Test the effect of different degrees of resolution (Δt integration step sizes) on results and run-time behaviour of our lovecycles model.

2.2.2 Entities and events: Discrete-event simulation

Discrete-event simulation views models as structured collections of objects bound into webs of relations and transformations. While continuous simulation abstracts from discontinuities, event-driven models assume that 'nothing relevant' happens between successive state transitions.

Although a wide variety of systems can be studied using discrete simulation techniques, the **queueing network paradigm** is the most popular. Here the aim is to explore the effects of capacity limitations and routing strategies on the flow of some class of transactions. The objects of interest are different kinds of resources, queues and workload items. Queueing theory can be used to analyze such models, but only if they have a relatively simple structure. Kleinrock (1976) gives a comprehensive survey of queueing models and some of their applications.

Inventory system simulations are another well-established class of simulation study. Banks and Malave (1984) give a good overview of their motivations and basic approach.

Over the last 25 years the important elements common to all discrete-event models have been identified and analyzed [Adams and Dogramaci (1979), Barnden (1981), Hooper and Reilly (1982), Nance (1979, 1981), Zeigler *et al.* (1979)]. Figure 2.3 shows five components of any discrete-event simulation.

1. Tools for discrete-event simulation must provide facilities for **model structuring** and execution. The most natural view is that of a collection of interacting objects moving through sequences of actions. *Objects* can be represented by counters, flags, simple variables or attribute vectors called entities. *Relationships* can be encoded in arrays, records or list structures. *Actions* leading to state transitions are typically described through functions and procedures and their dynamic interactions.
2. All dynamic models need a clock for **time management**. Discrete-event models typically use an asynchronous, so-called 'next-event' strategy for advancing model time. Here the clock's value jumps to the time of the next event on the monitor's agenda, once it has been established that no further actions can be performed at the current time instant. This method should be contrasted with the simple, synchronous 'next-epoch' technique used for time slicing in the previous chapter.
3. Since practically all interesting phenomena are too complex to be modelled in all their details, **random processes** must be used to represent less essential aspects. Inputs to so-called trace-driven models can be deterministic, using externally determined events, but most often models are internally driven by samples taken from some stored probability or frequency distributions.
4. One must always remember that simulation is an experimental technique. It can only be usefully employed if the data generated by simulation runs can be easily gathered and analyzed. Facilities for **statistical instrumentation and reporting** are therefore essential. They should provide for means, minima, maxima, frequency tables (histograms) and time plots (chronological graphs).
5. Sequencing state transitions is the responsibility of an execution monitor, often called a **run-time control system** (RTCS), **scheduler** or **simulation executive**. It needs an agenda, called a **sequencing set** (SQS), **event list**, **noticeboard** or **scheduling list** to keep track of pending events. Such event lists can be organized in a number of ways. Arrays are the most straightforward way for their implementation. Unfortunately, they are

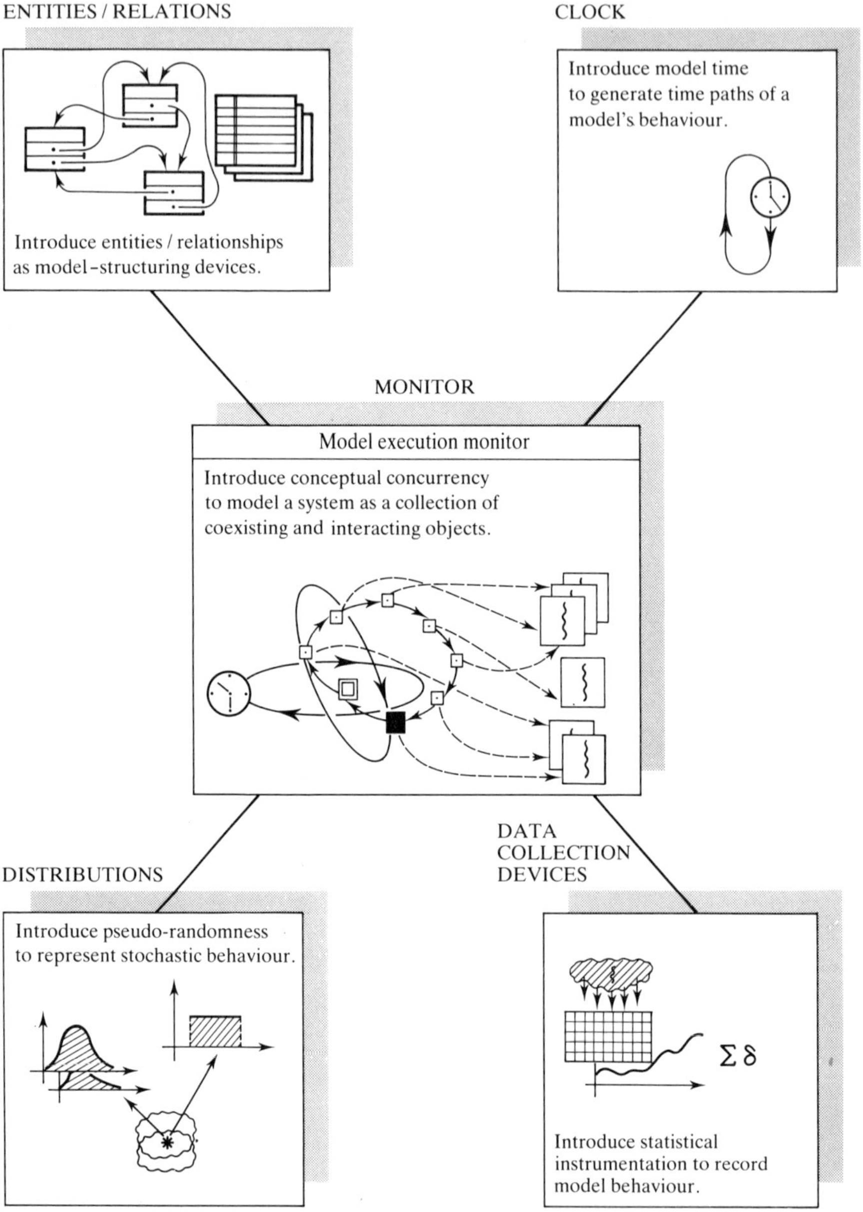

Figure 2.3 Discrete-event simulation: The main components

insufficiently flexible, particularly since it is difficult to predict the maximum number of pending events prior to a model's execution. Some programming tools of course do not offer any choice, since they fail to provide more sophisticated data structures.

Most event-list implementations are designed so that the next imminent event can very quickly be determined and removed from the agenda. Placing a notice for some future event usually takes more time, since a search for its proper position is typically involved in this. Simple linked lists are the most common event-list organization if pointer-based structures can be used; but trees, heaps and other more sophisticated data structures have been proposed and experimented with. Vaucher and Duval (1975), Franta and Maly (1977, 1978), Engelbrecht-Wiggans and Maxwell (1978), Ulrich (1978), McCormack and Sargent (1979), and Marsden (1984) describe some of these proposals.

The sequencing strategy used by a model monitor depends on the way descriptions of a model's behaviour are modularized. Activity, event and process-based world views may be distinguished:

1. **Events** define a set of simultaneous state changes. Between events the model's state remains unchanged; nothing relevant happens. In an airport model these would include plane and passenger arrivals and departures, start and finish of boarding, as well as start and finish of landing and takeoff.
2. A **process** is a sequence of events (nest of activities) ordered on time. These events must be logically connected, involving the same entity. All actions associated with this entity's actions are grouped into a single textual module, often referred to as its life cycle. The airport example could contain both plane and passenger processes, each defined by a description of the characteristics of its class.
3. An **activity** is a conceptual closure of some time-consuming action performed by an entity. It is typically guarded by a set of conditions under which it may start, and finishes after a specified time period. Descriptions of a plane's ground time, actions associated with landing and takeoff, as well as passengers' boarding activities would fall into this category.

These different perceptions of how to describe a model's behaviour have given rise to four paradigmatic sequencing schemes:

1. **Event scheduling** binds actions associated with individual events into single-entry modules, called **event routines**. The monitor selects events for execution, processing a time-ordered agenda of so-called event notices. **Event notices** consist of at least a time cell and a reference to an event routine. Event execution can schedule new events, creating an event notice and placing it at the appropriate position in the agenda. The clock is always updated to the time of the next imminent event – the one at the top of the agenda.
2. The **process interaction** approach focuses on the flow of entities through a model. This strategy views systems as webs of concurrent, interacting processes. The life cycle of each class of entities is described by a process class. **Process classes** can have multiple entries and exits, at

which a process interacts with its environment. The monitor uses an agenda, like the one just described, to keep track of pending tasks. However, apart from recording activation time and process identity, the monitor must now also remember the state in which a process was last suspended. Each process therefore owns a so-called **local sequence control** (LSC) as part of its state descriptor.

3. If a process also remembers the values of all private (local) variables, an **object-oriented programming style** can be employed. Here a module must encapsulate all relevant aspects (structural and behavioural) of an entity: its attributes, actions, and life cycle. Communication between objects is allowed only through well-defined interfaces, provided by those messages an object is willing to respond to. These are implemented through local functions and procedures.
4. In the **activity scanning** approach a model contains a list of activity procedures, each of which has two entry points. The first entry is guarded by a condition specifying when an activity may start and summarizes all associated actions. The second entry defines all finishing actions. Model execution is cyclic, with the monitor scanning all activities in a model, testing whether they can be started. This is done each time a change of state occurs. The clock is advanced only if a whole cycle leaves the model's state unchanged; in that case, it is set to the smallest value of all activities' finishing times. The monitor then executes the finishing actions of all activities that can finish at that point in time. A new start cycle begins thereafter.

   This **three-phase structure** is obviously computationally expensive. It is an example of what is called **conditional sequencing**, in which the programmer only needs to describe the conditions under which particular (start) events may occur. It is then up to the model monitor to make sure they are triggered at the right time. Methods for time-dependent event scheduling, as described for the event-, process- and object-oriented world views, are all so-called **imperative sequencing** strategies.

It is time to return to Manuel's lovepotion shop, which is to be modelled as a simple queueing scenario (**M/M/1**: exponential arrivals, exponential service, single server). In later sections this model will be used to demonstrate a number of different programming styles and languages.

Manuel treats lovesickness with potions he prepared the previous midnight. For simplicity, let us assume that there is always an unlimited supply of these. Figure 2.4 shows his shop as a so-called open queueing model.

Queueing networks can be characterized in terms of a number of prototypical entities. There are **sources** (symbolized by clouds) that generate workload items called **transactions**, the transactions themselves and **sinks** (clouds) through which transactions disappear. There are also **resources** or server units (denoted by flagged circles) performing some service on transactions, and **queues** that may form in front of these. Waiting lines occur

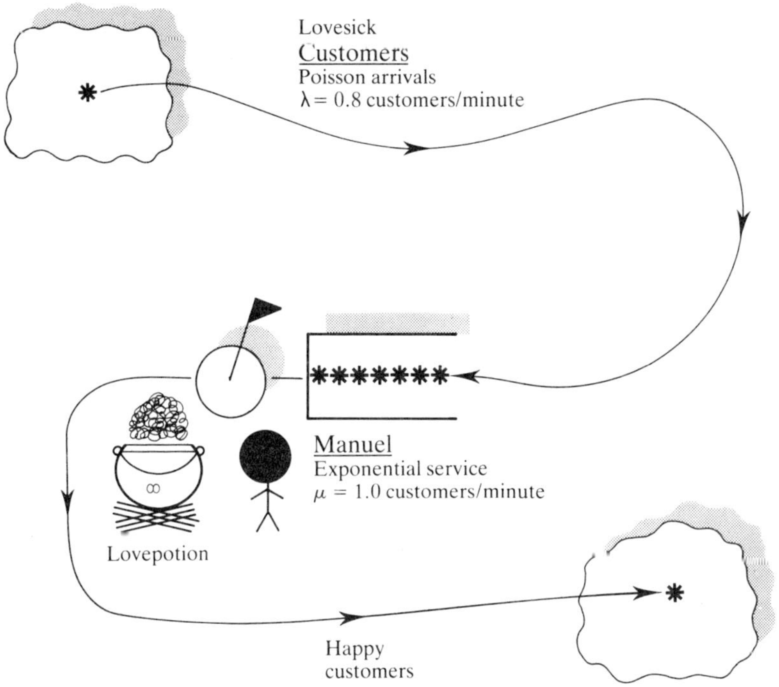

Figure 2.4 Manuel's shop: A queueing model ($M/M/1$)

because demand for services may at times exceed a server's limited capacity. The length of these queues will fluctuate, but it will be unbounded if the transactions' rate of arrival is consistently higher than the server's service rate.

Open queueing models assume that transactions are transient workload items; they appear from a source, traverse through the model and disappear at a sink. Such scenarios are, for instance, characteristic of bank or post office simulations, where customers appear, obtain service (eventually) and disappear again.

In so-called **closed queueing models** this is not the case. Here individual transactions can be permanent parts of a model, in the same way queues and servers are. Sources and sinks are not required. A good example of a closed queueing network is a computer simulation, where messages are generated by terminals, processed by a CPU and then return to their origins. The frame of an old message (transaction) can then be reused for a new one.

Assume that while business is booming (during an outbreak of lovesickness) Manuel's customers arrive at a rate of 0.8 per minute. Manuel serves (diagnoses, hands over potions, pockets donations) at a rate of 1.0 customers per minute. Experience shows that during such peak periods a queue will form. Always eager to please his clientele, Manuel wants to know the

Observed variables	• *Customers' flowtime* • *Queue length* • *Server's utilization* • Source's customer generation rate . . .
Initial state	• *Server idle, shop empty* • Reset statistics after a specified presimulation period • Start the model in steady state (if it is known before the simulation) . . .
Input schedule	• *Arrival and service rate distributions are sampled* • Arrival and service times are read from a file . . .
Terminal state	• *Run for a specified number of customer departures* • Run for a specified length of model time • Run the model until an equilibrium (steady state) is reached . . .
Data collection	• *Customers' flowtime (mean, variance, frequency tables, ...)* • *Queue length (mean, variance, max., frequency tables, exceeding a specified threshold, ...)* • *Server's utilization* • Number of service activities . . .

Figure 2.5 Experimental frames for queueing analyses of Manuel's shop (the italicized settings are assumed for the example programs)

average and maximum length of this queue, and also the time customers spend in his shop (waiting and being served). He hopes that this will help him decide whether or not he should occasionally employ a young pixie as an assistant.

Figure 2.5 shows some possible frames for experimenting with this model. The results, summarized in Figure 2.6, were obtained through a simulation for a presimulation sample of 1000 and a final observation interval of another 1000 customer departures.

Statistics were reset after 1000 samples were taken

Monitor statistics reset at 1225.68
Neg. exp. generator statistics reset at 1225.68
Neg. exp. generator statistics reset at 1225.68
Source statistics reset at 1225.68
Sink statistics reset at 1225.68
Queue statistics reset at 1225.68
Server statistics reset at 1225.68
Histogram statistics reset at 1225.68

1000 samples were taken since this monitor was last started.
The current clock time is 2453.55
Since last reset 997 transactions were created by this source.
Since last reset 1000 transactions were removed through this sink.

Manuel's performance:

Utilization of this server was 77.97%, since last reset.
The capacity of this server is 1 unit.
0 of these are currently in use.

Queue statistics:

Since its statistics were last reset, 997 items entered and 1000 left the queue.
Its current length is 3

The queue's main characteristics were:

Mean length = 3.01
Max. length = 19
Mean delay = 3.70

Customers' time in the shop:

Until 2453.55, 1000 observations were made.

This tally's characteristics were as follows:

Mean : 4.66
Sigma : 4.76
Min. obs. : 0.01
Max. obs. : 23.14

Overflow was: 0
Underflow was: 0

```
 0.000
        ************************************************
        ************************************************   0.44%
        ************************************************
 2.500
        *************************
        *************************   0.23%
        *************************
 5.000
        *************
        *************   0.11%
        *************
 7.500
        *********
        *********   0.08%
        *********
10.000
        *****
        *****   0.04%
        *****
12.500
        ****
        ****   0.03%
        ****
15.000
        ***
        ***   0.02%
        ***
17.500
        ***
        ***   0.02%
        ***
20.000
        *
        *   0.01%
        *
22.500

          0.00%

25.000
```

Figure 2.6 Manuel's shop: A queueing analysis

Exercise

1. Analyze some systems of your choice (car wash, bank branch, ferry terminal, time-sharing computer system, department store, saw mill) in terms of potential purposes for modelling and their relevant entities, properties, relationships, events, activities and processes.

A. Event routines

Event routines describe transformations at specified time instances, triggered under supervision of a monitor entity. SIMSCRIPT [Kiviat *et al.* (1968)], GASP [Pritsker (1974)] and SIMPAS [Bryant (1980)] are just some of the many simulation programming systems supporting this world view.

In most general-purpose programming languages event-oriented simulators are the simplest to implement. They are also quite efficient in terms of processing time, particularly for models with few event interactions. But this has to be paid for by placing quite some responsibility on the programmers' shoulders, because they must explicitly code all relevant scheduling relationships into the model's logic.

For Manuel's shop, four classes of events are used: *HelloEvent* and *ByeEvent* describe customer arrivals and departures, while *StartEvent* and *FinishEvent* define what happens when Manuel starts and finishes serving. Figure 2.7 shows their interactions and Figure 2.8 shows the monitor's data structures.

For such a simple model fewer event routines could have been used. *hello/start*, *finish/start* and *finish/bye* are called **bound events**. They can be integrated into single modules because their actions can be combined so that there is never any time delay between them. A start event, for instance, will

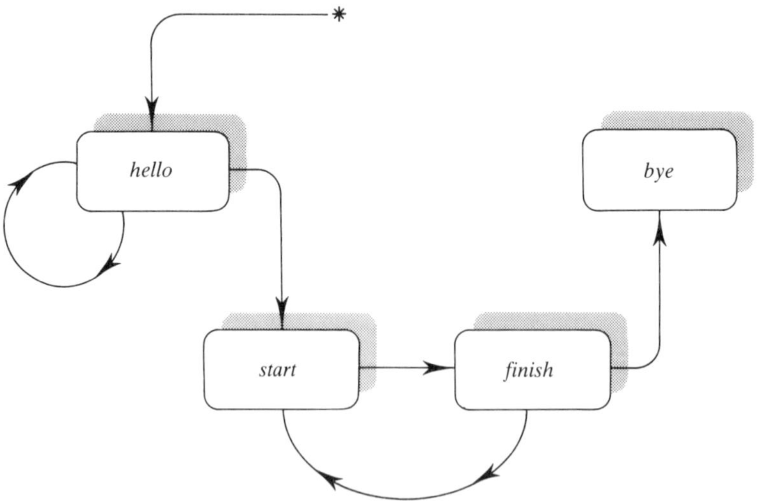

Figure 2.7 Scheduling relationships in an event-oriented model

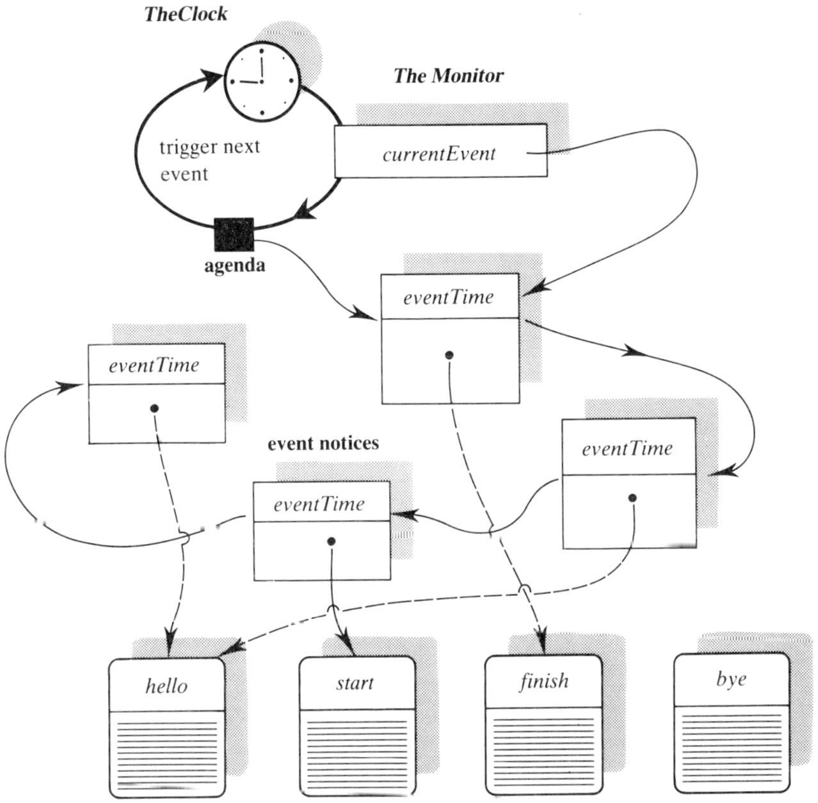

Figure 2.8 Execution structure of event-oriented models

immediately be caused by *hello* (if Manuel is idle), or can be immediately triggered by a finish event (if somebody else is waiting). There are no other ways a start event can ever occur. For this reason *hello* and *finish* form the minimal set of event routines needed. One can, however, argue that our solution, which requires four separate classes of events, leads to a conceptually cleaner structure.

Program 2.3 shows the event-oriented model in PASCAL. Apart from entity classes clock, monitor, transaction, source, server, queue and sink, we must introduce event notices. Note that in this model we chose to represent both the agenda and the queue as linked lists. Event notices and customers are truly dynamic data structures and we do not want to commit ourselves to some largely arbitrary limit on the number of events and customers our model can handle.

Event notices state the time at which an event is scheduled to occur, the class of event and any parameters the event may require (here a customer reference). Their *nextEvent* slot links them into an event list, ordered on ascending values of time. Customers are represented by transaction objects.

```
program ManuelsShop (INPUT, OUTPUT);
{the event-oriented version !}

  const
    MaxHistoClasses      =      20;
    MaxPlotPts           =      50;
    NoDelay              =     0.0;
    AlongNumber          =  MAXINT;
    PreSimulationSamples =     500;
    ExperimentSamples    =    1000;

  type
    anEventType = (hello, start, finish, bye); {each model must enumerate}
                                                {the names of its events  }
    PosIntType  = 0..MAXINT;

    {insert type declarations from Monte Carlo        toolbox}
    {insert type declarations from statistics         toolbox}
    {insert type declarations from discrete-event     toolbox}
    {insert type declarations from queueing-network   toolbox}

  var
    TheClock          : anAsynchClockType;
    TheMonitor        : anEventSchedulingMonitorType;
    Arrivals, Service : aNegExpGenType;
    Source            : aSourceType;
    Sink              : aSinkType;
    Q                 : aQueueType;
    Manuel            : aServerType;
    FlowTime          : aHistogramType;

    {insert functions & procedures from Monte Carlo        toolbox}
    {insert functions & procedures from statistics         toolbox}
    {insert functions & procedures from Discrete Event     toolbox}
    {insert functions & procedures from Queueing Network   toolbox}

  procedure ReadAndInitialize;
  {initiates the model's entities !}
    begin
      InitAsynchClock             (TheClock);
      InitEventSchedulingMonitor  (TheMonitor, AlongNumber, ExperimentSamples);
      InitNegExpGen               (Arrivals, 12345, 0.8 );
      InitNegExpGen               (Service , 67891, 1.0 );
      InitSource                  (Source               );
      InitSink                    (Sink                 );
      InitQueue                   (Q                    );
      InitServer                  (Manuel, 1            );
      InitHistogram               (FlowTime, 0, 25, 10 );
      ScheduleEvent               (TheMonitor, NoDelay, hello, nil);
    end;
```

Program 2.3 *An event-oriented analysis of Manuel's shop*

```
procedure ResetStatistics;
  begin
    WRITE      ('Statistics were reset after ', HowManySamplesQ (TheMonitor):5);
    WRITELN (' samples were taken!');
    WRITELN;
    ResetMonitorStats   (TheMonitor, AlongNumber, ExperimentSamples);
    ResetNegExpStats    (Arrivals  );
    ResetNegExpStats    (Service   );
    ResetSourceStats    (Source    );
    ResetSinkStats      (Sink      );
    ResetQueueStats     (Q         );
    ResetServerStats    (Manuel    );
    ResetHistogramStats (FlowTime);
    WRITELN;
  end;

procedure WriteReport;
{reports the results of a simulation run}
  begin
    ReportOnMonitor (TheMonitor);
    ReportOnSource  (Source      );
    ReportOnSink    (Sink        ); WRITELN;
    WRITELN ('Manuel''s performance: ');
    ReportOnServer (Manuel);          WRITELN;
    WRITELN ('Queue Statistics:');
    ReportOnQueue (Q);                WRITELN;
    WRITELN ('Customers'' Time in the Shop: ');
    ShowHistogram (FlowTime);
  end; {of WriteReport procedure}


{all EVENT ROUTINES are called EVENT prefixed by the name of the event}

procedure HelloEvent;
{creates incoming customers and puts them in a queue in front of Manuel}
  var
    romeo: aTAPtrType;
  begin
    {plan next arrival}
    ScheduleEvent (TheMonitor, NegExp (Arrivals), hello, nil);
    {create a customer}
    romeo := CreateNewTA (Source);
    {put him into the Q and schedule a start event}
    FileintoFIFOq (Q, romeo);
    ScheduleEvent (TheMonitor, NoDelay, start, nil);
  end; {of HelloEvent}

procedure StartEvent;
{starts a service activity if Manuel is idle}
  var
    romeo: aTAPtrType;
  begin
    if (EnoughServerUnitsAvailableP (Manuel,1)) and (Q.length > 0) then
```

(cont.)

```
        begin
          SeizeServer (Manuel, 1);
          {turn attention to waiting customer}
          romeo := TakeFirstFromQ (Q);
          {diagnose degree of lovesickness, mix and hand over remedy;
           accept donation}
          ScheduleEvent (TheMonitor, NegExp (Service), finish, romeo);
        end;
    end; {of StartEvent}

  procedure FinishEvent (romeo: aTAPtrType);
  {complete romeo's service activity and check whether other customers}
  {are waiting for Manuel}
    begin
      ReleaseServer (Manuel, 1);
      ScheduleEvent(TheMonitor, NoDelay, bye, romeo);
      ScheduleEvent(TheMonitor, NoDelay, start, nil);
    end; {of FinishEvent}

  procedure ByeEvent (romeo: aTAPtrType);
  {interviews romeo about queueing delays and gets rid of him}
    begin
      UpdateHistogram (FlowTime, FlowTimeQ (romeo));
      RemoveTA (Sink, romeo);
      RecordAnotherSample;
    end; {of ByeEvent}

  procedure RunModelForSamples (var TheMonitor: anEventSchedulingMonitorType);
    begin
      {Simulation loop:                                  }
      {Observe the specified number of customers depart!}

      while MoreSamplesToTakeP (TheMonitor) and MoreEventsP (TheMonitor) do
        with TheMonitor do
          begin
            GetNextEventNotice (TheMonitor);
            case KindOfEventQ (currentEvent) of
              hello  : HelloEvent;
              start  : StartEvent;
              finish : FinishEvent (TAreferenceQ (currentEvent));
              bye    : ByeEvent    (TAreferenceQ (currentEvent));
            end; {case}
            FlushOutEvent (currentEvent);
          end; {simulation loop}
    end;

  begin {***** Body of MAIN program *****}
    ReadAndInitialize;
    RunModelForSamples (TheMonitor);
    ResetStatistics;
    RunModelForSamples (TheMonitor);
    WriteReport;
  end.
```

Program 2.3 (cont.)

Their *nextInLine* attribute allows them to join queues – but only one queue at a time.

This model uses features from three of the toolboxes in Appendix 1. The stochastic components are specified by distribution objects for sampling customer arrivals and service times. One of the statistics of interest is the customers' mean flowtime. This is recorded by the sink, who interviews each customer as he or she leaves and stores this information in a tally object.

These requirements can be met by predefined objects from the Monte Carlo simulation (*aNegExpGenType*) and data collection toolboxes (*aTallyType*) together with their associated procedures. Note that a tally is the appropriate data collector to use because the only information of interest is the time each leaving customer spent in the shop; the interval between departures is of no importance. It would be taken into account by time-weighing data collectors (*measures*).

Some basic utilities for handling event-based models are provided through the discrete-event toolbox summarized in Table 2.4. We need an asynchronous clock and an event-scheduling monitor, both as globally accessible data structures.

The monitor owns an agenda of event notices and responds to scheduling messages. Event notices are created and manipulated by the *ScheduleEvent*, *GetNextEventNotice* and *FlushOutEvent* procedures. *MoreTimeToSimulateP*, *MoreEventsP* and *MoreSamplesToTakeP* are Boolean functions. They can be used to test the conditions for a simulation's termination. The *KindOfEventQ* function can be applied to an event notice to determine the type of event it encodes.

Here PASCAL's strong typing forces the toolbox to assume that the programmer has declared all event names used in a model as an enumerated type (*anEventType*).

RecordAnotherSample is a procedure that increments the monitor's *samplesTaken* slot. This could of course be done directly through an assignment statement, but our commitment to an object-oriented programming style causes us to refrain from manipulating an entity's internal attributes in such a way.

At any time we can ask about the number of completed samples, using the *HowManySamplesQ* function.

Simulation models, even relatively small ones, can quickly grow into quite complicated programs. Good design, coding and documentation are vital to cope with this complexity. However, things do have a tendency to go wrong at inconvenient times! Unfortunately, simulation programs, particularly those involving random numbers, list processing and event scheduling, can be notoriously difficult to debug. Any serious simulation tool must therefore provide as good a debugging environment as possible. This is not easy to implement in a PASCAL-based system, or one based on procedural extensions to most other general-purpose programming languages. Our toolboxes only hint at what could be provided. They offer some rudimentary features for taking snapshots of a model's state and partial traces of

Table 2.4 The event-based library

Object	: *anEventType*;	{defined by the model}
Object	: *anEventPtrType*;	{to an EventNotice}
Object	: *aTAPtrType*;	{to a (passive) TArecord}
Object	: *anEventNoticeType*;	
Methods	: **public** *KindOfEventQ (anEventPtrType): anEventType*; *TAreferenceQ (anEventPtrType): aTAPtrType*; *FlushOutEvent (anEventPtrType)*;	

Object	: *anAsynchClockType*;
Methods	: **public** *InitAsynchClock (anAsynchClockType)*; *SetClock (anAsynchClockType)*; {*in ACTIVITY_PROCS.LIB*}

Object	: *anEventListType*;
Methods	: **public** *ScheduleEvent (anEventSchedulingMonitorType, aRealDelay, anEventType, aTAPtrType)*; *GetNextEventNotice (anEventSchedulingMonitorType)*; *MoreEventsP (anEventSchedulingMonitorType): BOOLEAN*; *ShowAgenda (anEventSchedulingMonitorType)*;

Object	: *anEventSchedulingMonitorType*;
Methods	: **private** *HowManySamplesQ (anEventSchedulingMonitorType): INTEGER*; **public** *InitEventSchedulingMonitor (anEventSchedulingMonitorType, simLength, sampleSize)*; *TraceMonitor (anEventSchedulingMonitor)*; *UntraceMonitor (anEventSchedulingMonitor)*; *MoreTimeToSimulateP (anEventSchedulingMonitorType): BOOLEAN*; *MoreSamplesToTakeP (anEventSchedulingMonitorType): BOOLEAN*; *RecordAnotherSample*; *ResetMonitorStats (anEventSchedulingMonitorType, simLength, sampleSize)*; *ReportOnMonitor (anEventSchedulingMonitorType)*;

its execution. The *ShowAgenda* procedure can be called whenever the contents of the monitor's event list is of interest.

Simulations with stochastic components are sampling experiments, which must observe some basic principles of statistical design. It is usually a model's steady state that is of interest – not so much the state transformations involved in reaching it. Unfortunately it is difficult to tell just when a steady state has been reached. A few methods for model initialization and output analysis have been developed, founded on sound theoretical bases. **Regenerative simulations** [Crane and Lemoine (1977), Iglehart (1978)] and other strategies have been proposed and tested. Law and Kelton (1982) give a good overview.

Most often, however, quite pragmatic techniques are used for model initialization. A so-called **presimulation period** is a popular approach. A model is started in an empty state and run for a certain number of samples or some specified time interval – the presimulation period. After that all statistics are reset and data are then gathered during the continuation of the simulation experiment. Behind this strategy lies the assumption that the model will surely have reached a steady state if the presimulation period was long enough. Although this method has the attraction of simplicity, it is, unfortunately, very difficult to tell just how long is long enough. More formal methods with a stronger theoretical base should be used where possible to detect steady-state conditions.

The simulation toolboxes provide a number of methods to reset statistics counters (*ResetCount*, *ResetTally*, *ResetMeasure*, *ResetHistogram*, *ResetChronograph*). These procedures take the name of an appropriately typed object as arguments. It would be very convenient to have a global reset command instead of resetting all statistics counters individually. This would require that such objects were implicitly filed into a list at the time they are created, which could then be traversed, resetting all of them. In PASCAL it is not easy to have lists of objects of different types. Such a command would therefore be somewhat difficult to provide. SIMULA's class/subclass feature permits much more elegant list-processing facilities through which a global reset can be implemented. It has actually been provided for DEMOS scenarios.

The model of Manuel's shop was described in terms of terminology taken from the queueing network paradigm. A great number of real-life problems can be formulated as some kind of resource allocation or scheduling model, leading quite naturally to a queueing network representation. The queueing network toolbox offers support for such queueing scenarios. Table 2.5 shows the five predefined classes of objects: sources, sinks, transactions, servers and queues. They can represent both material and machine-oriented scenarios, although the predefined operations fit primarily a material-oriented modelling style.

This distinction between machine and material orientation goes back to job-shop simulations, one of the earliest applications of operations-research

Table 2.5 The queueing network library

Object	: *aSourceType*;
Methods :	**private** *HowManyCreationsQ* (*aSourceType*); **public** *InitSource* (*aSourceType*); *ResetSourceStats* (*aSourceType*); *CreateNewTA* (*aSourceType*); *TraceTAgeneration* (*aSourceType*); *UntraceTAgeneration* (*aSourceType*); *ReportOnSource* (*aSourceType*);

Object	: *aSinkType*;
Methods :	**private** *How ManyDeparturesQ* (*aSinkType*): *INTEGER*; **public** *InitSink* (*aSinkType*); *ResetSinkStats* (*aSinkType*); *RemoveTA* (*aSinkType*); *TraceTAdepartures* (*aSinkType*); *UntraceTAdepartures* (*aSinkType*); *ReportOnSink* (*aSinkType*);

Object	: *aTArecordType*;
Methods :	**private** *FlowTimeQ* (*aTAPtrType*): *REAL*;

models. Here jobs are routed through a network of machines requesting different operations to be performed. Routing strategies are controlled by so-called priority rules, which govern the way in which jobs are picked out of queues. The FIFO (first in, first out) rule is the simplest of a large number of such queueing disciplines. Coffman (1976) gives a good summary and analyses of job-shop scheduling.

In a **material-oriented model** the predominant entities, from whose point of view the model's behaviour is defined, are material items, commonly called transactions. Server units (machines) are passive entities, to be acquired, held and released again. Transactions route themselves from one machine to the next, following a sequence of actions defined by their life cycles.

Table 2.5 (cont.)

Object : *aServerType*;
Accessible Instance Variables : *capacity*, *state* {*idle*/*busy*};

Methods : **private**
ServerUtilizationQ (*aServerType*): *BOOLEAN*;

public
InitServer (*aServerType*, *capacity*);
ResetServerStats (*aServerType*);
SeizeServer (*aServerType*, *howMuch*);
ReleaseServer (*aServerType*, *howMuch*);
EnoughServerUnitsAvailableP (*aServerType*, *units*): *BOOLEAN*;
TraceService (*aServerType*);
UntraceService (*aServerType*);
ReportOnServer (*aServerType*);

Object : *aQueueType*;
Accessible Instance Variables : *length*;

Methods : **private**
MeanQlengthQ (*aQueueType*): *REAL*;
MaxQlengthQ (*aQueueType*): *INTEGER*;
MeanQdelayQ (*aQueueType*): *REAL*;

public
InitQueue (*aQueueType*);
ResetQueueStats (*aQueueType*);
FileIntoFIFOq (*aQueueType*, *aTAPtrType*);
TakeFirstFromQ (*aQueueType*): *aTAPtrType*;
TraceQops (*aQueueType*);
UntraceQops (*aQueueType*);
ReportOnQueue (*aQueueType*);

Machine-oriented models reverse this viewpoint. Server entities (machines) are now seen as active and dominating. They provide a modelling framework in which transactions are obtained, operated on, put into queues and finally removed. It is a good exercise to augment our toolbox with procedures to support machine-oriented simulations.

Both sources and sinks are needed for open queueing networks. Source objects are used to create and initialize a transaction's data structures. Note that, in line with what was said earlier about the importance of good debugging support, simple tracing procedures (*Trace*..., *Untrace*...) are included for all the predefined entity classes. The only statistic kept by sources is a creation count, which can be queried through the function *HowManyCreationsQ*. Sink objects are used to remove transactions from a model. A departure count is kept and reported through *HowManyDeparturesQ*.

At any time transactions can be queried about the length of time they have spent in a model (*FlowTimeQ*). This feature is typically used to record a transaction's model residence time immediately prior to its departure.

Server entities must be initialized to a given capacity. Single servers have a capacity of one. They can be seized if at least the requested number of capacity units are available. They must be released again at the end of a service. There is a function that checks whether there are enough server units available to fill a request (*EnoughServerUnitsAvailableP*). The main statistic kept for servers is their utilization. *ServerUtilizationQ* is used to compute it.

The queue data type in this toolbox is implemented by a linked list and procedures for treating it as a FIFO queue are provided. PASCAL's strong typing gets somewhat in the way again, forcing a commitment to the type of potential queue members. Queues are therefore solely transaction queues, and only one class of transaction is admitted into the models. As already mentioned, transactions have only one predefined link field, which prohibits simultaneous membership in more than one queue. By patching the toolbox's declarations and procedures these restrictions can quite easily be removed by more complicated models.

Measures of interest are typically a queue's mean and maximum lengths as well as the average time items spent in a waiting state. Summaries for any of these classes of objects can be generated by calls on appropriately named procedures (*ReportOn*. . .).

The facilities provided by the toolboxes make the event-oriented PASCAL program itself (Program 2.3) quite straightforward. The necessary entities (clock, monitor, arrival and service distributions, source, sink, Manuel, queue, histogram) are declared and instantiated. A monitor is set up to simulate the departure of 1000 transactions, with a presimulation sample of 200. The distribution objects are supplied with seeds and the appropriate mean rates (!), and Manuel's diagnostic capacity is set to one customer at a time. There are four event routines, one for each class of event.

Each time it is scheduled, *hello* creates a new customer, schedules an incarnation of itself to occur some time in the future and files the customer into the queue. It also attempts to start a service, by scheduling a start event. A start event can be triggered only if Manuel is idle and if there is at least one customer waiting for him. If a service activity can start, Manuel is seized, the first customer is taken out of the queue and a finish event is scheduled to take place after the number of time units needed for the service. This is predicted by sampling the *service* distribution. *finish* releases Manuel and, since other customers may be waiting for him, schedules an attempt to start a new service immediately. This will of course be successful only if there are any customers waiting. Otherwise Manuel will idle lazily in the sun until a customer arrives. The *hello* event will make sure he is eventually scheduled again.

bye causes customers to disappear from the shop, using the sink. Before this happens the histogram of transactions' flowtimes is updated and the completion of another sample is noted.

The procedure *RunModelForSamples*... and the MAIN program again exhibit a typical simulation loop:

while *MoreSamplesToTakeP* (*TheMonitor*) **and**
    MoreEventsP (*TheMonitor*) **do**
    {*or MoreTimeToSimulateP*}
  (1) *GetNextEventNotice* (*TheMonitor*);
  (2) **case** *KindOfEventQ* (*TheMonitor.currentEvent*) **of**
      ...
      {*a case list of event names associated with calls on event procedures*}
      ...
    end;
  (3) *FlushOutEvent* (*TheMonitor.currentEvent*);

Within this loop the monitor obtains the next event notice and removes it from the head of the agenda. The kind of event scheduled by the current event notice is determined and the appropriate procedure is executed. The event notice can then be destroyed.

This loop is guarded by the number of samples and the length of the monitor's agenda. The simulation therefore stops if enough customers have left the shop, or if there is nothing more to do. In our model the event list can never get empty because each time a *hello* event occurs it schedules its own reappearance.

Note that the *finish* and *bye* events need to know the identity of the customers they are serving. Since PASCAL procedures cannot remember any of their local variables from one invocation to the next, a customer reference must be communicated through a parameter of this event's event notice. In event-based models this kind of information must frequently be passed between events. Event-notice entities therefore have a predefined slot (*thisTA*) in which it can be saved. *ResetStatistics* is used to flush all statistics after a presimulation period, and procedure *WriteReport* reports the number of samples, queue statistics, and the average time customers spent in the shop.

Event-based simulations are quite straightforward to implement in a general-purpose programming language, although appropriate data types like records and lists permit much cleaner model and program structures. A major disadvantage of event-oriented programming styles is that the model's logic tends to be fragmented, with many scheduling commands scattered throughout the program. This stems from the inability to include time delays in event procedures. Passage of time can be modelled only through scheduling. The resulting fragmentation of control makes models hard to read and difficult to understand, particularly if they grow large, with many event routines and many potential interactions between them. Debugging complex event-based models can be a nightmare, even if good facilities for snapshots and traces are provided.

B. Process classes

Fragmentation of model logic can be avoided by using so-called life-cycle descriptions. Causal relationships between sequences of events are now emphasized in place of simultaneous state changes.

Patterns of actions that are logically connected in the mind of an analyst can be grouped into a single piece of code: a process class. This constrains complexity, localizing the dynamic aspects of a system under study. The model's logic becomes much easier to follow, at the expense of a somewhat more complicated implementation. Process-oriented frameworks have proved especially appropriate for modelling complex, highly interactive systems. SIMPL/1 [IBM (1972)], ASPOL [McDougall (1975)] and GPSS [Gordon (1979)] are representative for some of the languages supporting this view.

Figure 2.9 shows a process-based implementation of the scenario. It is still material oriented, with model dynamics described from a customer's viewpoint. Although the model contains only one class of process, this

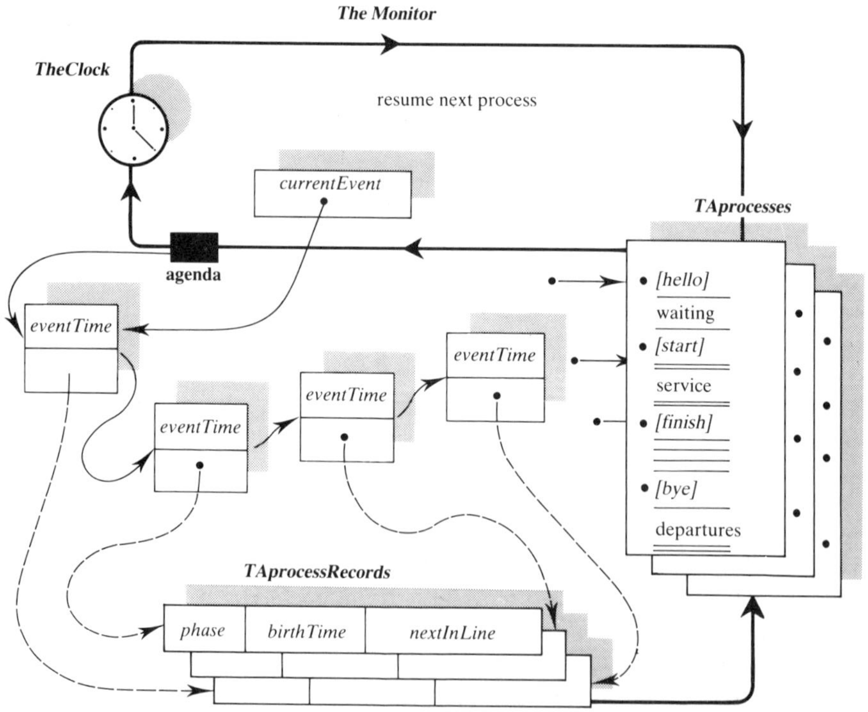

Figure 2.9 A process-based representation and its run-time control structures

Table 2.6 The process-based library

Object	: *aTAprocessClass* (*self*: *aTAPtrType*); {*FORWARD declaration*}
Object	: *aTAPtrType*; {*to a TAprocessType*}
Object	: *aTAprocessType*;
Methods :	public *CreateAndStartTAprocess* (*anInitEventPhase*, *aDelay*); *FlowToNextEvent* (*aTAPtrType*, *aDelay*); *ResumeNextTAprocess*;

approach can easily be generalized to more than one; coexisting, co-operating and competing for resources.

Since transactions are pointer-based structures we can have any number of customer instances.

Process-oriented frames group all actions of a class of entities into one module; here it is a *TAprocessClass* procedure. In this scenario a customer is both a data structure and a sequence of actions performed by that customer. A customer can be in any of four different phases: *hello*, *start*, *finish*, *bye*. These are phases in the life cycle of a process at which it interacts with the executive and other processes.

Much of the earlier discussion of event-based models is still valid. Table 2.6 shows the necessary modifications to the discrete-event toolbox. A singly linked list of event notices is again used as an agenda. Instead of having a reference to an event routine, event notices must now point to a process; and the global data structure defining the state of the process must remember its next re-activation point. The monitor cycle is also quite similar:

while *MoreSamplesToTakeP* (*TheMonitor*) **and**
MoreEventsP (*TheMonitor*) **do**
{*or MoreTimeToSimulateP* (*TheMonitor*)}
(1) *ResumeNextTAProcess*;

The PASCAL code in Program 2.4 shows this structure clearly. *start* and *bye* phases could be bound to *hello* and *finish*, yielding processes with only two entry points. Procedure *TAprocessClass* describes a customer's life cycle. *self* is a reference to a process instance's local data structures. Unfortunately a transaction's state cannot be included in the same module as its behaviour pattern, since PASCAL procedures cannot remember data from one invocation to the next. This information has to be stored in a separate activation record of *TAprocessType*.

The body of each *TAprocess* is a case statement, simulating a multi-entry procedure. The local sequence control (LSC), which indicates the current phase of a process, is stored in the *phase* slot of an activation record and can be accessed via a link between the process (life cycle) and its state (*self*).

```
program ManuelsShop (INPUT, OUTPUT);
{the process-oriented version !}

  const
    MaxHistoClasses =        10;
    MaxPlotPts      =        50;
    NoDelay         =       0.0;
    AlongNumber     = MAXINT;

  type
    anEventType = (hello, start, finish, bye);

    {insert type declarations from Monte Carlo          toolbox}
    {insert type declarations from statistics           toolbox}
    {insert type declarations from discrete-event       toolbox  -
     replacing aTArecordType with aTAprocessType}
    {insert type declarations from process-interaction  toolbox}
    {insert type declarations from queueing-network     toolbox}

  var
    LastPhase, FirstPhase : anEventType;
    TheClock              : anAsynchClockType;
    TheMonitor            : anEventSchedulingMonitorType;
    Arrivals, Service     : aNegExpGenType;
    Source                : aSourceType;
    Sink                  : aSinkType;
    Manuel                : aServerType;
    Q                     : aQueueType;
    FlowTime              : aHistogramType;

    {insert functions & procedures from Monte Carlo          toolbox}
    {insert functions & procedures from statistics           toolbox}
    {insert functions & procedures from discrete-event       toolbox}
    {insert functions & procedures from process-interaction  toolbox}
    {insert functions & procedures from queueing-network     toolbox}

  procedure ReadAndInitialize;
  {instantiates the model's entities}
    begin
      FirstPhase := hello;
      LastPhase  := bye;
      InitAsynchClock              (TheClock);
      InitEventSchedulingMonitor   (TheMonitor, aLongNumber, 1000);
      InitNegExpGen                (Arrivals, 12345, 0.8);
      InitNegExpGen                (Service , 67891, 1.0);
      InitSource                   (Source);
      InitSink                     (Sink);
      InitQueue                    (Q);
      InitServer                   (Manuel, 1);
      InitHistogram                (FlowTime, 0, 25, 10 );
      CreateAndStartTAprocess      (FirstPhase, NoDelay);
    end; {ReadAndInitialize procedure}
```

Program 2.4 *A process-based analysis of Manuel's shop*

```
procedure WriteReport;
{reports the results of a simulation run}
  begin
    ReportOnMonitor (TheMonitor);                          WRITELN;
    WRITELN ('Report on RN-generators ');
    WRITELN (' ============= ');                           WRITELN;
    ReportOnNegExpGen (Arrivals);
    ReportOnNegExpGen (Service );                          WRITELN;
    ReportOnSource       (Source);
    ReportOnSink         (Sink   );                        WRITELN;
    WRITELN ('Report on Manuel');
    WRITELN (' +++++++++ ');                               WRITELN;
    ReportOnServer (Manuel);                               WRITELN;
    ReportOnQueue(Q);                                      WRITELN;
    WRITELN ('The time customers spent in the shop: ');
    WRITELN (' ++++++++++++++++++++  ');                   WRITELN;
    ShowHistogram (FlowTime);
  end; {of WriteReport procedure}

procedure TAprocessClass; {self is aTAPtrType parameter; see forward}
                          {declaration in PROCESS_PROCS.LID      }

  procedure nextTriesToGrabServer (var q: aQueueType; var server:
                                   aServerType);
    var
      romeo: aTAPtrType;
    begin
      if (q.length > 0) and (EnoughServerUnitsAvailableP (server,1)) then
        begin
          romeo := TakeFirstFromQ (q);
          SeizeServer (server, 1);
          FlowToNextEvent (romeo, NoDelay);
        end; {if}
    end; {local nextTriesToGrabServer procedure}

  begin
    case self↑.phase of
      hello:
        begin
          {plan arrival of the next customer}
          CreateAndStartTAprocess (FirstPhase, NegExp (Arrivals));
          {queue for Manuel}
          FileIntoFIFOq (Q, self);
          nextTriesToGrabServer (Q, Manuel);
        end;
      start:
        {service activity has been successfully started}
        FlowToNextEvent (self, NegExp (Service));
      finish:
        begin
          {finish service activity}
          ReleaseServer (Manuel, 1);
```

(cont.)

```
            {signal Manuel's availability to others waiting in Q}
            NextTriesToGrabServer (Q, Manuel);
            FlowToNextEvent (self, NoDelay);
         end;
      bye:
         begin
            {record statistics}
            UpdateHistogram (FlowTime, FlowTimeQ (self));
            {remove customer from the shop}
            RemoveTA (Sink, self);
            RecordAnotherSample;
         end;
   end; {case}
end; {TAprocessClass procedure}

begin { ***** MAIN program *****}

   ReadAndInitialize;

   {Simulation loop: observe the specified number of customers depart}

   while MoreSamplesToTakeP (TheMonitor) and MoreEventsP
   (TheMonitor) do
      ResumeNextTAprocess;

   WriteReport;

end.
```

Program 2.4 (cont.)

Each customer steps through the same sequence of actions as in the event-based model. Upon arrival he or she schedules creation of a successor and queues for Manuel's services. Procedure *nextTriesToGrabServer* seizes Manuel, if he is available, and advances the customer's state. The customer then passes through the *start* state, advancing to the *finish* state after a delay of some randomly sampled time units. In the *finish* state Manuel is released. This is signalled to waiting customers by the *nextTriesToGrabServer* procedure. The final state is *bye*, where a histogram is updated by this customer's flowtime, the sample count is incremented and the process departs through the sink.

Procedure *FlowToNextEvent* is used to sequence processes, advancing them to their next state. It uses the *ScheduleEvent* procedure in the discrete-event simulation toolbox to create an appropriate event notice and file it into the monitor's agenda. For this it needs two parameters: a reference to the process state (*self*) and the requested delay between the current state and the next one.

Using this framework in PASCAL is somewhat clumsy, because a pseudo-coroutine structure must be simulated. Process-oriented modelling styles are best supported by languages that provide conceptual concurrency (such as MODULA, SIMULA). PASCAL-based models with process orientation must specify the names of the first and last phases of processes and enumerate all of them as *anEventType*. This information is used by *FlowToNextEvent* to check and correctly update the state of the process. (*FirstPhase* is not strictly required, but is included for reasons of symmetry.)

A new *aTAPtrType* replaces the definition given for event-based models. It must now refer to the activation record of a *TAprocess*, which is therefore used in place of the *TArecord* type. The monitor remains the same, since the basic principles of scheduling through event notices still apply. Only the scheduling procedures have to be changed.

A *TAprocess* is created by allocating and initializing an activation record and scheduling the event associated with the first phase of its execution. Procedure *CreateAndStartTAprocess* takes care of this. Procedure *FlowToNextEvent* has already been mentioned. After advancing the state of the process it schedules the next process phase. Procedure *ResumeNextTAProcess* is now used by the monitor in place of *GetNextEventNotice*. It obtains the event notice at the head of the agenda and re-activates the associated process.

This simple framework can be modified to cope with more than one class of processes. This is left as an exercise for the reader.

Program 2.4 again declares clock, monitor, distribution, source, sink, server, queue and histogram entities, initiated by the procedure *ReadAndInitialize*. The *TAprocessClass* is defined as described above.

The simulation loop steps through the agenda, advancing states and executing all processes for which active phases are scheduled. The clock is advanced in the classical next-event fashion – always to the time of the next imminent event.

The data structure for event notices is still the same as in the event-based approach. There are slots for the time of an event and the transaction involved in it. The 'class of event' slot now holds the current phase of the process representing this transaction's life cycle. It is stored as part of the transaction's state and automatically advanced by the *FlowToNextEvent* procedure. This procedure must be used to schedule process phases, not only across processes but also from one process phase to the next. The necessity of explicit sequencing commands within a process is part of the price to be paid for PASCAL's lack of a coroutine module.

Object orientation [Robson (1981), Rentsch (1982)] is a variant of process-based models which extends this framework. It integrates descriptions of data structures, procedures and process dynamics characterizing some class of objects into a single textual module. The basic idea was pioneered by SIMULA and is now widely recognized as a powerful vehicle for building sophisticated programming systems.

Languages like SMALLTALK, ACT, GLISP and others describe all computations as interactions between classes of objects. The object-oriented approach is particularly attractive if property inheritance is catered for. This permits easy, conceptually clean extensions to highly problem-oriented tools. DEMOS and DISCO (see Chapter 3) are good examples of the power of this approach.

Event- and process-based world views have been in the mainstream of simulation programming for some decades, with the advantages of process-oriented sequencing only recently becoming more widely recognized.

Implementation and efficiency problems are less important than they were some years ago. Most implementors of new or new versions of discrete-event simulation languages strive to offer process-oriented frameworks. GPSS, SIMULA, SIMPL/1 and ASPOL always were process-based systems; GASP has led to SLAM and SIMSCRIPT has been augmented by adding processes and resources. The process style of simulation will eventually replace other world views, as soon as newer programming languages, supporting some kind of genuine or pseudo- (coroutine) concurrency are used to host simulation environments. MODULA, EDISON, ADA, SMALLTALK and others are among this new generation.

C. Activity procedures

There is a third category of programming style, which lies somewhat outside the mainstream of simulation programming. These are so-called activity-oriented models. While activity-oriented languages like CSL and SIMON have been quite widely used in England since the early 1960s, this world view never gained any significant foothold in other countries. Unlike event-based models, activity orientation offers the convenience of not having to worry about explicit interactions between events.

Activity descriptions try to capture the notion of causally connected start and finish events, clustering descriptions of state transitions at the start and finish of some time-consuming activity. Instead of having to schedule them procedurally from wherever in the model they may be triggered, the programmer may declare conditions under which such clusters of changes will occur. Activities are therefore associated with a set of conditions, stating when an activity of a given class may start, and a time cell that specifies when it may finish. This approach is called **conditional sequencing** because the sequence of transformations that such a description gives rise to is determined by repetitive scans of temporal and other conditions.

Control is usually implemented by event-driven inspection of indicators (time cells and conditions). A model executive cycles through all activities, asking whether they are able to start or finish. Willing activities are executed. If no further activity can occur at this time instant, the clock is advanced to the time of the next imminent event and the whole cycle is repeated. ECSL [Clementson (1966)] and SIMON [Hills (1965)] are among the programming systems based on this idea.

Program 2.5 views Manuel's shop through an activity-oriented scenario. There are two classes of activities – *arrival* and *service* – which move transactions between states (waiting to enter the shop, entering the shop, waiting to be served, being served). In most activity-oriented systems, which are commonly based on FORTRAN, arrays are used to implement queues and agenda. This forces a binding of the maximum number of activity instances and queue members at compile time.

Our model uses linked lists provided by the discrete-event toolbox. In line with most activity-based systems a closed queueing network viewpoint is adopted and customer records (transactions) are moved between waiting states associated with the two activities (*arrival*: outside, waiting to enter the shop; *service*: in the queue, waiting to be served). Since customer records are now permanent, rather than transient parts of the model, we must keep a fixed number (*MaxCustomers*) of frames around. When not in use they are put *outside*.

It is also assumed that the average number of customers in the shop is to be measured. This is an appropriate application for the measure data type from the statistics toolbox. Measures keep a time integral of some numerically encoded state, weighing each value by the time period through which it endured. The state to be observed is the number of customers in the shop. It is therefore necessary to update the measure each time this value changes, either through a customer arrival (+1) or through a departure (−1). At the end of a simulation experiment the average will be reported.

An arrival activity, when started, creates a customer by taking an empty frame from outside, updates the measure of customers in the shop and puts the customer at the tail of the queue. It then causes the arrival activity to delay itself for some time, whereafter it can be started again.

A service activity is started whenever both an idle server and waiting customers exist, and finishes after a service period, putting the frames of leaving customers back outside.

An overview of the model's structure is given by Figure 2.10. This kind of representation is called an **activity** or **life-cycle diagram** [Hutchinson (1975a)]. It symbolizes states as circles and activities by boxes.

Initially there are 100 customer frames waiting outside. Manuel is idle and the queue is empty. A customer frame must be available to start an arrival activity. The finishing time of the arriving customer is sampled from an exponential distribution. In order to start, a service activity needs an idle Manuel and a non-empty queue. Service time is also exponential.

For this model we use the discrete-event toolbox modified and augmented by a number of specialized features listed in Table 2.7. The event-scheduling monitor is now replaced by an activity scanning monitor, implementing the **three-phase structure**. Apart from the usual attributes of a monitor, this entity owns a linked list of activity notices (agenda). Note that PASCAL does not permit us to link activities directly into a list. They can only be represented by procedures, and pointer-based procedure references are not possible. Therefore activity notices must be introduced, preserving the symmetry with the event-scheduling approach.

In contrast to the event-based strategy only one set of activity notices will be created at the start of a simulation. In our simple example we need two classes of activities (*arrival* and *service*) and there is only one instance of

```
program ManuelsShop (INPUT, OUTPUT);
{the activity-oriented version !}

  const
    MaxHistoClasses =        10;
    MaxPlotPts      =       100;
    MaxCustomers    =       100;
    AlongNumber     = MAXINT;

  type
    {we need a dummy "anEventType" to compile the discrete-event toolbox}
    anEventType    = (startArrival, finishArrival, startService, finishService);
    anActivityType = (arrival, service);

    {insert type declarations from Monte Carlo        toolbox}
    {insert type declarations from statistics         toolbox}
    {insert type declarations from discrete-event     toolbox}
    {insert type declarations from activity-scan      toolbox}
    {insert type declarations from queueing-network   toolbox}

  var
    Romeo                   : aTAPtrType;
    TheClock                : anAsynchClockType;
    TheMonitor              : anActivityScanMonitorType;
    ArrivalDist, ServiceDist : aNegExpGenType;
    Outside, Q              : aQueueType;
    Manuel                  : aServerType;
    NoOfCustInShop          : aMeasureType;

    {insert functions & procedures from Monte Carlo        toolbox}
    {insert functions & procedures from statistics         toolbox}
    {insert functions & procedures from discrete-event     toolbox}
    {insert functions & procedures from activity-scan      toolbox}
    {insert functions & procedures from queueing-network   toolbox}

  procedure ReadAndInitialize;
  {instantiate the model's entities}
    var
      index: 1..MaxCustomers;
    begin
      InitAsynchClock          (TheClock);
      InitActivityScanMonitor  (TheMonitor, AlongNumber, 1000);
      InitNegExpGen            (ArrivalDist, 12345, 0.8);
      InitNegExpGen            (ServiceDist, 67891, 1.0);
      InitQueue                (OutSide);
      InitQueue                (Q);
      InitServer               (Manuel, 1);
      InitMeasure              (NoOfCustInShop);
      {Let <MaxCustomers > 'empty' TA-frames wait outside the shop}
      for index := 1 to MaxCustomers do
        begin
          new (Romeo);
          FileIntoFIFOq (Outside, Romeo);
        end;
```

Program 2.5 *An activity-based analysis of Manuel's shop*

```
    InitAndStartActivity (arrival);
    InitAndStartActivity (service);
  end; {of ReadAndInitialize procedure}

procedure WriteReport;
{reports the results of a simulation run}
  begin
    WRITE     ('RN-generators:');
    WRITELN (' +++++++ ');                WRITELN;
    ReportOnNegExpGen (ArrivalDist);      WRITELN;
    ReportOnNegExpGen (ServiceDist);      WRITELN;
    WRITE     ('Manuel''s Statistics:');
    WRITELN (' +++++++++ |  ');           WRITELN;
    ReportOnServer       (Manuel);        WRITELN;
    WRITE     ('Queue Statistics ');
    WRITELN ('(Outside and Q) :');        WRITELN;
    ReportOnQueue (Outside);              WRITELN;
    ReportOnQueue (Q);                    WRITELN;
    WRITE     ('No. of Cust. in Shop:');
    WRITELN (' ********************* '), WRITELN;
    ReportOnMeasure      (NoOfCustInShop);
  end; {of WriteReport procedure}

procedure ArrivalActivityClass (self: anActivityPtrType;
                                state: anActivityPhaseType);
{enters arriving customers; sleeps between arrivals}
  var
    romeo: aTAPtrType;
  begin
    case state of
      start:
        begin
          {let a customer in}
          romeo := TakeFirstFromQ (Outside);
          UpdateMeasure (NoOfCustInShop, +1);
          FileIntoFIFOq (Q, romeo);
          SetFinishTime (self, NegExp (ArrivalDist), nil);
        end;
      finish: ;
    end; {case}
  end; {ArrivalActivityClass procedure}

procedure ServiceActivityClass (self: anActivityPtrType;
                                state: anActivityPhaseType);
{Processes waiting customers}
  var
    romeo: aTAPtrType;
  begin
    case state of
      start:
        begin
          {turn attention to waiting customer}
          romeo := TakeFirstFromQ (Q);
```

(cont.)

```
            self↑.thisTa := romeo;
            SeizeServer (Manuel, 1);
            {diagnose, mix and hand over potion}
            SetFinishTime (self, NegExp (ServiceDist), romeo);
          end;
        finish:
          begin
            ReleaseServer (Manuel, 1);
            romeo := self↑.thisTA;
            {record a customer's experiences}
            UpdateMeasure (NoOfCustInShop, -1);
            RecordAnotherSample;
            {customer departs; put "empty" frame back outside}
            FileIntoFIFOq (Outside, romeo);
          end;
      end; {case}
    end; {ServiceActivityClass procedure}

  procedure StartActivities;
    var
      nextActivity: anActivityPtrType;
    begin
      while not EndOfAgendaP do
        begin
          nextActivity := GetNextActivityQ (TheMonitor);
          if (PhaseOfActivityQ (nextActivity) = start) and
            (OutSide.length > 0)   then
            begin
              if (ClassOfActivityQ (nextActivity) = arrival) then
                begin
                  ActivityStarted;
                  ArrivalActivityClass (nextActivity, start);
                  AdvanceActivityPhase (nextActivity);
                end; {if}
              if (ClassOfActivityQ (nextActivity) = service) then
                if (Q.length > 0) and (EnoughServerUnitsAvailableP (Manuel, 1))
                then
                  begin
                    ActivityStarted;
                    ServiceActivityClass (nextActivity, start);
                    AdvanceActivityPhase (nextActivity);
                  end; {inner if}
            end; {outer if}
        AdvanceActivityScan;
        end; {while}
    end; {procedure StartActivities}

  procedure FinishActivities;
    var
      nextActivity: anActivityPtrType;
    begin
      ResetActivityScan;
```

Program 2.5 (cont.)

```
      while not EndOfAgendaP do
        begin
        nextActivity := GetNextActivityQ (TheMonitor);
          if (PhaseOfActivityQ       (nextActivity) = finish) and
            (FinishTimeOfActivityQ (nextActivity) = TheClock.NOW) then
            begin
              if (ClassOfActivityQ (nextActivity) = arrival) then
                begin
                  ArrivalActivityClass    (nextActivity, finish);
                  AdvanceActivityPhase (nextActivity);
                end; {if}
              if (ClassOfActivityQ (nextActivity) = service) then
                begin
                  ServiceActivityClass    (nextActivity, finish);
                  AdvanceActivityPhase (nextActivity);
                end; {inner if}
            end; {outer if}
          AdvanceActivityScan;
        end; {while}
  end; {procedure FinishActivities}

begin {***** MAIN program*****}

  ReadAndInitialize;

  {Simulation loop: observe the specified number of departing customers}

  while StillMoreSamplesP (TheMonitor) do
    begin
      {PHASE I :try to start all activities which can proceed NOW !}
      repeat
        {reset scan to start of activity list}
        StartNewScan;
        {start any activity which can go NOW}
        StartActivities;
      until ScanWithoutAnyStartsP;

      {PHASE II :update "The.Clock" to soonest finishing time!}
      SetClock (TheClock);

      {PHASE III :trigger all activities which finish at this time      !}

      { NOTE: 1)   finishing can only depend on the clock time.     }
      { ++++ 2)   there will be at least one activity able to finish }
      {           at this time(from phase II)                        }

      FinishActivities;

      {and round and round and round...and roundabout,we go!}

    end; {of simulation loop}

  WriteReport;

end.
```

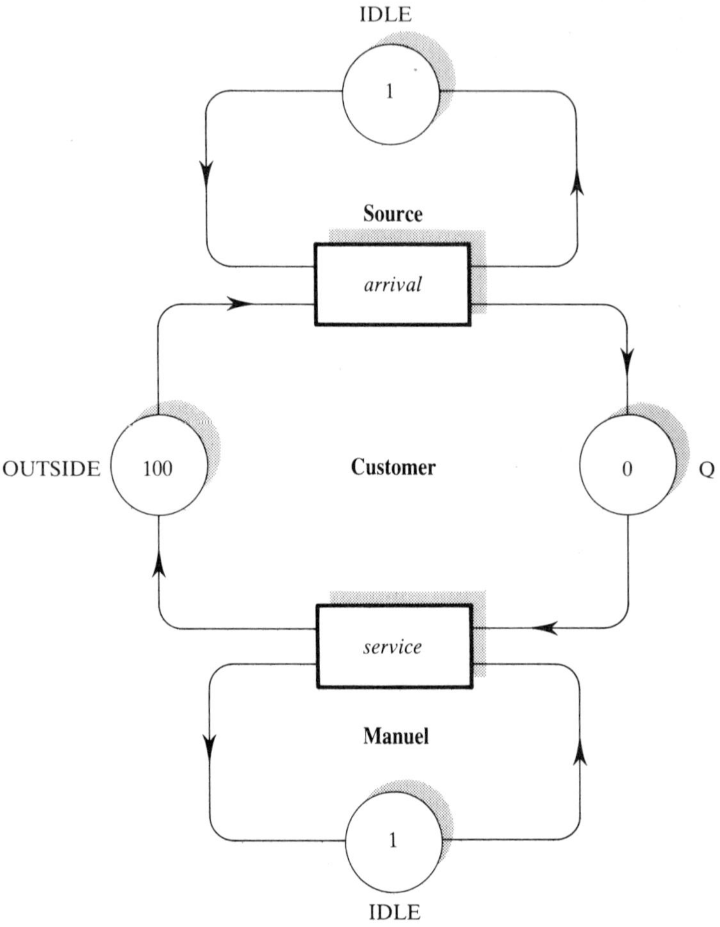

Figure 2.10 Activity cycles in Manuel's shop

each of them. It would be quite easy to permit dynamic creation and deletion of activity routines. Writing the necessary procedures is a good exercise.

Activity notices store information about the identity of the activity they denote, its class, the phase it will be in when next triggered (start or finish) and the time at which it will finish. If an activity has not yet started, a long number should be stored in the finish-time slot to prevent the monitor from trying to finish it. There is also a slot for passing a reference to the transaction currently processed by this activity. It is used to remember a customer's identity from start to finish. Note also that it would be appropriate to store a description of starting conditions here.

Unfortunately PASCAL permits neither unevaluated expressions nor function references as fields of record structures. All our activity-based

Table 2.7 The activity-based library

Object : *anActivityType*; {*list of activity classes in the model*}
Object : *anActivityPhaseType*; {*start and finish*}
Object : *anActivityPtrType*; {*to an activitynotice*}
Object : *anActivityNoticeType*;

Methods : **public**
InitAndStartActivity (*anActivityType*);
AdvanceActivityPhase (*anActivityPtrType*);
SetFinishTime (*anActivityPtrType*, *aDelay*, *aTAptrType*);
PhaseOfActivityQ (*anActivityPtrType*): *anActivityPhaseType*;
ClassOfActivityQ (*anActivityPtrType*): *anActivityType*;
FinishTimeOfActivityQ (*anActivityPtrType*): *REAL*;

Object : *anActivityListType*;

Methods : **public**
EndOfAgendaP: *BOOLEAN*;
GetNextActivityQ (*anActivityScanMonitorType*): *anActivityPtrType*;

Object : *anActivityScanMonitorType*;

Methods : **public**
InitActivityScanMonitor (*anActivityScanMonitorType*, *simLength*, *sampleSize*);
AdvanceActivityScan;
ResetActivityScan;
ActivityStarted;
StartNewScan;
ScanWithoutAnyStartsP: *BOOLEAN*;
StillMoreSamplesP (*anActivityScanMonitorType*): *BOOLEAN*;
StillMoreTimeP (*anActivityScanMonitorType*): *BOOLEAN*;

---

models must therefore contain a *StartActivities* procedure, with 'in-line' code for testing all activities' starting conditions. Programs should also supply a similar *FinishActivities* procedure.

The toolbox extension described by Table 2.7 offers a number of pre-defined utilities, which simplify the task of setting up activity-based models. There are methods for instantiation and initialization of activity instances and the activity-scan monitor.

The reason for having a monitor different from the event-scheduling one can again be traced back to PASCAL's strong typing rules. The agenda has to be an activity list instead of an event list, and the current activity is a pointer to an activity instead of an event. Other than that, the two monitors

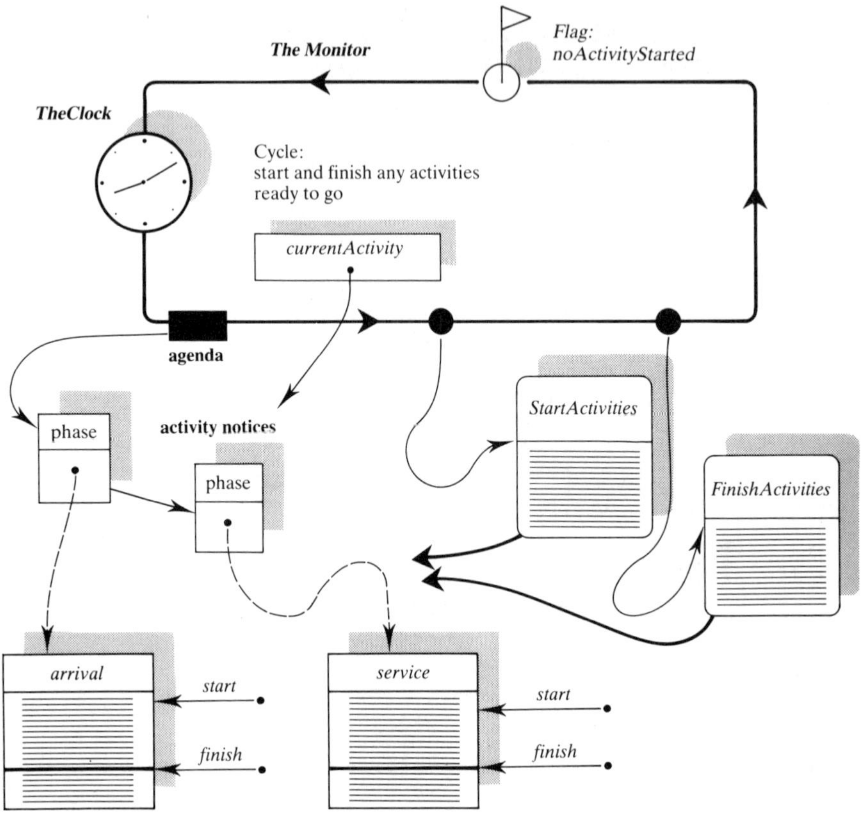

Figure 2.11 Execution structure of activity-oriented models

are identical. A generator steps through an agenda of activities. It also causes a side effect, advancing the global monitor's *currentActivity* pointer. Once the agenda is exhausted it returns **false**, setting *Current Activity* to **nil**.

There is also a procedure to set the finish-time slot of an activity and to advance the clock to the smallest finish time across all activities on the agenda.

Figure 2.11 summarizes the model's structure. It is assumed that an activity's start may depend on any combination of conditions, temporal or otherwise. Finishing, on the other hand, may depend only on model time. Since activities may be influenced by each other's actions, the scan of start conditions has to be repeated until no activity is able to start throughout a whole cycle (*ScanWithoutAnyStartsP*). Otherwise, for example, the starting conditions of a service activity waiting for a customer may not be tried again, even though the finish phase of another service may have produced a state that could now trigger it.

In this simple model such repeated scans will never be strictly necessary.

There is only one instance of each activity and the ordering of activity startups (arrivals before service) in the *StartActivities* procedure ensures that a single scan is 'safe'. Cyclic scans may, however, be necessary in the more general case, where complex interactions in the model may prohibit such a simple solution.

It also runs against the grain of an activity-oriented world view to force the programmer to consider side effects of the order in which startup routines trigger activities. The programmer must not be expected to give clues about activities' possible causal connections and independencies, even though this means more work for the monitor during model interpretation.

Procedure *StartNewScan* sets the *noActivityStarted* flag of the monitor, and function *ScanWithoutAnyStartsP* can be used to test its status.

The program for this activity-based scenario (Program 2.5) is again more complicated than the event-based one. This should, as for the process-oriented program, not be interpreted as a property inherent in that world view, but rather as a reflection on the limitations of PASCAL as a simulation tool. Like most older general purpose programming languages it is really well equipped only to deal with event-oriented simulations.

The constant *MaxCustomers* defines the number of customer frames initially waiting outside Manuel's shop. The necessity for such a limit is caused by the decision to use a closed queueing network. As noted above, *AlongNumber* ensures that only activities that have already started will be tested during the monitor's finishing cycle.

The now familiar classes of entities are declared, instantiated and initialized. Note that procedure *ReadAndInitialize* has to fill the *outside* queue with 100 customer frames. The rest of the model defines an *arrival* and *service* activity, as well as the mandatory *StartActivity* and *FinishActivity* procedures.

When an *ArrivalActivityClass* is started, it generates customer arrivals by taking empty customer records from the list *outside*, filing them into the queue in front of Manuel. It also updates the measure of customers in the shop, reporting the change in state. The *SetFinishTime* procedure is used to set the activity's finish time to the time of the next arrival, delayed by a sample from the *arrivals* distribution. The case label associated with the finish phase is empty; thus, no actions are performed. Its only effect is to prepare that activity for restart.

The start phase of the *ServiceActivityClass* procedure turns attention to a waiting customer by taking him or her out of the queue and seizing Manuel. The customer's finish time is set to the current clock time plus some time delay supplied by the service distribution. When the *ServiceActivityClass* finishes, it releases Manuel and records the customer's departure. A customer departure is accomplished by filing the transaction into the *outside* list, ready for reuse by a new customer. The change in the number of customers in Manuel's shop again leads to an update of the *NoOfCustInShop* measure and, via procedure *RecordAnotherSample*, to an incrementation of the monitor's *samplesTaken* slot.

A procedure must also be written to test which activity can start under which conditions. An arrival activity can always be started if *start* is the value of its *nextPhase* slot. No other condition controls the triggering of this activity. The start of a service requires at least one waiting customer and an idle Manuel.

Procedure *StartActivities* also advances the next state of a started activity from *start* to *finish* and signals this by calling the *ActivityStarted* procedure.

Procedure *FinishActivities* tests the time slots of all activities on the agenda whose next state will be a finish phase. If any can finish at this clock time, the appropriate block of code is executed and its next phase is set to *start*.

The main simulation loop is again guarded by the number of samples. It implements a strategy known as the **three-phase structure**. This strategy, truc to its name, goes through three distinct phases for each clock update.

while *StillMoreSamplesP* (*TheMonitor*) **do**
  {*or StillMoreTimeP* (*TheMonitor*)}
  (1) Phase I: **repeat**
      (1.1) *StartNewScan*;
      (1.2) *StartActivities*
    until *ScanWithoutAnyStartsP*;
  (2) Phase II: *SetClock* (*TheClock*); {*to smallest finishing time on agenda*}
  (3) Phase III: *FinishActivities*; {*all which finish at this time*}

Phase 1 tries to start all activities that can possibly proceed at this state and time. In our program this is the responsibility of procedure *StartActivities*. The monitor has to cycle until no further starts can be triggered during one complete scan of all activities. As long as there is at least one state change the whole scan must be repeated. The reason for this lies in the fact that there might be a previously unsuccessful activity whose conditions could now allow it to be started. A predefined Boolean function, *ScanWithoutAnyStartsP*, is therefore used to guard the scan.

Phase 2 is entered upon completion of Phase 1. Here the clock is set to the earliest of finishing times across all activities on the agenda. Procedure *SetClock* performs this task.

Phase 3 then finishes all activities scheduled to finish at this clock time. Because of the next-event style in which the clock was advanced, there must be at least one ready to go. Procedure *FinishActivities* implements this phase.

It should be obvious that such a pattern-driven sequencing strategy can become quite expensive in computational terms. If there are many interactions between activities, waiting on conditions set by others, then Phase 1 can require a great number of passes over the agenda, repetitively scanning each activity's starting condition. This is wasteful. A state change should ideally lead only to tests and restarts of those activities that could possibly be affected by it. The great number of unnecessary scans is the price that has to be paid for the convenience of declarative (conditional) scheduling. The programmer is not asked to detail the way in which events interact. This information is explicit in the scheduling commands used in event-based models, and in the clustering of action sequences into process classes in the case of process-based models. Although a preprocessor could possibly derive this knowledge from an analysis of a model's structure, this is not normally done. Simulation monitors using the activity-scan strategy are therefore typically involved in a lot of repetitive cycling – a brute force approach to model execution.

The basic idea of **declarative sequencing** is not limited to activity-based models. It can and has been used in process-based frameworks as well. GPSS (see Chapter 3) is a good example. There this strategy is referred to as 'conditional sequencing' and involves transactions whose progress may be delayed by waiting for some condition to occur (a server becoming available, a switch being reset). The programmer only specifies the condition and the model interpreter has to make sure it detects it at the right stage in a model's execution.

The inefficiency of this kind of sequencing is reflected in the notoriously slow speed of most GPSS processors. Much of this can be attributed to the overhead of repetitive scans across the model's entities and state indicators. A much better solution is offered by DEMOS (see Chapter 3), whose CONDQ primitive also allows processes to wait on conditions, but they do so passively, allowing the programmer to explicitly reactivate them by signals from any place in the model where their conditions may be affected.

The signalling required to implement such passive waiting does not put too much of a burden on the programmer's shoulders, indeed it may well lead to a superior model structure through a more conscious appreciation of entity interactions. For typical models it is an order of magnitude more efficient than the simple-minded 'active-waiting' approach embodied in GPSS and the three-phase structure used for the activity-scan strategy.

Although it is inherently inefficient, particularly for models with frequent interactions between activities, activity scanning does allow the programmer to localize control knowledge in starting conditions and finishing times for each activity. This makes models readable, easy to design, modify and extend. The strategy is related to so-called **production systems** used in artificial intelligence programming. Production systems have been well researched [see Davis and King (1977) and Nilsson (1980)] and have recently successfully been used to build so-called expert systems. Their primary advantage over other knowledge-representation schemas is the independence of chunks of knowledge, which can therefore be changed, deleted or added to a working system without having to consider and recode interactions with other modules. Production rules can be triggered by the current state of a model of a given miniworld. Conflicts must of course be resolved somehow, if more than one production may fire during one scanning cycle. The order of rule sequencing may well make a difference to the final result. A number of strategies for conflict resolution have been explored and are reported in Davis and King (1977).

Activity-based simulators typically use a first-detected, first-triggered rule for activity activations, with repeated scans of the agenda taking care of interactions. This is a simple strategy but not always a safe one.

Exercises

1. Try to write a global procedure that can be used to reset *all* the statistics counters in a model in one sweep. You will have to attach an additional list to the model monitor and you must provide a data type for reference to data collectors. Because of PASCAL's typing rules you will also have to define all data collector types as variants of the same record type. The initialization procedures for data collectors will have to be modified so that they link into this list automatically.

2. Experiment with different data structures (arrays, rings, trees, etc.) for queues and agendas. You will want to test these objects with models using different scheduling patterns.

3. Extend our toolbox so that it can cope with models involving more than one class of process.

4. How could conditional sequencing be implemented in event-based models?

2.2.3 Time events, state events and processes: Combined discrete/continuous simulation

Combined simulation tries to integrate both the discrete-event and the continuous-change approach to the study of process behaviour. For some applications this is a natural view of system dynamics, although for a long time user communities of these world views were almost completely disjoint. A combined-simulation model can reflect either a process that progresses in an essentially continuous fashion, with discrete events modelling occasional discontinuities; or it can be driven by an event-based scheduler, with some regular subprocesses represented continuously.

Two different kinds of events are used to interface the two modelling frameworks. **Time events** are scheduled by a discrete-event monitor – for instance, in order to change a continuous state variable's state at some given point in time. **State events** may be triggered by a time-slicing monitor when a state variable's level crosses a specified threshold, which may lead to certain actions (event) in a model's discrete-event part.

A continuous process embedded in an essentially discrete-event framework will now be considered. The alternative will be explored in Chapter 3.

For this the scenario of Manuel's shop needs extending. So far it has been assumed that an unlimited supply of freshly heated potion is always available, and we have abstracted from the cauldron's actual contents.

Unfortunately Manuel ran into difficulties validating this model. It always seemed to give predictions that proved to be too optimistic. After empirically confirming that the customer arrival- and service-time distributions and their parameters are essentially correct, Manuel concluded that there must be some other factor that kept him from performing according to expectations.

As it turns out, the time he spends refilling his cauldron has been overlooked, and during this time he cannot attend to waiting customers. Filling the cauldron typically takes Manuel about 3 minutes, with a standard deviation of 0.1 minute. This measure follows a normal distribution.

Another unaccounted influence has also been identified. To preserve the potion's potency it must be kept at quite a high temperature, causing rapid evaporation. Since this proceeds at a regular rate measured at 6 litres an hour (0.1 litre per minute), the level of potion in the cauldron is naturally modelled by a continuous process. Of course, this level is also lowered each time a customer buys a potion. This is represented by a normal distribution, with a mean of 0.5 and a sigma of 0.1 litre.

Apart from the familiar discrete-event framework of Figure 2.6, we must now consider a state event. To ensure that he never runs out of potion, Manuel wants to be immediately informed whenever the cauldron's level drops below 2 litres (it holds 30 litres). For this purpose he has installed a wooden semaphore, to be set by a demon watching the cauldron. Manuel always checks the semaphore's state before serving a customer. If it is set, he immediately refills the cauldron and resets the semaphore. It will still be assumed that an unlimited supply of cold potion is available.

An implementation of this scenario uses all six toolboxes: since the model has stochastic components the Monte Carlo method utilities are needed; data-collection utilities are required for statistical instrumentation; the quasi-continuous simulation utilities are needed for modelling the cauldron and the associated evaporation process; and the discrete-event and queueing-network simulation toolboxes support coding the discrete-event part of the model. A few additional procedures are necessary for communicating between continuous- and discrete-model components, and these are listed in Table 2.8. There is a procedure (*ChangeState*) to cause discrete changes in a continuous-state variable's level and one to evaluate conditions associated with continuous processes (*WaitUntil*). This implementation of *WaitUntil* is unfortunately not what is really required. Ideally it should be possible to attach state events to continuous processes via some kind of

Table 2.8 The combined library

Object :	{various previously declared classes}
Methods :	**public** *ChangeState* (*aContVarType*, *someRealQty*); *WaitUnitlP* (*condition*): *BOOLEAN*; *TimeEventNOWp* (*anEventSchedulingMonitorType*): *BOOLEAN*;

```
program ManuelsShop (INPUT, OUTPUT);
{the combined discrete/continuous version !}

  const
    MaxHistoClasses =      100;
    MaxPlotPts      =       50;
    LevelSymbol     =      '*';
    NoDelay         =      0.0;
    EvaporationRate =     - 0.1; {litres per minute}
    AlongNumber     = MAXINT;

  type
    anEventType = (hello, start, finish, bye, signalFill, finishFill);

    {insert type declarations from Monte Carlo              toolbox}
    {insert type declarations from statistics               toolbox}
    {insert type declarations from continuous-simulation    toolbox}
    {insert type declarations from discrete-event           toolbox}
    {insert type declarations from queueing-network         toolbox}

  var
    TheClock          : aSynchClockType;
    TheMonitor        : anEventSchedulingMonitorType;
    Arrivals, Service : aNegExpGenType;
    Filling , Serving : aNormalGenType;
    Source            : aSourceType;
    Sink              : aSinkType;
    Q                 : aQueueType;
    Manuel            : aServerType;
    Cauldron          : aContVarType;
    NoOfFills         : aCounterType;
    FlowTime          : aHistogramType;
    PotionLevel       : aChronographType;
    LowPotionSignal   : boolean; {set whenever the potion level drops below}
                                 {a threshold(2 litres).Manuel checks this   }
                                 {signal before he starts any service       !}

    {insert functions & procedures from Monte Carlo              toolbox}
    {insert functions & procedures from statistics               toolbox}
    {insert functions & procedures from continuous-simulation    toolbox}
    {insert functions & procedures from discrete-event           toolbox}
    {insert functions & procedures from queueing-network         toolbox}
    {insert functions & procedures from combined D/C             toolbox}

  procedure ReadAndInitialize;
  {instantiates all the model's entities}
    begin
      InitSynchClock              (TheClock, 0.1);
      InitEventSchedulingMonitor (TheMonitor, aLongNumber, 500);
      InitNegExpGen  (Arrivals, 12345, 0.8);
      InitNegExpGen  (Service,  67891, 2.0);
      InitNormalGen  (Filling,  23457, 2.0, 0.2);
      InitNormalGen  (Serving,  89123, 0.5, 0.1);
```

Program 2.6 *A combined discrete/continuous analysis of Manuel's shop*

```
    InitSource      (Source);
    InitSink        (Sink   );
    InitQueue       (Q      );
    InitServer      (Manuel,     1);
    InitContVar     (Cauldron, 3.0);
    InitCounter     (NoOfFills    );
    InitHistogram   (FlowTime, 0, 40, 10);
    InitGraph       (PotionLevel, 4.0, 100, 50, 50);
    LowPotionSignal := false;
    ScheduleEvent (TheMonitor, NoDelay, hello, nil);
  end; {of ReadAndInitialize procedure}

procedure WriteReport;
{reports the results of a simulation run !}
  begin
    WRITELN ('RN-generators:');
    WRITELN ('  ++++++   ');                                    WRITELN;
    ReportOnNegExpGen (Arrivals  );                             WRITELN;
    ReportOnNegExpGen (Service   );                             WRITELN;
    ReportOnSource         (Source    );                        WRITELN;
    ReportOnSink           (Sink      );                        WRITELN;
    WRITELN ('Manuel''s Statistics:');
    WRITELN ('  +++++++++  ');                                  WRITELN;
    ReportOnServer         (Manuel  );                          WRITELN;
    ReportOnQueue          (Q       );                          WRITELN;
    WRITELN ('Cauldron Refills:');
    WRITELN ('  +++++++   ');                                   WRITELN;
    ReportOnCounter        (NoOfFills);                         WRITELN;
                                                                WRITELN;
    WRITELN ('The time customers spent in the shop:');
    WRITELN ('  ********************************  ');           WRITELN;
    ShowHistogram (FlowTime);                                   WRITELN;
    WRITELN ('The Graph of the level of potion against time:');
    WRITELN ('  ***************************************  ');    WRITELN;
    ShowGraph (PotionLevel);
  end; {of WriteReport procedure}

procedure HelloEvent;
{creates incoming customers and puts them in a queue in front of Manuel}
  var
    romeo: aTAPtrType;
  begin
    {plan next arrival}
    ScheduleEvent (TheMonitor, NegExp (Arrivals), hello, nil);
    {create a customer}
    romeo := CreateNewTA (Source);
    {put him into the Q and schedule a start event}
    FileIntoFIFOq (Q, Romeo);
    ScheduleEvent (TheMonitor, NoDelay, start, nil);
  end;{of HelloEvent}
```

(cont.)

```
procedure StartEvent;
{starts a service activity if Manuel is idle}
  var
    romeo: aTAPtrType;
  begin
    if LowPotionSignal then
      begin
        if (EnoughServerUnitsAvailableP (Manuel, 1)) and (Q.length > 0) then
          begin
            SeizeServer (Manuel, 1);
            ScheduleEvent (TheMonitor, Normal (Filling), finishFill, nil);
          end;
      end
    else
      if (EnoughServerUnitsAvailableP (Manuel, 1)) and (Q.length > 0) then
        begin
          SeizeServer (Manuel, 1);
          {turn attention to waiting customer}
          romeo := TakeFirstFromQ (Q);
          {diagnose degree of lovesickness, mix and hand over remedy; accept
           donation}
          ScheduleEvent (TheMonitor, NegExp (Service), finish, romeo);
        end;
  end; {of StartEvent}

procedure FinishEvent (romeo: aTAPtrType);
{complete romeo's service activity and check whether other customers}
{are waiting for Manuel}
  begin
    ChangeState (Cauldron, - Normal (Serving));
    ReleaseServer  (Manuel, 1);
    ScheduleEvent (TheMonitor, NoDelay, bye, romeo);
    ScheduleEvent (TheMonitor, NoDelay, start, nil     );
  end; {of FinishEvent}

procedure ByeEvent (romeo: aTAPtrType);
{interviews romeo about queueinngg delays and gets rid of him}
  begin
    UpdateHistogram (FlowTime, FlowTimeQ (romeo));
    RemoveTA (Sink, romeo);
    RecordAnotherSample;
  end; {of ByeEvent}

procedure SignalFillEvent;
{This event signals that the cauldron will need filling.        }
{This will be attended to by Manuel (StartEvent) as soon}
{as he is able. Until then the evaporation will proceed          }
{at its normal rate                                              }
  begin
    LowPotionSignal := true;
  end; {of SignalFillEvent procedure}
```

Program 2.6 (cont.)

```
procedure FinishFillEvent;
  begin
    ChangeState (Cauldron, 3.0 - Cauldron.level);
    LowPotionSignal := false;
    UpdateCounter (NoOfFills);
    ReleaseServer  (Manuel, 1);
    ScheduleEvent (TheMonitor, NoDelay, start, nil);
  end; {of FinishFillEvent procedure}

begin{*****Body of MAIN program*****}

  ReadAndInitialize;

  {Simulation loop: Observe the specified number of customer depart !}

  while AnotherTickP (TheClock) and MoreSamplesToTakeP (TheMonitor) do
    begin
      {update and record levels of all continuous variables}
      If (( ROUND (TheClock.NOW * 10) mod 10) = 0) and
          (TheClock.NOW < 100) then
          StorePoint (PotionLevel, Cauldron.level, TheClock.NOW,
          LevelSymbol);
      UpdateState (Cauldron.level, EvaporationRate);
      {watch out for possible state events}
      if WaitUntilP ( (Cauldron.level < 2) and (not LowPotionSignal) ) then
        ScheduleEvent(TheMonitor, NoDelay, signalFill, nil);
      {now check whether any time events are pending}
      while TimeEventNOWp (TheMonitor) do
        with TheMonitor do
          begin
            GetNextEventNotice (TheMonitor);
            case KindOfEventQ (currentEvent) of
              hello      : HelloEvent;
              start      : StartEvent;
              finish     : FinishEvent (TAreferenceQ (currentEvent));
              bye        : ByeEvent    (TAreferenceQ (currentEvent));
              signalFill : SignalFillEvent;
              finishFill : FinishFillEvent ;
            end; {case}
            FlushOutEvent (currentEvent);
          end; {with}
    end; {simulation loop}

  WriteReport;

end.
```

'**demon**' procedure. Such data- or pattern-driven invocation is not possible in PASCAL, and the alternative of simple-minded scans through all the model's state variables, in the spirit of an activity-based model, is not particularly attractive. A more primitive version of *WaitUntil* is therefore chosen, which requires that this procedure is explicitly called from wherever the attached condition may possibly be changed.

The combined discrete–continuous scenario implemented by Program 2.6 is almost identical to the event-based version of Manuel's shop (Program 2.3). Assuming that a chronological graph is to be used to plot the potion level, the plot symbol and rate of evaporation are defined as symbolic constants. Two new events are also introduced: starting and finishing refilling of the cauldron. New entities are the cauldron, represented by a continuous-state variable, a count recording the number of refills (*NoOfFills*) and a chronological graph to plot the potion level against time. The semaphore attached to the cauldron (*LowPotionSignal*) must also be declared. Procedure *ReadAndInitialize* instantiates these entities and binds them to appropriate initial values. Scheduling a *hello* event triggers the simulation.

Note that the event-scheduling monitor is used, but a synchronous clock is needed for time slicing the progress of the quasi-continuous processes. When necessary, the procedure implementing the new *StartFill* event will set the semaphore to signal a low potion level. It also attracts Manuel's attention, if he is currently idle. If successful, it schedules a *FinishFill* event, delayed by the required time interval. If Manuel is busy, then this event relies on the *StartService* event to check the state of the semaphore (still set) and tries again to trigger a *StartFillEvent* before Manuel attends to a waiting customer; *StartEvent* is patched accordingly for this. Procedure *FinishFillEvent* will set the cauldron's level to 30 litres and reset the semaphore. A *StartEvent* is also scheduled without delay. Manuel can now serve his customers again, if there is anybody still waiting.

In combined discrete–continuous programs the main simulation loop may be guarded by both the number of epochs (time slices) and the number of samples recorded by the event-scheduling monitor:

while *AnotherTickP* **and** *MoreSamplesToTakeP* **do**

(1) integrate all continuous variables (see CS shell);
(2) check if discrete events may occur now (see DES shell);
(3) record the state of the model;

Since the simulation is again sample-driven, the temporal termination condition will never apply; a long number is stored in the monitor's *simTime* slot.

The clock time slices at the requested resolution, each time updating the cauldron's state. This is the same approximation to a truly continuous simulation that was used before. The new state is always recorded in the chronological graph. This loop also watches to see whether the cauldron level falls below the specified threshold. It signals a state event (*StartFill*) if it does.

Random-number generators
****

502 samples were taken by this neg. exp. distribution. The sample mean was: 1.24
501 samples were taken by this neg. exp. distribution. The sample mean was: 0.51

Since last reset 502 transactions were created by this source.
Since last reset 500 transactions were removed through this sink.

Manuel's statistics:

The capacity of this server is 1 unit.
1 of these is currently in use.
Utilization of this server was 90.84%, since last reset.

Queue statistics:

Since its statistics were last reset, 502 items entered and 501 left the queue.
Its current length is 1

The queue's main characteristics were:

Mean length = 8.62
Max. length = 27
Mean delay = 10.73

Cauldron refills:

Value of this counter at 623.69: 157

Customers' time in the shop:

Until 623.69, 500 observations were made.

This tally's characteristics were as follows:

Mean : 11.20
Sigma : 8.24
Min. obs. : 0.02
Max. obs. : 31.99

Overflow was: 0
Underflow was: 0

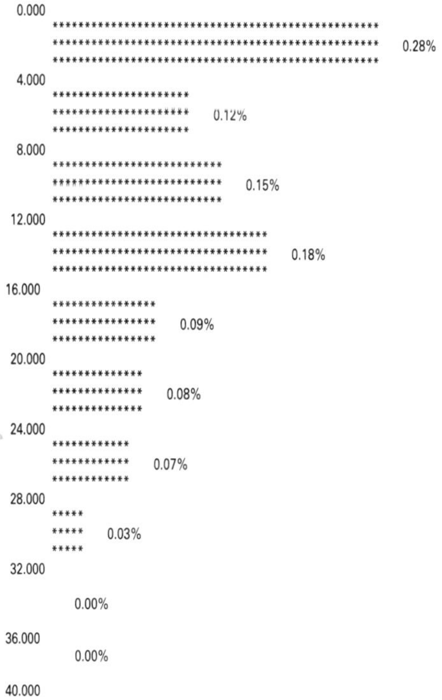

Figure 2.12 Manuel's shop with refills: The results (cont.)

The graph of the level of potion against time
****

105 plot points were collected.

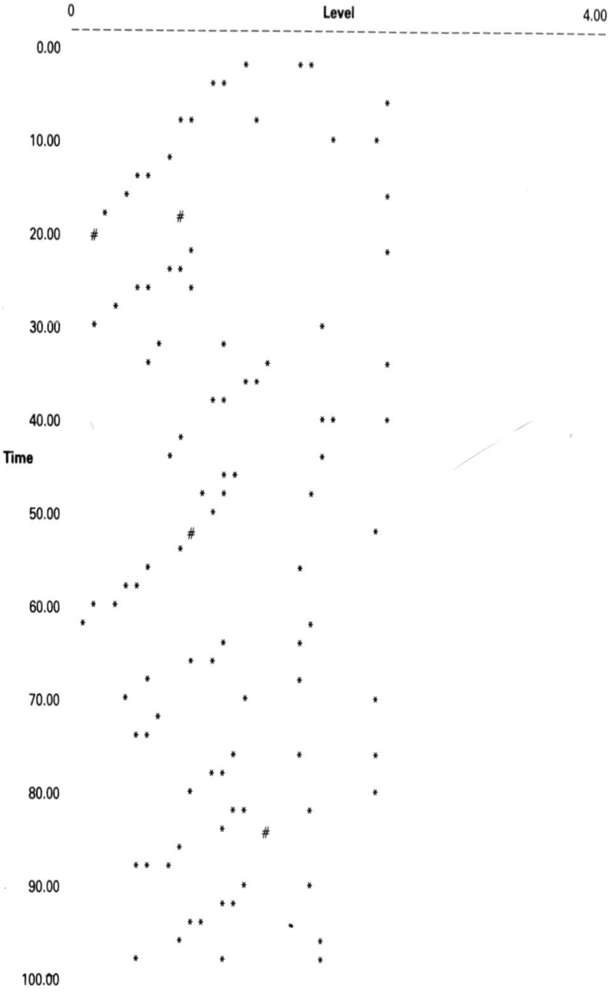

Figure 2.12 (cont.)

Finally, the event-scheduling agenda is checked for any discrete events due to occur at this time. These are processed in the same fashion as for other event-based models.

This cycle will recur until the number of samples terminates the loop.

Chapter 3
Simulation Programming Systems

"NOW, WITH THE NEW MATH..."

3.1 History and family lines

Over the past three decades a great variety of simulation programming languages, catering for different modelling styles, have been designed and implemented. An annual catalogue of simulation software is published by the Simulation Council [SCS (1984, 1985)].

MONTECODE [Kelley and Buxton (1962)] was an early attempt to design a procedure package for simple **distribution sampling**. GERTS [Moore and Clayton (1976)] is a more recent system, used to describe and simulate stochastic planning networks. Nowadays distribution sampling is well supported by discrete-event simulation software.

Continuous-simulation languages evolved from attempts to simulate analogue computers. Figure 3.1 shows the pedigrees of some of the best-known systems. The CSSL report [Strauss *et al.* (1967)], commissioned by the Simulation Council, was a major milestone. Since then a number of software systems have been based on its proposals. Most implementations are still FORTRAN-based [see Cellier (1982) and Karplus (1982)]. Continuous-simulation languages typically offer a range of different integration schemes. Most use expression-based scenarios (differential equations) with a model monitor sorting equations into the right order. Macros, special provisions for dealing with stiff differential equations and graphical displays are sometimes supported. Because of the number-crunching nature of numerical solutions to differential equations, run-time efficiency and accuracy are overriding concerns. Most research is consequently targeted on these areas.

Figure 3.2 traces some of the more important families of **discrete-event simulation languages**. Solid lines stress development paths, whereas dotted lines show major cross-pollinations. Reviews and comparisons of popular systems are given by Tocher (1965), Dahl (1968), Kiviat (1971), Unger (1978) and Kreutzer (1983). The article by Dahl is still particularly well worth reading.

The GASP, SIMSCRIPT, GPSS and SIMULA families have the longest traditions, originating in FORTRAN IV and ALGOL 60. PL/1-based (SIMPL/1) and PASCAL-based (SIMPAS, SIMONE, etc.) languages have been more recently developed. Over the last few years ADA has also often been proposed as a host for discrete-event simulators.

Early applications evolved from using general-purpose programming languages to subroutine packages of simulation-oriented procedures. Later interpreted, precompiled and compiled systems provided better packaging of common concepts and utilities.

There have been disjoint development paths of self-contained, special-purpose discrete-event simulation languages within the USA and Europe.

The material-oriented approach to queueing scenarios [GPSS: Gordon (1979)] and event-oriented modelling frameworks [SIMSCRIPT: Kiviat *et al.* (1968)] developed in North America. These systems were originally built as FORTRAN extensions. At the same time scenarios based on activity

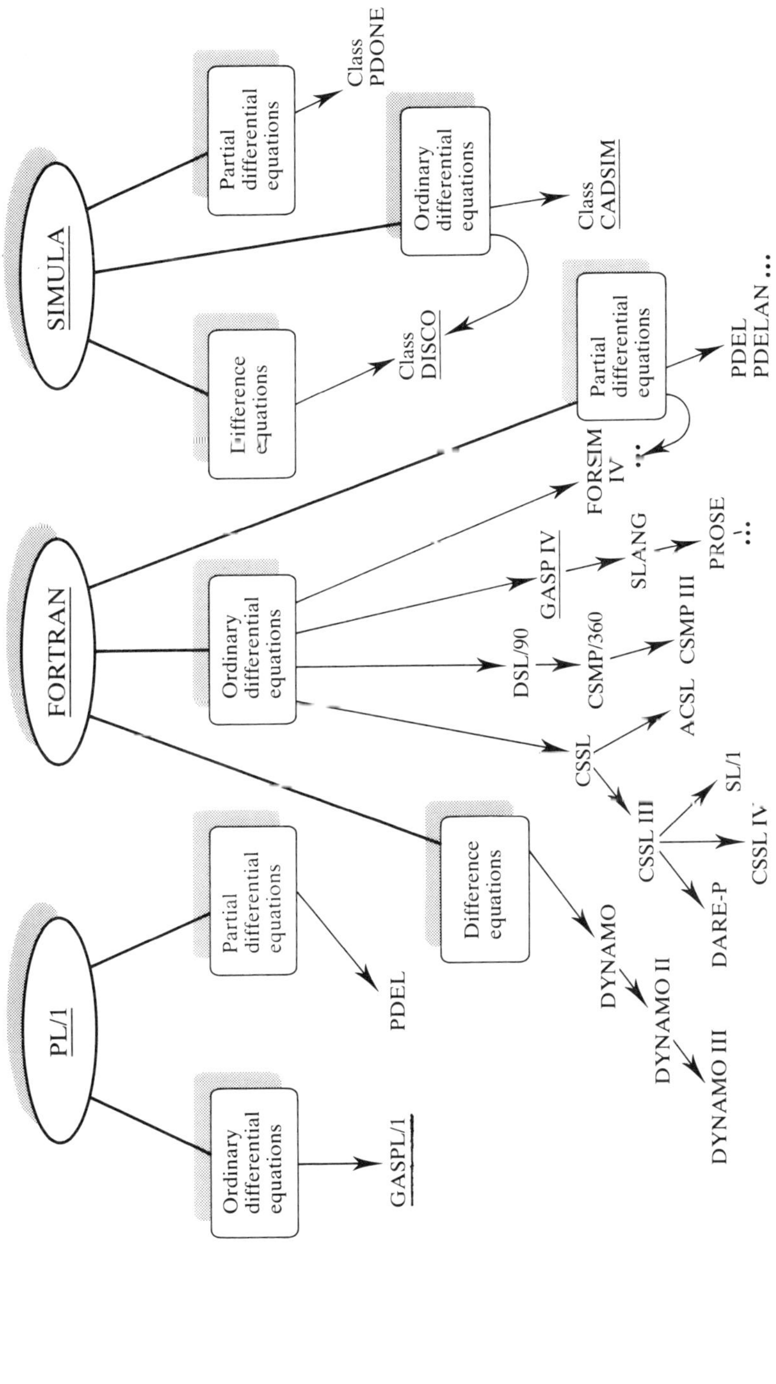

Figure 3.1 Family trees of some continuous-system simulation languages (combined discrete/continuous languages are underlined)

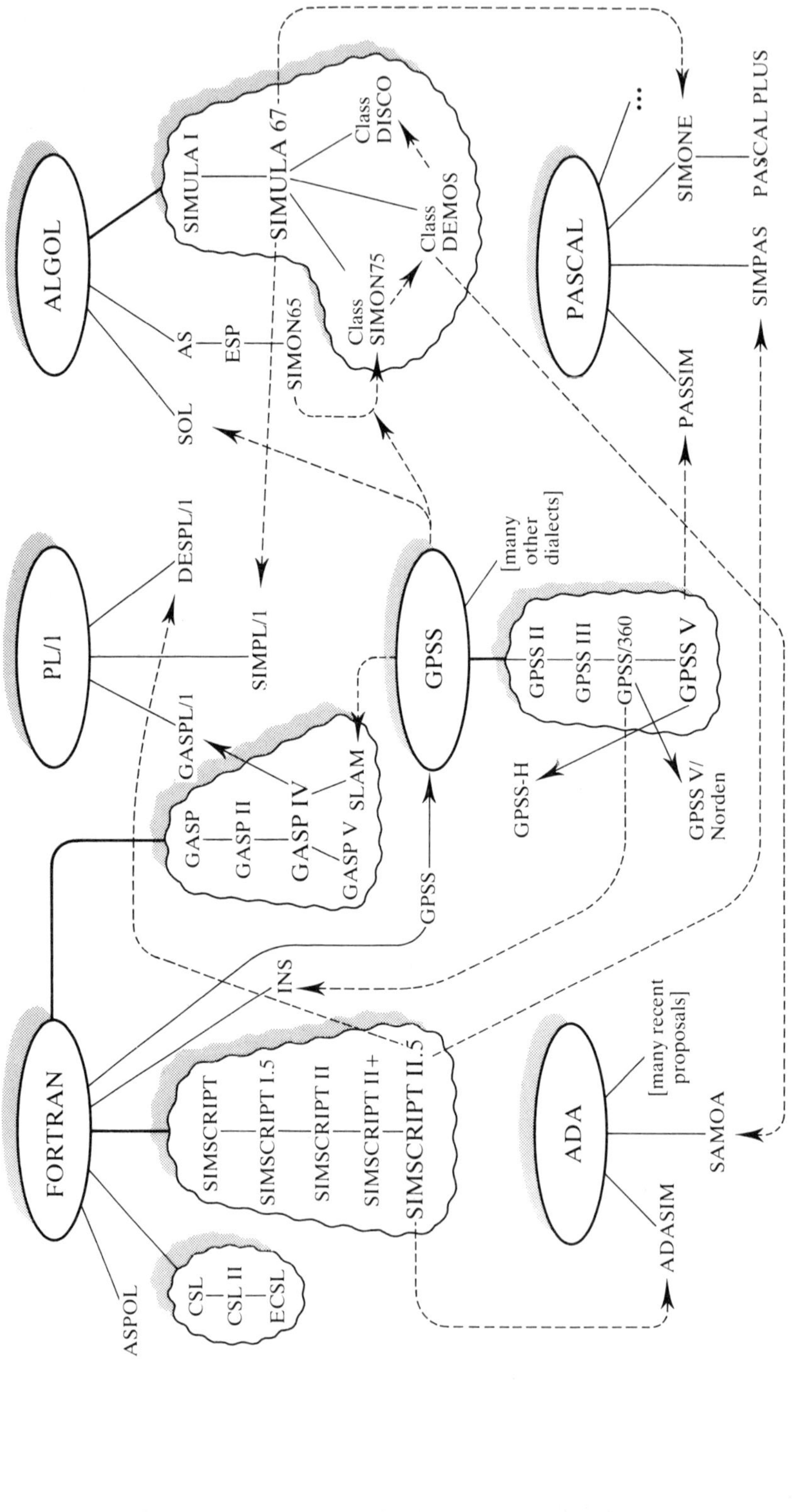

Figure 3.2 The discrete-event simulation language landscape

cycles, and activity [CSL: Clementson (1966)] and process-based sequencing schemes were designed and implemented by researchers in Europe and the UK, where ALGOL 60 was more popular.

Process-oriented techniques were already implicit in GPSS. An ALGOL-based generalization had been proposed by SOL [Knuth and McNeley (1966)]. SIMULA I [Dahl and Nygaard (1966)] was influenced by SOL, but it also contributed the idea of classes, which are integrated descriptions of object structure and behaviour; this was a logical extension of the blocks, procedures and own variables of ALGOL 60.

The IFIP conference on Simulation Programming Languages in 1968 [Buxton (1968)] was a major milestone in the history of discrete-event simulation. Since then the development of new systems has slowed, but there has been some convergence of the most important language families. Some notable developments since 1968 are:

1. At the conference it was repeatedly emphasized that a simulation language like SIMULA or SIMSCRIPT would be well suited for any kind of system description. Apart from special capabilities for modelling stochastic processes and multi-level descriptions of model dynamics, these languages offer all features of general-purpose programming languages. Newer versions of simulation languages are now competing with other general-purpose programming systems, offering simulation capabilities as an added bonus (level 5 of SIMSCRIPT; CLASS SIMULATION of SIMULA).
2. Process or object-oriented world views have been widely accepted as the best frames for model structuring and are supported by most major simulation languages, such as GPSS, SIMULA, SIMPL/1, and extensions of SIMSCRIPT II.5 [Russel (1976)] and GASP IV [Washam and Pritsker (1976)].
3. Some languages offer features for combined discrete–continuous modelling, a development that was long retarded by a lack of communication between user communities (primarily engineers at the continuous side of the fence, and mathematicians and computer scientists at the discrete side). Oren (1973) and Cellier (1979a, 1979b) give comprehensive surveys of relevant concepts and tools. GASP IV [Pritsker (1974)], DISCO [Helsgaun (1980)], SLAM [Pritsker and Pegden (1979)] and CADSIM [Sim (1975)] are examples of the latter.
4. Investigation of interactive programming environments and graphical interfaces [Alemparte *et al.* (1975); Hoffman and Carlson (1981); Milner *et al.* (1982); Hoover and Perry (1983)] provide some interesting perspectives for future simulation laboratories. As hardware prices continue to fall this has become an increasingly active area of research.
5. The importance of model and program structuring and software reliability is receiving increasing attention from all simulation-language designers.

6. Methodologies for establishing model credibility and experimental design and evaluation are currently largely unsupported by simulation software. More guidance could be given, especially in interactive environments. A body of well-established techniques is available, although mostly scattered through journals and conference proceedings. Kleinjen (1974/75, 1979) are useful items in this literature.

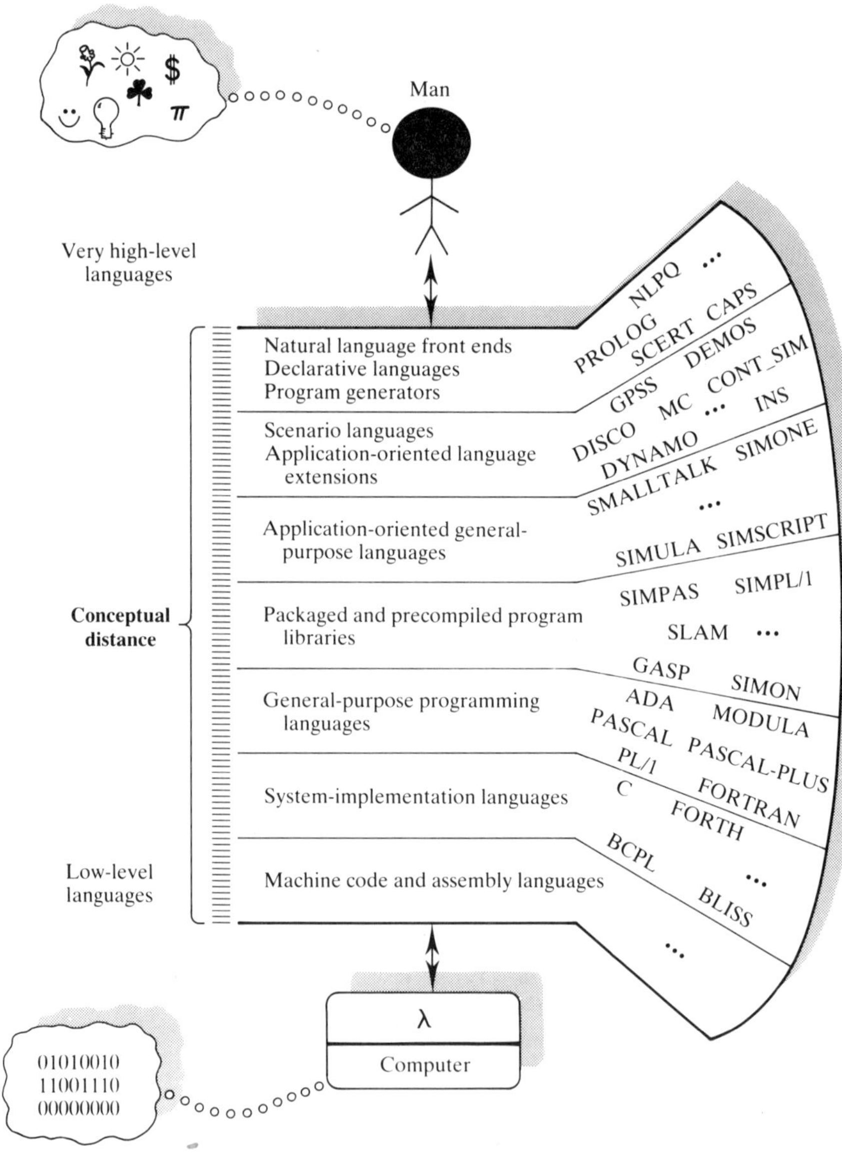

Figure 3.3 Levels of man–computer communication

Today we find a great number of programming systems that can be classified according to their levels and frames for representing model behaviour (world views).

Raising the semantic level of a language leads to improvements in ease of learning and use, speed of model development and reduced error rates. High-level simulators can almost directly carry familiar conceptual frameworks, giving considerable guidance to users. This has to be paid for by a decrease in flexibility and generality, processing and storage efficiencies.

Relatively low-level tools (like PASCAL or SIMULA) are almost universal. They can be applied across a wide range of modelling tasks. The usefulness of higher-level systems like GPSS, on the other hand, quickly deteriorates once they are moved outside their intended area of application.

Usually four levels of software are distinguished (Figure 3.3):

1. Modelling with general-purpose programming languages (FORTRAN, PL/1, PASCAL, etc.).
2. Modelling with procedure-oriented simulation software. This includes self-contained systems as well as subroutine packages or preprocessors on top of some host languages (MIMIC, GASP; SIMPAS, SIMPL/1; SIMSCRIPT, ECSL, SIMULA, etc.).
3. Modelling with descriptive scenarios (GERTS; CSMP, CSSL, ACSL, PDEL, DYNAMO; SLAM, GPSS, INS, DEMOS, DISCO, etc.).
4. Modelling with model generators (DRAFT, CAPS, etc.) and declarative systems – a field for specialized software packages, such as computer system simulation (SCERT, SAM, CASE, etc.).

In this chapter we will briefly discuss some systems bound at each of these levels.

3.2 Using general-purpose programming languages

In principle any programming language can be used for simulation modelling, although this amounts to reinvention of the wheel and debugging it all over again can be very time consuming. Prepackaged objects and processes can ease the modelling effort considerably. Some utilities (such as random-variate generators), model instrumentation and reporting aids are relatively easily implemented. The primitives for dealing with model time, concurrency, the creation of dynamic objects and the handling of dynamic relationships are, however, less well supported by most of the widely used programming systems.

The event-oriented framework is the easiest world view to implement. It does not, however, offer a very user-friendly modelling environment. As shown in Chapter 2, process-, activity- and object-oriented frames require more sophisticated schemes for storage allocation and sequencing, which cannot so readily be implemented in one of the usual general-purpose programming systems.

3.3 Providing custom-made data types, operators and control structures for system simulation

3.3.1 Procedure libraries

Most distribution sampling (MONTECODE, GERTS) and many continuous-simulation systems are implemented as FORTRAN-based procedure packages.

As you may have noticed from the toolboxes in the previous chapter, prepackaged procedures for event scheduling and execution can be quite clumsy to use. A number of other PASCAL-based simulation libraries are reported in the literature. Hac (1982), Cavouras (1983), Vaucher (1984b), and Frantz and Trott (1984) describe some of these. FORTRAN-based discrete-event simulators like GASP, SIMON [Hills (1965)] and SLAM are among the most popular library packages for simulation programming.

SLAM ([Pritsker and Pegden (1979)], [Pritsker (1983)] offers support for queueing-network scenarios.

GASP IV [Pritsker (1974, 1979)] is a package of FORTRAN IV or PL/1 [GASPL/1: Pritsker and Young (1975)] subroutines related to system simulation. It carries an event-oriented framework. A monitor routine, called GASP, operates on a model representation and scheduling records stored in COMMON arrays. Features provided by GASP are similar to those defined for the event-based PASCAL programs in Chapter 2. Arrays and procedures are the only structuring devices. Since it is hosted by FORTRAN, GASP is restricted to static storage management, which requires compile-time dimensioning of all data structures (that is, size of agenda, maximum number of transient entities, maximum queue lengths, etc.).

All FORTRAN-based systems suffer from the necessity of having to use short names (up to six characters only). Although at first this may only appear a minor inconvenience, it does, in fact, prove to be a very severe handicap for writing readable and well-structured models. Since good clustering and structuring devices (records, lists, processes) are not available, even simple models are already quite difficult to decipher. More complicated ones also tend to suffer from 'name pollution', since almost all data structures turn out to be global, and distinct and meaningful names are difficult to invent, particularly within the restriction of six character positions.

These problems grow worse at an exponential rate as model complexity increases. One is often faced with a tangled mesh (mess) of variables representing entities and links between them, event routines and scheduling references. Such models are very difficult to understand, communicate, debug and verify, and their credibility can be difficult to defend. This is a good example of the way in which pragmatic decisions in early programming languages, solely motivated by machine architectures and efficiencies relevant at the time they were first designed, violate basic principles of good, habitable user interfaces. The only advantage that can be attributed to

FORTRAN-based simulation systems is their wide availability and portability.

Although discussions of the pros and cons of programming styles and languages are always in danger of inspiring an almost religious fervour in both the faithful and heretics, it can safely be said that FORTRAN is by now an outdated tool. It has served its purpose well; but it was designed to cater for an environment characterized by batch operation, expensive processor and memory resources, and an absence of any guidelines for the design of good cognitive tools. The computing scene is now radically different from what it was in the late 1950s. The use of FORTRAN for building sophisticated models and programs is now far too tedious, clumsy and unreliable to be recommended.

Using a more modern, conceptually richer and reliable general-purpose programming language like PASCAL can alleviate some of the problems, even though it falls short of being a truly useful tool for implementing process- or object-oriented frameworks. Attempts to overcome PASCAL's shortcomings in this respect either have to extend the language or turn out to be clumsy to use [see PASSIM: Uyeno (1980), Horton and Turner (1986) and the process-oriented PASCAL model]. Extending PASCAL means patching the compiler. This has been done by Kaubisch *et al.* (1976) and Perrot *et al.* (1980) {SIMONE}, Welsh (1979) {PASCAL PLUS}, Lindstrom and Skansholm (1981), Kriz and Sandmayr (1980) and others, but it largely destroys portability. SIMULA, MODULA and ADA are far better suited as base languages for simulation programming environments, while PL/1 [Hac (1984)] and some more exotic languages, such as SETL [Franta and Maly (1973)], ALGOL 68 [Shearn (1975)] and ALGOL-W [Stirling (1978)], have also served as hosts.

3.3.2 Language interpreters

Very few interpreted systems for simulation have been proposed and implemented. Most versions of GPSS are implemented by interpreters, with a so-called presimulation phase, mapping models into a set of fixed-size tables. Systems hosted in interactive environments like BASIC, APL, SMALLTALK [Goldberg and Robson (1983)] or PROLOG [Clocksin and Mellish (1981)] are naturally interpretative.

Interpreters are relatively easy to program, using an appropriate base language like LISP [Hasemer (1984)]. LISP-based interpreter languages like FRL [Roberts and Goldstein (1977)], KRL [Bobrow and Winograd (1977)], PEARL [Deering *et al.* (1981)], ACT [Theriault (1982)] and others have traditionally been used to explore so-called miniworlds in the context of artificial intelligence research, a domain closely related to the use of simulation in the exploratory sense.

Advantages gained by interpretative execution are good, interactive programming environments, as well as superior execution monitoring, error tracing and correction facilities. The price to pay is a slower speed of execution. GPSS/H, a compiled GPSS version, reports a 500–600% improvement in run–time efficiency [Henriksen (1975), Reitman (1982)] over typical interpreter-based implementations.

There is a basic trade-off between user and machine efficiency. While it must be compiled into low-level representations to gain speed, knowledge must be kept at a high level to be understandable and easy to change. Although this is less important for its application to theory exploration, the speed penalty involved in interpretative model exploration may well be unacceptably high for the processor-intensive use of simulation as a tool for numeric predictions. While sufficiently powerful work stations (such as modern LISP machines) may somewhat diminish the importance of these considerations, a potential compromise is offered by SMALLTALK and many advanced LISP systems, where interpretative execution can be employed during development and testing, while modules can be compiled once they have stabilized.

3.3.3 Simulation-language translators

A. Precompilers

Precompilers are the easiest way to construct user-friendly modelling environments quickly and without too much implementation effort, particularly if appropriate tools are used. Many continuous-simulation systems are implemented by preprocessors into FORTRAN (CSMP, CSSL, ACSL, PDEL). SIMPL/1 [IBM (1972)] is a process-oriented, discrete-event simulator precompiled into PL/1.

Program 3.1 shows a SIMPAS model of Manuel's shop. SIMPAS [Bryant (1981a, 1981b)] is a precompiler written in PASCAL. It translates a PASCAL program using simulation-oriented data types and commands into standard PASCAL. The simulation facilities of SIMPAS were patterned after those of SIMSCRIPT. In the program there is an event-oriented model monitor, which offers statements for starting a simulation experiment (**start simulation**), stopping a simulation (**schedule main**. . .) and for event sequencing (a variety of **schedule** statements).

A status indicator can be used to signal contingencies to the main program. An agenda of event notices is kept in the classical next-event fashion and there is a clock called **time**. Classes of events are described in **event** routines, which are basically PASCAL procedures. Event routines are automatically associated with event notices, so that they can be scheduled.

There are **queue** and **queue member** data types. Variables declared as queue members are transient objects, instances of which can be dynamically created and destroyed. Various **insert** and **remove** statements are provided to manipulate queues, for which certain statistics (mean length, max. length, waiting time, etc.) are automatically computed.

An extensive library of random sampling procedures is provided and can be included on demand. There are ten predefined random-number streams, with preinitialized seeds.

Further statistical instrumentation is available through so-called 'watched variables', whose values are observed by the model monitor. Since PASCAL is a strongly typed language, they must be declared as either Boolean, integer or real. One can request frequency-counted (tallied) or time-weighted (accumulated) statistics. This feature corresponds to our notion of tally and measure objects in Chapter 2. Therefore **a_boolean** is a Boolean variable whose values are observed and weighted by the time period for which they accumulated. In the case of a server's *status* variable (0 for idle, 1 for busy) it produces a utilization statistic. In our program the *FlowTimes* variable tallies all occurrences of the customers' transition times (**t_real**), before leaving the shop. A **display** command can be used to print some standard statistics (mean, variance, etc.) for queues and watched variables. PASCAL's standard input/output must be used for more extensive or better formatted reports.

To aid error detection, selective or complete (**trace_all**) event traces and dumps (**dmp_evset**) can be requested [see Figure 3.4 (a)]. Queue objects must be initialized and watched variables must be cleared before a model is started.

SIMPAS also offers a number of additional features not used in this example.

Note that the **forward** declarations are required for the definition of events in their 'natural' order. PASCAL normally insists on declaration before reference, which means that the *hello* event would have to be declared last, with *bye*, *finish* and *start* preceding it.

The main disadvantage of precompilers such as this one is the additional execution time required for the precompilation phase. The system is, however, easy to use, assuming that a good PASCAL compiler is available. Run-time error messages unfortunately refer to the PASCAL rather than the SIMPAS source code.

B. Compilers

SIMSCRIPT [Kiviat *et al.* (1968)] and SIMULA [Birtwistle *et al.* (1973)] are probably the most widely used, best-documented and best-supported self-contained compiled languages with discrete-event simulation capabilities. Both languages were designed to be user-extensible, so that new features for specialized applications could easily be added. SIMULA's class concept has proved to be an excellent vehicle for building application-oriented contexts, while level 7 of SIMSCRIPT was never released [Markowitz (1979a, 1979b)].

SIMSCRIPT was originally a FORTRAN-based language. Although it has since outgrown most of FORTRAN's restrictions, some of this heritage can still be noticed. Some weird and potentially dangerous features are still part of the language, presumably in order to preserve compatibility with older versions.

Variables can be implicitly typed ('*I* to *N*' rule) or declared. Comparison operators are FORTRAN-like (.**lt**.: less than). **integer** and **real** are the only data types. A **define** facility provides macros that can be used to make models more readable.

```
program ManuelsShop (INPUT, OUTPUT);

  const
    NormalTermination = 0;

  type
    aCustomer = queue member ↑
                   ShopEntryTime : real;
                end;

    aSource   = record
                   lambda : real;
                end;

    aQueue    = queue of aCustomer;

    aServer   = record
                   busy : a_boolean;
                   mu   : real;
                end;

  var
    testing     : char;
    status      : integer;
    source      : aSource;
    Manuel      : aServer;
    Q           : aQueue;
    FlowTimes   : t_real;
    samples,
    SampleSize  : integer;

  include expo;

  event hello;                         forward;
  event StartService;                  forward;
  event finish (romeo: aCustomer);     forward;

    procedure ReadAndInitialize;
      begin
      samples      := 0;
      readln (SampleSize);
      readln (source.lambda, Manuel.mu);
      clear Manuel.busy;
      Manuel.busy := false;
      initialize Q;
      clear FlowTimes;
      schedule hello now;
      writeln ('Do you want to run in debugging mode?');
      read (testing);
      if testing = 'y' then
        trace_all := true
      else
        trace_all := false;
    end {of ReadAndInitialize procedure};
```

Program 3.1 *An event-oriented queueing analysis of Manuel's shop (in SIMPAS)*

```
   event hello;
     var
       romeo: aCustomer;
     begin
       create romeo;
       romeo ↑ .ShopEntryTime := time;
       insert romeo in Q;
       schedule StartService now;
       schedule hello delay expo (source.lambda, 1);
       if testing = 'y' then   dmp_evset;
     end {of hello event};

   event StartService;
     var
       st     : real;
       romeo : aCustomer;
     begin
     if (not manuel.busy) and (not q.empty) then
       begin
         manuel.busy := true;
         remove the first romeo from Q;
         st := expo (Manuel.mu, 2);
         schedule finish (romeo) delay st;
       end {if};
     end {of StartService event};

   event finish;
     begin
       FlowTimes := time - romeo ↑ .ShopEntryTime;
       destroy romeo;
       manuel.busy := false;
       samples := samples + 1;
       if samples > SampleSize then
         schedule main (NormalTermination) now
       else
         schedule StartService now;
     end {of finish event};

begin { ********** Body of main Program ********** }
  ReadAndInitialize;
  start simulation (status);
  if status = NormalTermination then
    begin
      WRITE   ( <heading> );
      WRITELN; WRITELN ('Server utilization   :',manuel.busy.mean:10);
      WRITELN; WRITELN ('Flowtimes of customers');
      display FlowTimes;
      WRITELN; WRITELN ('Statistics for QUEUE in front of Manuel');
      display Q.size;
    end
  else
    WRITE ('Error # ', status: 1);
  end {of Model}.
end
```

```
OK, seg #simmanuel                                                    (a)

INPUT
500
0.8  1.0

Do you want to run in debugging mode ? y

-------------------------------------------------------------------------
hello        (id  1) occurs at time                      0.0000000E+00
last scheduled/rescheduled at time                       0.0000000E+00
at line   39    in program/module                          manuelsshop
-------------------------------------------------------------------------

-------------------------------------------------------------------------
startservice (id  2) scheduled for time                  0.0000000E+00
last scheduled/rescheduled at time                       0.0000000E+00
at line   60    in program/module                          manuelsshop
-------------------------------------------------------------------------

-------------------------------------------------------------------------
hello        (id  3) scheduled for time                  7.1929644E-01
last scheduled/rescheduled at time                       0.0000000E+00
at line   61    in program/module                          manuelsshop
-------------------------------------------------------------------------

-------------------------------------------------------------------------
   EVENT SET DUMP                                 TIME= 0.0000000E+00
-------------------------------------------------------------------------
startservice (id  2) scheduled for time                  0.0000000E+00
last scheduled/rescheduled at time                       0.0000000E+00
at line   60    in program/module                          manuelsshop
-------------------------------------------------------------------------

-------------------------------------------------------------------------
hello        (id  3) scheduled for time                  7.1929644E-01
last scheduled/rescheduled at time                       0.0000000E+00
at line   61    in program/module                          manuelsshop
-------------------------------------------------------------------------
                          END EVENT SET DUMP
-------------------------------------------------------------------------
```

Figure 3.4 Event trace and analysis of queueing delays in Manuel's shop: (a) with debugging option; (b) after termination

SIMSCRIPT offers flexible report-generation statements, which are pleasant to use. It is available on a wide range of computer equipment. Newer versions of the language have added primitives for process-oriented [Russel (1976)] and combined discrete–continuous [Delfosse (1976)] modelling. A more interactive version, which runs on personal work stations (IBM, PC, for example) has recently been announced.

SIMSCRIPT supports event-oriented frameworks similar to SIMPAS. It was designed to have a readable syntax. This accounts for the number of 'noise' words (**a**, **an**, **in**, **and**, . . .) which can optionally be used in com-

```
------------------------------------------------------------------
startservice (id  2) occurs at time                 0.0000000E+00
last scheduled/rescheduled at time                  0.0000000E+00
at line   60   in program/module                    manuelsshop
------------------------------------------------------------------

------------------------------------------------------------------
finish       (id  4) scheduled for time             1.2735295E+00
last scheduled/rescheduled at time                  0.0000000E+00
at line   75   in program/module                    manuelsshop
------------------------------------------------------------------

------------------------------------------------------------------
hello        (id  3) occurs at time                 7.1929644E-01
last scheduled/rescheduled at time                  0.0000000E+00
at line   61   in program/module                    manuelsshop
------------------------------------------------------------------
```

(b)

```
OK, seg #simmanuel

INPUT
500
0.8  1.0

Do you want to run in debugging mode ? n

Manuel's Lovepotion shop
**********************

Simulation stops at: 6.302E+02

Server utilization: 7.37E-01

Flowtimes of customers
nobs=  501 max.= 1.50853E+01 mean= 2.45697E+00 var= 6.31325E+00

Statistics for queue in front of Manuel
nobs= 1015 max.= 14          mean= 1.32556E+00 var= 4.73890E+00
```

mands. SIMSCRIPT subscribes to a so-called **entity/attributes/sets/events** world view. Entities are used to model real-life objects, characterized by a number of attributes. Sets are provided for entity–entity groupings and relationships. Events and scheduling commands define a model's dynamic behaviour.

Every SIMSCRIPT program consists of a **preamble**, a **main** program, and a number of **event** and other **routines**. The **preamble** defines all global data structures: event notices, permanent entities, temporary entities and variables. While permanent entities become permanent parts of a model,

```
preamble

event notices include hello, start, finish, bye

permanent entities
every source has a lambda
every manuel has a status, mu, utilization, and owns a queue

temporary entities
every customer has a ShopEntryTime and may belong to a queue

define samples, SampleSize and FlowTime as variables
define idle    to mean 0
define busy    to mean 1
define minutes to mean units

accumulate utilization (manuel) as the mean of status (manuel)
tally      MeanFlowTimes        as the mean of FlowTime

main
read lambda (source), mu (manuel), and SampleSize
let samples         = 0
let status (manuel) = idle
schedule a hello in negexp.f (lambda (source), 1) minutes

start simulation

perform aFinalReport now
stop
end

event hello
create a customer called romeo
let ShopEntryTime (romeo) = now
file romeo in the queue (manuel)
schedule a start now
if samples.lt. SampleSize then
  schedule a hello in negexp.f (lambda (source), 1) minutes
return
end
```

Program 3.2 *An event-oriented queueing analysis of Manuel's shop (in SIMSCRIPT)*

temporary entities can be dynamically created and destroyed. Entities or **the system** may own and be members in any number of lists. This flexibility is not available in our PASCAL toolboxes or in SIMULA. The **tally** and **accumulate** commands can be used for model instrumentation, apart from writing explicit procedures for data collection and compression.

Program 3.2 shows a SIMSCRIPT II.5 model of Manuel's shop. It is very similar to the SIMPAS program.

```
event start
if status (manuel) = idle and queue (manuel) is not empty then
   let status (manuel) = busy
   remove the first customer called romeo from the queue (manuel)
   schedule a finish (romeo) in negexp.f (mu (manuel), 2) minutes
regardless
return
end

event finish (romeo)
schedule a bye (romeo) now
let status (manuel) = idle
schedule a start now
return
end

event bye (romeo)
let FlowTime = now - ShopEntryTime (romeo)
destroy romeo
add 1 to samples
return
end

routine for aFinalReport
begin report
start new page
   begin heading
   print 2 lines as follows
                                         Manuel's Shop
                                         -------------
   end ''heading
skip 2 output lines
print 1 line with utilization (manuel) as follows
   Manuel's Utilization was **.**
print 1 line with MeanFlowTimes like this
   Customers spent an average of ***.* minutes in Manuel's shop.
end ''report
return
end
```

3.4 Building higher-level scenarios

Programming languages have traditionally been classified as higher or lower level, with the human programmer/analyst as the point of reference. Low-level languages are close to machine structures, reflecting architecture and operation of some particular processor–memory configuration. High-level languages are closer to a human's natural mode of thinking, reflecting architecture and operation of our cognitive system. Some benefits of high-level man–machine communication have already been briefly discussed in Chapter 1.

Programming languages have traditionally been bound close to a particular computational model, the von Neumann architecture. This is slowly beginning to change as a drop in hardware prices makes very high-level languages more attractive. Many of these fourth- or even fifth-generation systems are either program generators with a natural-language front end for some restricted task domain, or they claim to support a less procedural style of programming (for example, PROLOG).

Figure 3.3 shows various programming and simulation languages and the levels at which they are bound. Although many simulations have been written in very low-level languages, the benefits of using customized building blocks for simulation programming were recognized at quite an early stage. RAND's Job-Shop Simulator was one of the first parameter-driven program generators and GPSS was the first so-called scenario language.

Scenario languages offer predefined modules (data types and procedures) that simplify animations of some microworld. The queueing network paradigm is a good example of this. An analyst can program in terms of high-level concepts like sources, sinks, transactions, servers, queues and their transformations (create TA, seize server, enter queue, etc.), instead of coding in terms of low-level data structures, procedures and control-flow directives.

Scenarios are typically specified initially through some graphical notation like petri nets, system dynamics, block or activity-cycle diagrams. These are then mapped on to the concepts supported by the programming system. DYNAMO, DEMOS, GPSS and DISCO are only some of a number of such scenario languages that are frequently used for simulation-model building.

Even though they are high-level in the sense that they support problem-oriented modelling styles, they are not very high-level languages. This term is usually reserved for highly declarative systems, where a programmer only describes a solution's properties and expects the system to generate instances of it.

CAPS is a **model generator** that builds model descriptions in terms of activity cycles, assembled during a dialogue with its users. Another kind of declarative system is exemplified by non-interactive model generators for computer-configuration models, like SCERT, SAM and CASE [see Kreutzer (1977)]. NLPQ [Heidorn (1974)], a queueing network generator (GPSS) with an associated natural language interface, is representative of early attempts to provide more intelligent programming assistance.

3.4.1 An introduction to object-oriented programming and SIMULA

A. Principles of object-oriented programming

Object-oriented programming styles view programs as collections of active components (sometimes called **classes** or **actors**) and their patterns of interaction. Each component has a name and responds to a number of messages. **Messages** may retrieve values of private variables, change the value of some private variable, or trigger a computation causing some side effect (reading, printing, moving a cursor on the screen). The principle of strong localization of information in closed modules is strictly adhered to. Communication between objects takes place only through messages, which therefore define a protocol of all the requests an object will respond to. Other data structures, functions, procedures are private and can only be used from within an object.

Another characteristic of object-oriented programming is the use of some **inheritance mechanism**. Classes of objects may inherit properties and message patterns from other classes. Hierarchical structures are most commonly supported, where classes can be further specialized into subclasses, subsubclasses, and so on. Hewitt (1977), Rentsch (1982) and MacLennan (1983) briefly discuss the conceptual bases of this programming paradigm.

Object-oriented programming has proved to be a very elegant and productive programming style for dealing with complex models and software systems. It has so far predominantly been used for simulation programming (SIMULA classes) and experimental systems in artificial intelligence (actor languages). SMALLTALK is the paradigmatic object-oriented system. It has pioneered and popularized these ideas for a wider sector of the computing community, adding window-based interaction and powerful techniques for graphical animations – applications that are quite naturally expressed in an object-oriented framework.

To be able to host this programming style a language has to support some concept of classes or abstract data types, together with some mechanism for property inheritance. SIMULA can (but need not) be used in such a manner, if the programmer exerts some self-discipline. A number of LISP-based systems have been constructed to support this computational method as an extension of the 'frame' approach to knowledge representation [Minsky (1975)]. The strictly hierarchical property inheritance typified by SIMULA and SMALLTALK is too restrictive for some applications. SMALLTALK can quite easily be extended to support inheritance networks [Borning and Ingalls (1982)]. Other languages [GLISP: Novak (1983); Flavours: Weinreb and Moon (1981)] offer more flexibility.

Figure 3.5 sketches the structure of an object-oriented framework for queueing-network simulations. Although the basic idea is similar to the PASCAL toolbox, there is now a three-level hierarchy. All objects are specializations of classes *aProcessObject* and *aStructureObject*. Process objects can actively execute tasks, while structure objects are passive entities, responding to a set of predefined messages.

All objects are parameterized by their name and inherit *report* and *reset* procedures for printing and resetting their personal statistics. More specialized objects are defined as subclasses: *aDist*, *aCustomer*, *aSource*, *aQueue*, *aServer*, *aSink*, *anObserver*. These subclasses are further specialized (*anExpoDist*, *aNormalDist*, *aUnifDist*; *aFIFOqueue*; *aSingleServer*; *aCounter*, *aTally*).

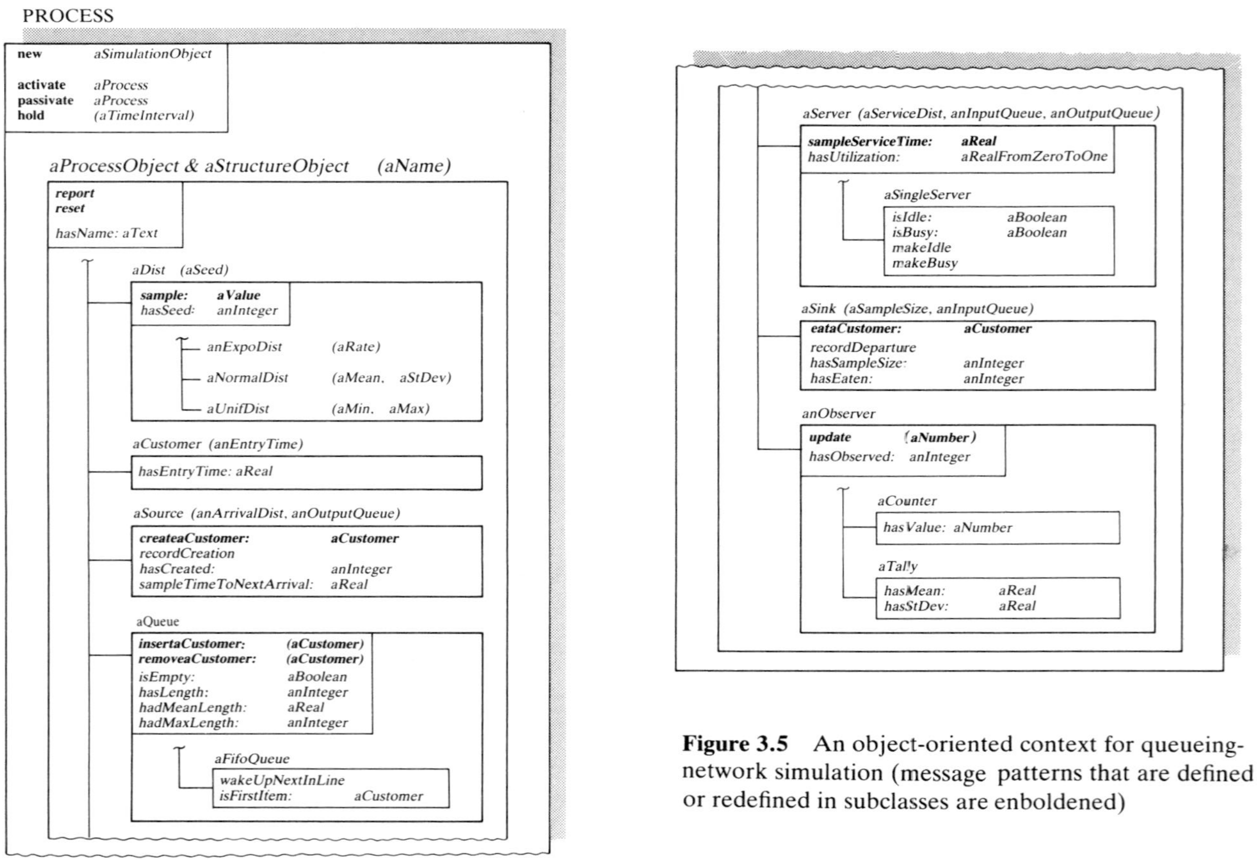

Figure 3.5 An object-oriented context for queueing-network simulation (message patterns that are defined or redefined in subclasses are enboldened)

All **distribution objects** have procedures to show their seeds and respond to sampling messages. There are subclasses for exponential, normal and uniform distributions.

All **customers** answer messages to report their *EntryTime*.

All **sources** know their inter-arrival-time distribution and the queue into which arriving customers are put (reference variables). They also respond to creation messages and enquiries about the number of customers they have already created.

Sinks are parameterized by *sampleSize* and *inputQueue*. They can be used to delete finished customers' records and count the number of completions.

Queues respond to messages to insert or remove customers, wake up the next waiting customer, produce a reference to the first queue item, query their current mean or maximum lengths, and answer whether or not they are empty. FIFO queues provide specific implementations for the insertion, deletion, and wake-up messages. Other queue types (such as priority queues) may define these differently.

Apart from knowing their name, how to report and reset their statistics (inherited from *aProcessObject*), all **server objects** know their input and output queues, and their service-time distribution. They respond to queries about their capacity, their service-time (sampling from the service time distribution) and utilization. A *SingleServer* is a specialization of this class. Its capacity is one workload item. It responds to messages to set and answer queries about its idle and busy states.

All **observer objects** keep an observation count. Counter objects are observers that are incremented by values passed to them. Tally objects are specializations of observers that can be asked to record observations (*update*) and compute their non-weighted mean and standard deviation.

Appendix 1 shows an implementation of this *MiniQNcontext* written in SIMULA. Since SIMULA forms the base of many of the software systems discussed in this chapter, we must briefly look at its most relevant features.

B. Programming in SIMULA

SIMULA is a programming language with simulation capabilities. Its earliest version, SIMULA I, was designed and implemented by Dahl and Nygaard in 1966 as a tool for discrete-event simulations. SIMULA 67, the current version, is a totally redesigned language. Birtwistle *et al.* (1973), Franta (1977) and Papazoglou *et al.* (1984b) are good introductions. Birtwistle *et al.* (1984) give a fairly complete bibliography.

SIMULA is a general-purpose programming tool that has been well standardized and implemented. It is quite widely used today, particularly in Europe and Australia. Even though it is now a relatively old language, some of its underlying ideas and features are still superior to those found in much more current programming systems. SIMULA has never gained any widespread use in commercial environments, but it has had a great influence on programming language design. Languages supporting the idea of data encapsulation (CLU, ALPHARD, MESA, CONCURRENT PASCAL) and so-called actor languages used by the artificial-intelligence community (PLASMA, ACT, SMALLTALK) are ultimately rooted in SIMULA.

Compilers for the language were originally distributed and supported by the Norwegian Computing Centre (NCC) in Oslo. It now hosts SIMULA A.S., a commercial company which acts as a co-ordinating body for matters associated with SIMULA development. The SIMULA Standards Group (SSG) and the SIMULA Development Group (SDG) have done an excellent job in keeping language implementations free from non-standard dialects and extensions of dubious value. A portable (S-PORT) version has recently become available and has led to a number of new implementations on a wide range of machines. Appendix 2 gives an overview of these. The Association of SIMULA Users (ASU) is an active user group, serving as an information-clearing house and publishing a quarterly newsletter.

A number of microcoded hardware designs for SIMULA machines [such as Papazoglou *et al.* (1984)] have been proposed, and a number of implementations on personal work stations have appeared. Although it would be a very attractive base for flexible and powerful modelling work benches, there is as yet no sophisticated programming environment like the SMALLTALK one (see Section 3.6).

SIMULA is an ALGOL-based language. A slightly modified ALGOL 60 forms the base of its programming features. This is extended by a flexible I/O system, text-handling facilities, classes, reference variables, coroutines and system-defined utilities for list processing and discrete-event simulation.

We will examine SIMULA's main features by studying the stable-marriage problem, set in a miniworld populated by such famous fantasy characters as *Kermit*, *Skooter*, *Animal*, *Linus*, *MissPiggy*, *Henrietta*, *Lucy* and *SnowWhite*.

The stable-marriage problem has been well described in the literature [Gale and Shapley (1962), McVitie and Wilson (1969, 1970), Wilson (1972), Itoga (1978, 1981, 1983)]. Our discussion is based on that of Allison (1983). The problem may be summarized as follows:

> Given a set of N males, a set of N females, and a list of preferences of marriage partners for each male and female, find a stable marriage for the whole group. A marriage constellation is said to be unstable if there are a male and a female who are not married to each other, but who both prefer each other to their current partners.

Gale and Shapley showed that at least one and possibly many solutions always exist and gave an algorithm to solve the problem. McVitie and Wilson (1969) published three ALGOL programs for its solution. One of these is a nice example of a *tournament* strategy. In this scenario males and females are considered as individual entities, each with a list of partner preferences and a reference to a current fiancée. The following procedure constructs a *male-optimal* solution.

Each male proposes in order of preference until he is accepted. He is then kept in suspense by his fiancée until either he is dropped in favour of a better choice or the tournament is over. If dropped, he continues his courting by asking for the hand of his next choice. Eventually he will be married to his final financée.

Each female accepts her more preferred suitor. She accepts the first proposal she gets, but later she may drop her fiancée whenever she receives a better offer.

In McVitie and Wilson's ALGOL-based solution all control information resides in the main program and procedure calls are used to alter it. The algorithm is more naturally formulated in an object-oriented style, using coroutines. Here each instance of a male or female independently controls its own destiny. There is a separate instance of class *Male* and class *Female* for each of them. Males loop over their preferences; to propose they must **resume** a female. A female initially accepts the first suitor's proposal. Thereafter she rejects the less favoured, using **resume** to pass control to a rejected male. The main program resumes each male in turn, until all are happily married.

Two versions will be given to solve this problem. Both have in the past been used to stage a famed marriage festival held on the plains of Morven, a rugged and inaccessible kingdom in the far north of DreamWorld.

Once they reach the age of 18 all of Morven's citizens take part in a big tournament, where they hunt for suitable brides and grooms. Tourists from all corners of DreamWorld come to witness this event, held on the seventh full moon of each year. In order to assure stable marriages the tournament is conducted according to the ritual reported here. This strategy will eventually lead to a male optimal solution, in line with Morven's basic social philosophy (to keep males happy).

A model of this algorithm is not a simulation as such, although the basic idea of taking a miniworld through a sequence of states is similar to one. Contrary to a true simulation model this procedure will always find a solution.

SIMULA is a good programming tool to describe such a scenario. Any possible world can be defined in terms of entities and relationships between them. Entities have certain properties and behaviour patterns. Relationships are represented by associations between entities or by membership in certain sets. In the miniworld under discussion males and females are such entities, owning properties like name, current fiancé(e) and partner preferences. In a conventional world males know how to propose and females know how to accept or reject a proposal. Relationships are formed by proposals, engagements, and by belonging to the male brotherhood or female sisterhood. To implement this conceptual view in a computer-based model it is necessary first to map entities on to structured objects and relationships on to associations between them. Computations will implement state-changing activities.

One particularly convenient world model leads to descriptions of objects in terms of slots for attributes and processes for their life cycles. Associations are defined by reference variables, arrays or list structures. Slots of objects can be filled by values, or by functions and procedures computing values or causing state changes whenever they are executed.

SIMULA views models as populated by communicating objects. Its basic structuring device is the idea of a class of objects, sharing a common structure and behaviour pattern. In this sense males and females can be perceived as instances of classes *Male* and *Female*. Although *Kermit*, *Skooter*, *Linus* and *Animal* are individuals with different names, fiancées and preferences, for the purpose of our model they can be described by the same set of characteristics. Only the values filling their slots are different. The same holds for *MissPiggy*, *Henrietta*, *Lucy* and *SnowWhite*, who also share the same properties, with different instantiations.

All males share a procedure for making proposals, and all females share knowledge on how to accept or reject a suitor. Our males also have the same simple life style! They propose to the next-best female on their list until the tournament is over. All females continue to consider proposals, accepting or rejecting them, until no further proposals are made.

A SIMULA **class** provides an encapsulation of all aspects of some type of object under a common name. It contains descriptions of an object's properties, the actions it can perform and the life cycle that instances of this class will follow, once they have been created. Any number of class instances can be dynamically created, with their attributes instantiated to appropriately chosen values. In our model we include a class description for male objects and one for female objects. At the start of a simulation the model instantiates *Kermit*, *Skooter*, *Linus* and *Animal* to males with different partner preferences. *MissPiggy*, *Henrietta*, *Lucy* and *SnowWhite* are similarly created as female objects and are supplied with distinct preferences for male companions.

A look at the attributes shows that most of them are identical for both males and females. Any person, regardless of sex, has a name, an association with a current fiancé(e) (possibly **none**) and a list of partner preferences. Males and females differ only in that males know how to propose and females know how to react to a proposal. It seems unreasonable to declare these identical attributes twice, once in class *Male* and again in class *Female*. SIMULA therefore provides a **prefixing** mechanism, through which properties may be inherited by an object.

All common attributes and life-cycle aspects can be extracted from the two classes (*Male* and *Female*) and bound into a new class called *Person*. It is specified that any person has a name, partner preferences and a current fiancé(e). Persons also know how to initialize these properties once they have been created. The *Male* and *Female* classes can be prefixed by class *Person*, which means that they automatically inherit a person's life style and all of its attributes.

Class descriptions are now simplified considerably. Persons are specialized to male persons by adding a procedure for making proposals and appending a loop of proposals to a person's basic life style. Female persons add a procedure to accept or reject proposals and their life cycle is extended to allow them to react to chains of proposals.

The idea of hierarchies of classes of objects has proved to be an extremely powerful vehicle for the description of complex systems in a well-structured and natural way. Together with a coroutine facility, which allows objects to step through a sequence of actions and communicate with other objects in a quasi-concurrent fashion, it almost ideally realizes the concept of closure via textual clustering of conceptually related information. This strategy permits the organization of descriptions of complex systems into almost independent modules, dividing them into intellectually manageable chunks with well-behaved interactions between them. The idea of problem solving by step-wise decomposition [Dahl *et al.* (1972)] is based on the same principle.

Objects are individual class instances, which are created dynamically:

```
Kermit:- New Male (...);
```

These objects permanently acquire all properties, functions, procedures and behaviour patterns defined for a class and all its superclasses. This sequence of superclasses

```
class Person;
  Person class Male;
  Male class Muppet;
  Muppet class FrogMuppet;
  ...
```

is called a **prefix chain** (*Male* is prefixed by *Person*, *Muppet* is prefixed by *Male*. . .). Property inheritance follows such prefix chains.

SIMULA's class mechanism is an extremely versatile structuring device. An object can be purely passive, without a life cycle. It then consists of data and/or procedure slots only, similar to a PASCAL record. An equivalent to PASCAL's dot notation is used to access its attributes:

```
class Frog;
  begin
    text aName;
    boolean isMuppet;
  end;
ref (Frog) Kermit;
Kermit:- new Frog;
Kermit.aName:- copy ("Kermit");
Kermit.isMuppet:= true;
```

Note the use of :−, a special kind of assignment symbol. SIMULA is one of the few languages that distinguishes between assignment of values (:=) and reference assignment (:−) at the syntactic level. Reference assignment is used to bind pointers to objects. Such objects can be instances of pre- or user-defined classes or strings of characters (text objects):

```
currentProposer:- this Male (1, 4);
currentProposer.aName:- copy ("Kermit");
```

The qualifier **this** refers to the currently executing instance of a class. **copy** is a standard procedure to allocate storage for text objects.

In contrast to PASCAL we can also fill slots by functions or procedures:

```
class Frog;
  begin
  . . .
    procedure WriteName;
      outtext (aName);
  end;
```

Kermit.WriteName; will now write Kermit's name into the current output file.

If a class consists of a life cycle only, without any attributes, its instances are procedures or coroutines. Coroutines can be resumed until they pass through their last **end** statement.

A class template may also consist of a name only, with no attributes or life cycle of its own. It then serves solely as a label for dynamically created data items.

Classes can be used to model active entities when they provide both properties and life cycle. The life cycle may contain some standard initializations and can be executed in a coroutine fashion:

```
class Male;
  begin
    text aName;
    ref (Female) array preferences (1: NoOfFemales);
    ref (Female) fiancee;
    procedure propose (myDelight);
                              ref (Female) myDelight;
      begin
        . . .
        resume (myDelight);
      end of propose procedure;
  fiancee:- preferences (1);
  begin
    integer index;
  for index:= 1 step 1 until NoOfFemales do
    propose (preferences (index));
      end of block;
  end of Male class;
```

Objects exist for as long as they have not terminated their life cycles or can still be referenced. SIMULA's run-time system keeps reference counts for each object. It reclaims a class instance's memory space once it becomes inaccessible. A **garbage-collection** program performs this task on demand. Normally the life cycles of all classes on a prefix chain are concatenated, but there are mechanisms to modify this simple scheme.

Classes can also have formal parameters, which must be filled with actual values at the time objects are created. Subclasses inherit all the parameters of their superclasses:

```
class Person (anIndex, noOfCandidates);
  integer anIndex, noOfCandidates;
  ...
  Person class Male;
  ...
Kermit:- new Male (1, 4);
```

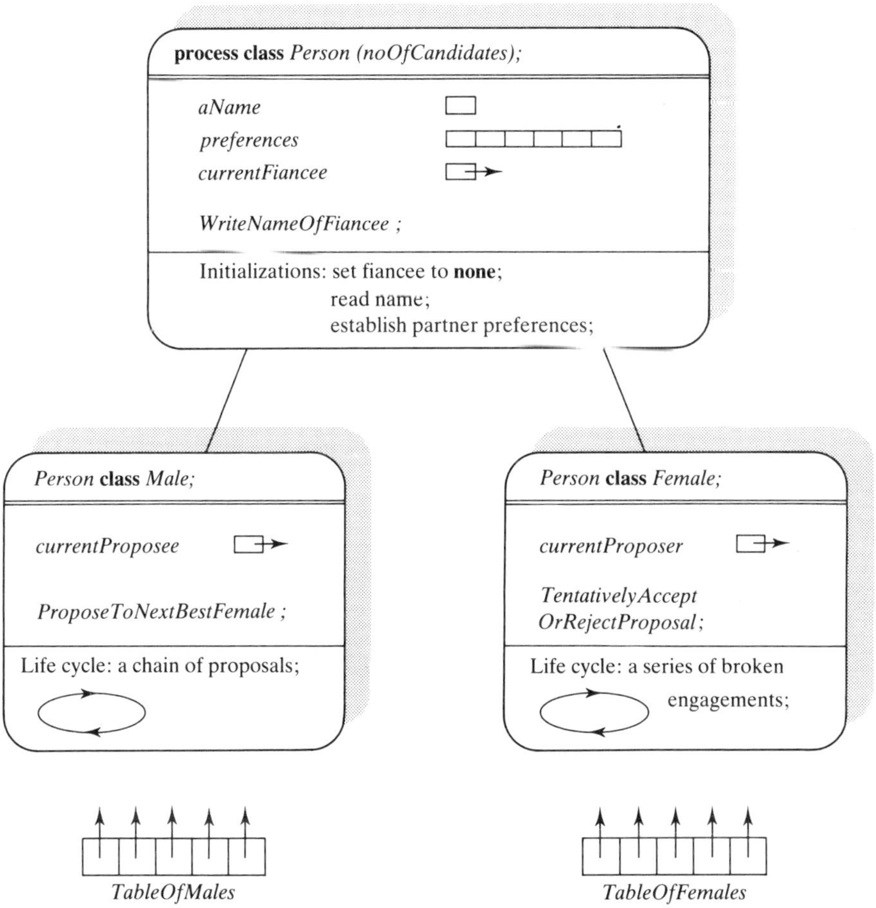

Figure 3.6 Class/subclass structure for the stable-marriage problem

```
begin
comment: A SIMULA Model of Morven's Marriage Tournament;

comment: We use coroutines for males and females. Data is read from a
         specified Data file, and results are written into a specified
         Results file;

  ref (infile) Data;

comment: The Data file should look like this:

         <first bride name>
         <a list of scores for each male in the tournament, establishing
           her preferences>
         <second bride name>
           ...
         <last bride name>
           ...
         <first groom name>
         <a list of scores for each female in the tournament>
         <second groom name>
           ...
         <last groom name>
             ...;

  ref (outfile) Results;
  integer     NoOfPotentialPartners;

NoOfPotentialPartners := inint;

comment: Set up Files and dimension the problem;

Data    :- new infile  ("CUPIDDATA");
Results :- new outfile ("CUPIDRESULTS");
Data    .open (blanks (80));
Results .open (blanks (80));

begin
comment: Model Block;

  ref (Male)   array TableOfMales   (1: NoOfPotentialPartners);
  ref (Female) array TableOfFemales (1: NoOfPotentialPartners);

  procedure ExchangeRings (bride, groom); ref (Female) bride;
                                          ref (Male  ) groom;
  begin
  inspect Results do
    begin
    outtext (bride.aName); outtext (" accepts a proposal of ");
    outtext (groom.aName); outimage;
    end of inspection;
  bride.currentFiancee :- groom;
  groom.currentFiancee :- bride;
  end of ExchangeRings procedure;
```

Program 3.3 *Stable marriage by coroutines (a SIMULA model)*

```
class Person (anIndex, noOfCandidates); integer anIndex, noOfCandidates;
begin
  text aName;
  integer array preferences (1 : noOfCandidates);
  ref (Person) currentFiancee;

  procedure WriteNameOfFiancee (aFile); ref (outfile) aFile;
    if currentFiancee =/= none then
      aFile.outtext (currentFiancee.aName)
    else
      aFile.outtext ("nobody (sigh !)");

comment: A Person's Body;

comment: initialize fiancee and preferences;

currentFiancee :- none;

Data.inimage;
aName :- copy (Data.image.strip);
Data.inimage;
  begin
    integer partnerIndex;
  for partnerIndex := 1 step 1 until NoOfPotentialPartners do
    preferences (partnerIndex) := Data.inint;
  end;

detach;   comment: Back to the main program;

end ***** of Person class *****;

Person class Male;
begin
  ref (Female) currentProposee;

  procedure ProposeToNextBestFemale (preferenceListIndex);
                                    integer preferenceListIndex;
  begin
  currentProposee :- TableOfFemales (preferences (preferenceListIndex));
  currentProposee.currentProposer :- this Male;
  Results.outtext (aName); Results.outtext (" proposes to ");
  Results.outtext (currentProposee.aName); Results.outimage;
  resume (currentProposee);
  end of ProposeToNextBestFemale procedure;

comment: A Male Person's Body;

comment: describe his lifecycle;
    begin
      integer preferenceListIndex;
    for preferenceListIndex := 1 step 1 until noOfCandidates do
      ProposeToNextBestFemale (preferenceListIndex);
    end of block;
```

(cont.)

```
outtext (aName); outtext(" resigns; disappointed !"); outimage;

end ***** of class Male *****;

Person class Female;
begin
    ref (Male) currentProposer;

    procedure TentativelyAcceptOrRejectProposal;
    begin
    if preferences (currentProposer.anIndex) <
      preferences (currentFiancee .anIndex) then
      begin
      comment: this guy is better than the one I am engaged to, so submit
               to him and send the old fiancee hunting for a new bride;
        ref (Male) yesterdaysHero;
      yesterdaysHero :- currentFiancee;
      Results.outtext (aName); Results.outtext (" breaks her engagement to ");
      Results.outtext (yesterdaysHero.aName); Results.outtext (" !");
      Results.outimage;
      ExchangeRings (this Female, currentProposer);
      resume (yesterdaysHero);
      end
    else
      inspect Results do
        begin
        comment: blissfully happy with the current engagement, so send
                 proposer away, disappointed;
        outtext (aName); outtext (" rejects "); outtext (currentProposer.aName);
        outtext ("'s proposal"); outimage;
        resume (currentProposer);
        end of inspection;
    end of TentativelyAcceptOrRejectProposal procedure;

comment: A Female Person's Body;

comment: always accept the first proposal;

ExchangeRings (this Female, currentProposer);
detach;     comment: return to main program to give other males a chance;

comment: her life cycle;
while true do
  TentativelyAcceptOrRejectProposal;

end ***** of class Female *****;

comment: ********** main body **********;

comment: Set up Brides and Grooms;

comment: It is a somewhat sexist model. Females are passive, waiting
         to be asked. Males are active, hunting for partners.
         This can easily be changed!;
```

Program 3.3 (cont.)

```
begin
    integer personindex;
for personIndex := 1 step 1 until NoOfPotentialPartners do
    TableOfFemales (personIndex) :- new Female (personIndex,
                                   NoOfPotentialPartners);
for personIndex := 1 step 1 until NoOfPotentialPartners do
    TableOfMales (personIndex)   :- new Male (personIndex,
                                   NoOfPotentialPartners);
for personIndex := 1 step 1 until NoOfPotentialPartners do
  resume (TableOfMales (personIndex));
end of Initialization block;

comment: Report on Results;

inspect Results do
  begin
    integer personIndex;
  outimage; outimage;
  outtext ("The results of this year's tournament are !"); outimage;
  outtext (" ====================== "); outimage;
  outimage;
  for personIndex := 1 step 1 until NoOfPotentialPartners do
    begin
    outtext ("The Bride of ");
    outtext (TableOfMales (personIndex).aName);
    outtext (" will be : ");
    TableOfMales (personIndex).WriteNameOfFiancee (Results);
    outtext (" ! ");
    outimage;
    end;
  outimage;
  for personIndex := 1 step 1 until NoOfPotentialPartners do
    begin
    outtext ("The Groom of ");
    outtext (TableOfFemales (personIndex).aName);
    outtext (" will be : ");
    TableOfFemales (personIndex).WriteNameOfFiancee (Results);
    outtext (" ! ");
    outimage;
    end;
  outimage; outimage;
  outtext (" ********** Congratulations ! ********** ");
  outimage;
  end of Reporting block;

end of Model block;

Data   .close;
Results.close;

end of Program;
```

Figure 3.6 summarizes our model's class structure, and Program 3.3 shows a corresponding SIMULA implementation.

The code for class *Person* demonstrates SIMULA's typical ALGOL style. As in PASCAL, parameters have to be typed, but the type declaration follows the parameter list. SIMULA supports three different parameter-passing strategies: by reference, by value and by name. Any object, including procedures and references to classes, can be passed as a parameter. Default modes are: 'call by value' for reals, integers, characters and booleans; and 'call by reference' for text, class, array and procedure objects. Call by reference corresponds to PASCAL's concept of **var** parameters.

'Call by name' passes parameters as unevaluated expressions, whose textual definition is substituted for each occurrence of the formal parameter. By explicitly overriding the standard defaults, parameters can be forced to be passed by value or name:

class *Person* (*aName*); **value** *aName*; **text** *aName*;

SIMULA offers a **text** data type whose variables may denote strings of arbitrary length. There is also a **character** data type, whose values must be single symbols. Characters are enclosed in single quotes, whereas text must be surrounded by double quotes:

aLuckyDigit:= *'7'*;
aSillyFrog:– **new** *Person* (*"Kermit"*);

The **array** data type follows ALGOL's conventions. ALGOL permits dynamic arrays, whose boundaries need only to be bound at the time a block is entered. Elements of arrays may be of any scalar type, including references to class instances. This last feature gives much more flexibility than is possible within PASCAL's relatively restrictive framework:

integer *NoOfFemales*;
NoOfFemales:= **inint**;

begin
  ref (*Female*) **array** *TableOfFemales* (*1*: *NoOfFemales*);
  . . .
end *of block*;

Arrays of arrays are not supported by SIMULA. One can instead define arrays of higher dimensionality.

Reference variables are similar to PASCAL pointers. They can only denote class instances and must be qualified by the name of a class:

ref (*Person*) *currentFiancee*;

During program execution they may be set to point to instances of this or any subclass of this class:

currentFiancee:– **new** *Person* (*3*, *4*);
currentFiancee:– **new** *Male* (*2*, *4*);

```
currentFiancee:- new Female (1, 4);
currentFiancee:- none;
```

As far as possible SIMULA tries to check for type violations at compile time. Although the language has been carefully designed to permit this, its references cannot always be made unambiguous and run-time checks have to be made instead.

Dot notation accesses an object's attributes from outside:

```
currentProposee.currentProposer:- this Male;
```

This is a uniform means of access to all of the aspects of a class; procedures can therefore be triggered in the same way:

```
ref (Male) Kermit;
...
Kermit:- new Male (1, 4);
...
Kermit.proposeToNextBestFemale (preferences (1));
```

There is also an **inspect** command, which can be used in a similar fashion to PASCAL's **with** statement.

SIMULA's model of input and output is centred on the idea of a file buffer called **image**. All input/output procedures operate through this buffer, either explicitly or implicitly. An **image** is a standard attribute of any file object known to the system. The standard default is **image** as a shorthand for **sysin.image** and **sysout.image**, where **sysin** and **sysout** are the standard input and output devices. These are normally associated with keyboard and terminal screen, although they can easily be reset:

```
ref (infile) Data;
...
Data:- new infile ("CupidData");
sysin:- Data;
```

There are a number of standard procedures to read from and write into a file buffer. The buffer is automatically filled for a read operation and its contents is automatically written to the file for a write operation. This can be explicitly requested through **inimage** and **outimage**. The functions and procedures **inint**, **inreal**, **outint**, **outreal** and **outfix** are for reading and writing numbers from and into the file buffer; **intext**, **inchar**, **outchar** and **outtext** read and write textual information:

```
ref (outfile) Results;
...
Results:- new outfile ("CupidResults");
Results.open (blanks (80));
Results.outtext ("Hear Ye, hear Ye, hear Ye!");
Results.outimage;
```

This defines a reference variable to a file called *Results*, instantiates it to a file object with external name *CupidResults*, opens the file, writes a text into its buffer and the buffer's contents to the file. The **open** procedure instantiates the file buffer to 80 initially empty locations.

In the example program *WriteNameOfFiancee* is a procedure-valued attribute, whose execution produces a side effect; writing a text into an output file:

```
Kermit.WriteNameOfFiancee;
```

class *Person* has both a structure (head) and a behaviour pattern (body). The statements in the body of the class are executed as soon as a new person is created. They initialize a person's name and partner preferences by reading from *Data*, our input file. Note that in SIMULA, as in ALGOL, comments are statements, terminated by a semicolon:

```
comment: A Person's Body;
```

The variable *partnerIndex* is used to initialize the *preferences* array. It is local to a block of only one **for** statement:

```
ref (infile) Data;
Data:- new infile ("CupidData");
Data.inimage;
begin
  integer partnerIndex;
for partnerIndex:= 1 step 1 until NoOfCandidates do
  preferences (partnerIndex):= Data.inint;
end;
```

Such blocks are a useful mechanism to restrict both scope and lifetime of auxiliary variables. They can be nested anywhere within a program and encapsulate declarations for local data structures and statement sequences. PASCAL's compound statement is not the same concept, since compound statements may not have any local declarations. The closest analogue to PASCAL would be a parameterless procedure.

Note also that a **for** loop's syntax differs slightly from the way it would be written in PASCAL. SIMULA's **for** statement has various options and is quite versatile. It can, for instance, run up and down with variable step width or it can step through lists of values:

```
for index:= NoOfPartners step -2 until 0 do...
for currentFiancee:- Kermit, Skooter, Animal do...
...
```

In our model the *currentFiancee* slot is initialized by a statement in a person's body:

```
currentFiancee:- none;
```

The reserved word **none** indicates that a reference variable currently does not point to any object. This statement is not strictly necessary, since SIMULA, unlike PASCAL, performs default initialization for all variables. All pointers and text variables are automatically set to **none**, all numeric variables to zero, all characters to blank and all Booleans to **false**. It is, however, good programming practice to document this through explicit initializations.

Males, in the stable-marriage model, are a special class of persons. They inherit all of a person's parameters, properties and behaviour patterns via the prefix:

Person **class** *Male*;...

Both classes and blocks can be prefixed in this fashion, providing strictly hierarchical property inheritance. All parameter lists, declarations and class bodies are concatenated; even for long prefix chains:

class *Object*;...
Object **class** *PhysObject* (*aWeight*);...
Object **class** *SpiritualObject* (*aSignificance*);...
PhysObject **class** *Animal*;...
Animal **class** *Pet* (*anOwner*, *aName*);...
Pet **class** *Cat* (*aBreed*, *aColour*);...
ref (*PhysObject*) *Thing*;
Thing:- **new** *Cat* (*15*, *"Wolfgang"*, *"Fritz"*, *"?"*, *"grey"*);

There is no restriction on the depth of inheritance trees grown by such prefix chains.

Males add a *currentProposee* slot to a person's attributes, which may be filled by any female. They also own methods for indexing into crowds of females and for proposing. To propose, they obtain a reference to the next elegible female, make this male known as a proposer and activate the female to ponder the proposal. The next female to apply to is found by taking the number of the next preference and indexing into a global table of female references kept by our program. The selected pointer refers to a female object whose body will be resumed, triggering her procedure for proposal evaluation.

SIMULA supports **coroutines** as one of its main control structures. Coroutines implement a more democratic relationship between caller and callee than the master–slave sequencing implied by conventional functions and procedures. Every object owns a **local sequence control** (LSC), which always remembers the last statement executed by a class instance. If resumed instead of called, it will continue on its life cycle where it last stopped executing (in order to resume another object).

After creation, **detach** can be used to make an object independent of and return control to its creator. It can then be resumed by anyone who obtains a reference to it. Specifically,

resume (*currentProposee*);

activates the female object a male is currently at the brink of proposing to, giving her a chance to consider the proposal. This male will wait in suspense until he is either rejected and resumed by her, or until the end of the tournament when he learns that he has won the hand of this fair lady.

A male's body adds some new statements to the initializations performed by each person. Males' life cycles are very simple: a stream of proposals is generated, waiting for the females' reactions, until either they are finally accepted or there are no more females to propose to. Because of the way the marriage tournament works, the last contingency should never occur.

Any SIMULA object can be active, passive or terminated. The terminated state is reached by falling through the last **end**. All of a terminated object's properties are still accessible as long as it can be referenced from somewhere. It can now, of course, no longer be resumed. If, in addition, it can no longer be accessed, its storage space may be reclaimed by the garbage collector.

The description of properties and actions local to class *Female* is similar to that of males:

```
while true do
  TentativelyAcceptOrRejectProposal;
```

implements an 'infinite' cycle of courtships. This will eventually terminate when no male in the model is resumed by his current fiancée, yielding the male-optimal solution. All males have been given a chance to propose in order of preferences and no female is willing to settle for less than she currently holds. All marriages are stable – for the moment anyway. Control now returns to the body of the main program.

SIMULA programs are blocks of statements, enclosed by **begin**... **end** brackets. There is no provision as such for naming programs, but it is good practice to do so through comments. Any SIMULA block may contain declarations, statements and other blocks. Scope rules are similar to PASCAL. Declarations may include **real**, **integer**, **character**, **boolean**, **ref**erence variables, **text**, **array**, **procedure** and **class**. The body of the outer block of a SIMULA program typically opens and closes files, creates class instances, initializes data structures and triggers objects, procedures and a final print-out of results.

Note that

```
inspect Data do
  begin
  ...
  end;
```

defines a scope for naming objects, similar to PASCAL's **with** statement. Since *Data* has been instantiated to an **infile**, all calls on input procedures (**inint**, etc.) within that block are automatically qualified by it (the same as

Data.**inint**;...). **inspect** can also be used to check the class membership of objects:

ref (*Person*) *currentFiancee*:
...
inspect *currentFiancee*
  when *Male* **do**...
  when *Female* **do**...
otherwise do...;

The statement

TableOfFemales (*personIndex*).*aName*:– **copy** (*Data*.**image**. *STRIP*);

demonstrates the flexibility of SIMULA's text handling. The left-hand side evaluates to a **text** variable (a *Person*'s name). Text variables are implemented as references to strings; that is why the symbol for reference assignment (:–) must be used. The right-hand side builds a string object holding the leftmost non-blank characters on the current input line of the *Data* file. *Data*.**image** refers to the *Data* file's buffer. **strip** is a standard procedure that can be used to strip all trailing blanks. The system procedure **copy** is needed to allocate storage for this string object, so that *aName* can be set to point to it.

The body of our example program also shows how new objects are created. The **new** function generates class instances. If we want to refer to them later we must point appropriately qualified reference variables to these:

ref (*Female*) **array** *TableOfFemales* (*1*: *NoOfPotentialPartners*);
...
TableOfFemales (*personIndex*):–
  new *Female* (*personIndex*, *NoOfPotentialPartners*);

Note that the two parameters of this female are inherited from her superclass. Class *Person* was defined with two formal parameters and class *Female* is a subclass of *Person*. Note also that in SIMULA **new** is a run-time command. Class instances can be dynamically created (and destroyed) at the time a model is executed. This is different from languages like ADA or MODULA, which provide concepts similar to classes (packets and modules). There **new** is executed when a program is compiled, which means that the number of instances is frozen at that time.

Figure 3.7 shows the results of a tournament for some sample data.

SIMULA can be classed as an extensible language, although there are no facilities for definition or redefinition of infix operators. The class concept allows us to define our own data and procedure abstractions and encapsulate them into hierarchically arranged contexts, building on concepts defined in subordinate classes. SIMULA's lowest level may be used as a base language for richer, higher-level programming environments. This is the level we have

```
Kermit proposes to SnowWhite
SnowWhite accepts a proposal of Kermit
Skooter proposes to Henrietta
Henrietta accepts a proposal of Skooter
Linus proposes to SnowWhite
SnowWhite breaks her engagement to Kermit !
SnowWhite accepts a proposal of Linus
Kermit proposes to Henrietta
Henrietta rejects Kermit's proposal
Kermit proposes to Lucy
Lucy accepts a proposal of Kermit
Animal proposes to SnowWhite
SnowWhite rejects Animal's proposal
Animal proposes to Henrietta
Henrietta rejects Animal's proposal
Animal proposes to Lucy
Lucy breaks her engagement to Kermit !
Lucy accepts a proposal of Animal
Kermit proposes to MissPiggy
MissPiggy accepts a proposal of Kermit

The results of this year's tournament are !
***********************************

The Bride of Kermit will be: MissPiggy !
The Bride of Skooter will be: Henrietta !
The Bride of Linus will be: SnowWhite !
The Bride of Animal will be: Lucy !

The Groom of MissPiggy will be: Kermit !
The Groom of Henrietta will be: Skooter !
The Groom of Lucy will be: Animal !
The Groom of SnowWhite will be: Linus !

*********************************** Congratulations ! ***********************************
```

```
INPUT
MissPiggy
8 1 3 4 2 5 6 7
Henrietta
8 3 2 4 5 6 7 2
Lucy
8 1 2 4 7 6 5 2
SnowWhite
8 1 3 4 5 2 6 7
Kermit
4 2 3 1
Skooter
2 3 4 1
Linus
4 2 1 3
Animal
4 2 1 3
Beauregard
4 3 2 1
BigBird
1 2 3 4
Grumpy
3 1 2 4
SuperMan
4 2 3 1
```

Figure 3.7 Stable marriage by coroutines: The results

discussed so far, although we have not mentioned all of its features. Most implementations support separate compilation of classes (**external class**. . .) and some offer mechanisms to protect internal attributes from external access (**hidden/protected**). Birtwistle *et al.* (1973) and Franta (1977) are still the best introductory textbooks.

A number of **complement languages** have been written to extend SIMULA's basic capabilities. Birtwistle *et al.* (1984) give an excellent tutorial on how this can be done. Special SIMULA contexts have been designed for writing interactive dialogues, graphics and simulation programming. Some of these will be exemplified in the following chapters.

Basic contexts for list processing and discrete-event simulation are part of any common-base SIMULA implementation. These are provided by system classes **simset** and **simulation**. Prefixing can make them available to any programmer who wants to use their features.

A block or class prefixed by **simset** or **simulation** inherits **simset's list processing facilities**.

```
simset begin
...
end;
```

All of SIMULA's utilities for discrete-event simulation are available within any class or block prefixed by **simulation**:

```
ref (Reality) MyWorld;
simulation class Reality;
  begin
  ...
  end;
MyWorld:- new Reality;
```

Class **simulation** is prefixed by **simset** because it uses linked lists to implement queues and agenda. **simset**'s list-handling primitives are therefore automatically accessible to any **simulation** user.

Class **simset** introduces linked lists as a basic data type. It defines so-called head and link objects. Class **head** represents lists of entities. These can be of any class.

```
ref (head) QofEligibleFemales;
...
QofEligibleFemales:- new head;
```

Instances of classes prefixed by **link** can be members in lists; although **simset** restricts such memberships to one list at a time. There are predefined procedures for filing link objects into and removing them from lists:

```
SnowWhite.into (QofEligibleFemales);
MissPiggy.out;
```

Objects can also be linked to other objects by **precede** or **follow**. Functions **first** and **last** are used to obtain references to the head or tail of a list:

```
currentProposee:- QofEligibleFemales.first;
```

Procedure **clear** flushes all members from a list; the Boolean function **empty** tests its status and **cardinal** returns the current number of members:

```
if QofEligibleFemales. empty then
  Despair;
  ...
remainingChances:= QofEligibleFemales.cardinal;
```

Class **simulation** offers all of **simset**'s features, a large collection of random-drawing procedures (**draw**, **randint**, **uniform**, **negexp**, **erlang**, **normal**, **poisson**), a clock (**time**), an agenda (**sqs**), a number of scheduling statements and procedures, as well as other utilities. There are also procedures (**discrete**, **linear**, **histd**), which sample frequency tables to approximate empirically determined frequency distributions.

The world view of the class **simulation** is process oriented. There is a single agenda, called the **sequencing set** (**sqs**), into which event-notice entities are filed. This is accomplished by a variety of scheduling commands.

Event notices have a time aspect, noting when the event is scheduled to occur, and a process reference.

Entities must be defined as **process** classes; only then are we allowed to schedule them:

> **process class** *Person* (...);
> ...
> **activate this** *Person*;

Processes are implemented through coroutines. Figure 3.8 [adopted from Franta (1977)] shows the various states they can be in, and the scheduling statements governing their transitions. A process can be 'active', executing its life cycle, or 'suspended', waiting to continue at some future (scheduled) time. It can also be 'passive', waiting to be scheduled. After it has exhausted all actions on its life cycle it will be 'terminated'.

Class **simulation** offers statements to **activate**, **reactivate**, **hold**, **passivate**, **wait** or **cancel** processes:

> **ref** (*Person*) *currentFiancee;*
> **real** *LengthOfCourtship*, *mean*, *sigma*;
> **integer** *rnSeed*;
> ...
> *currentFiancee*:- **this** *Female*;
> *LengthOfCourtship*:= **normal** (*mean*, *sigma*, *rnSeed*);
> **hold** (*LengthOfCourtship*);
> **activate** *currentFiancee*;

activate and **reactivate** can be modified by a number of clauses to specify time instants (**at** *t*) or delays (**delay** *t*). Detailed information about the order in which event notices should be filed can also be given (**prior**, **before**, **after**). This is often useful for tie-breaking between quasi-simultaneous events.

There is a predefined procedure to determine the time of a process' next scheduled phase (**evtime**). Processes can also be tested for being **idle** (passive) or **terminated**. **current** sets a reference to the currently active process and **nextev** refers to the next imminent event's event notice. The pseudo-variable **main** can be used to schedule the outer shell of a simulation program.

To demonstrate these features the rules of Morven's marriage festival will be modified to make it slightly more realistic. A time dimension and a stochastic element are both introduced. The model is now a true simulation,

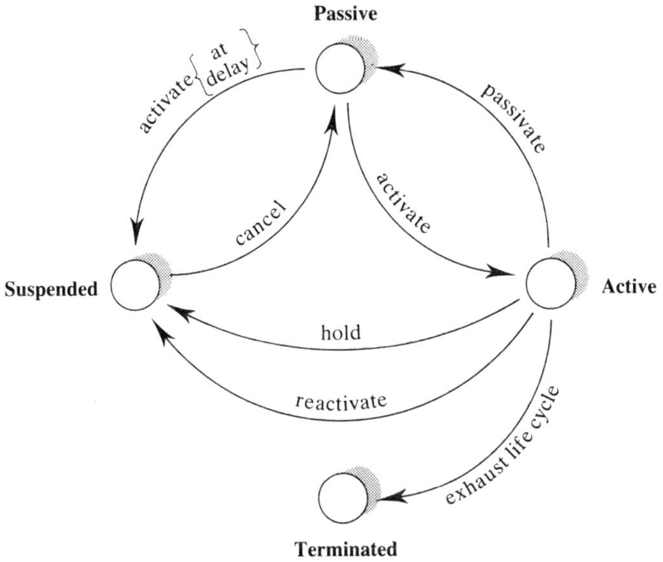

Figure 3.8 States and transitions of **process** objects

since the strategy (even though it is still iterative and systematic) can no longer guarantee to find a stable marriage for all partners.

Females are still lined up in an array, but males now have to queue in front of them, in order of arrival. They will try to propose to their next most preferred choice, but they may have to wait for the end of a previous courtship. To give females a chance to consider a proposal, the life cycles of pairs are suspended for some random time interval a courtship is going to take. This time is sampled from a normal distribution with a mean of five and a sigma of two minutes. Females then make up their mind on whether to accept or reject a proposal. Thereafter a rejected male again joins the tail of the queue in front of his next most preferred eligible female. The female will continue to remain open to suggestions.

There is now another complicating factor. In Morven it is socially quite humiliating to stay unattached for too long, particularly for females. During the tournament females are therefore quite nervous, afraid to be stranded without a partner at the end of the festival. This phenomenon has often been referred to as the 'old maid syndrome'. It causes females to finalize their choices after a certain time period has elapsed. The length of this period depends very much on personality, but it has generally been found to range from 30 to 120 minutes. Since no more is known a uniform distribution will be used to represent it.

This time the tournament cannot go on until all males have been claimed. It will finish when all females have panicked (121 minutes).

The modified class hierarchy for this model is shown in Figure 3.9. Since the new strategy must cope with unequal numbers of males and females it is

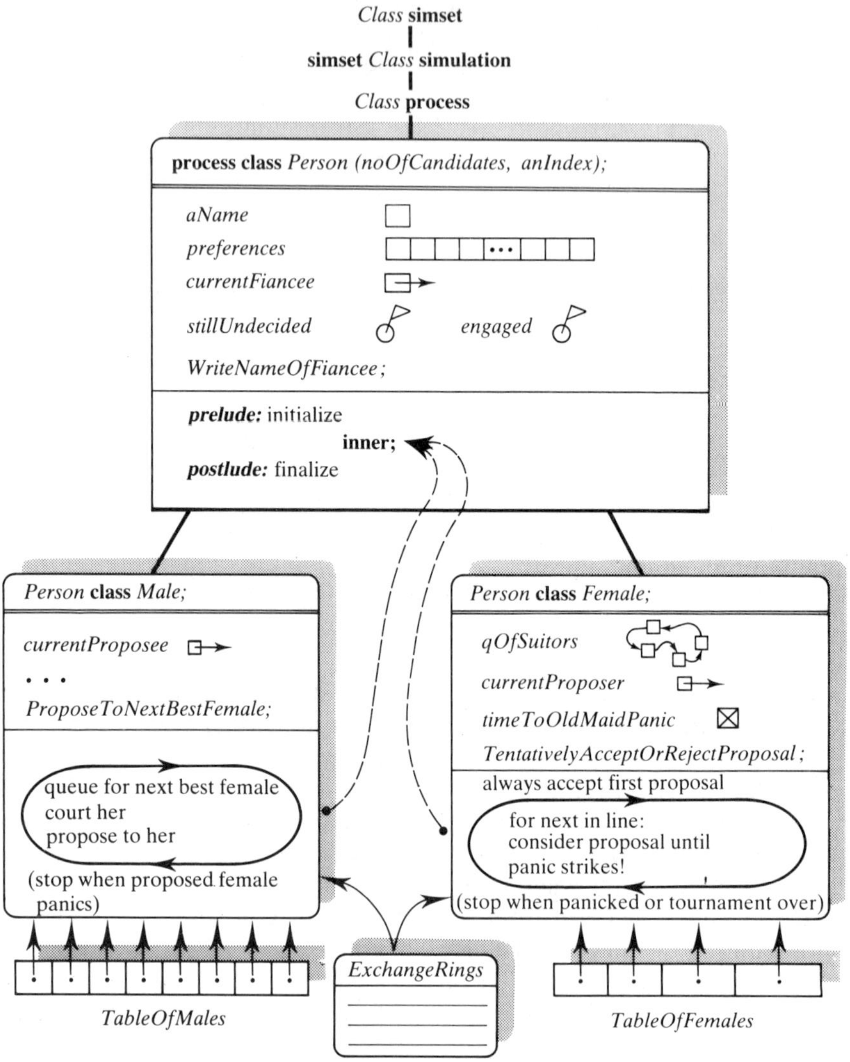

Figure 3.9 Stochastic stable marriages: Class/subclass structure

tried on a sample of eight males and four females; Figure 3.10 shows our results.

Apart from some additional variables we notice that the body of class *Person* (Program 3.4) now contains one block of statements for initializing data structures (*prelude*) and another one to report the result of the tournament for each male or female (*postlude*).

Life cycles of classes and their subclasses are normally concatenated. This can sometimes be inconvenient. In our example, we would like each person to execute the *prelude* first, then their male or female life cycles, and

```
Kermit proposes to SnowWhite at 4.34
SnowWhite accepts a proposal of Kermit at 4.34
Skooter proposes to Henrietta at 4.95
Henrietta accepts a proposal of Skooter at 4.95
BigBird proposes to MissPiggy at 5.08
MissPiggy accepts a proposal of BigBird at 5.08
MissPiggy panics and settles on BigBird at 5.08 !

***** BigBird has successfully courted MissPiggy !

Grumpy proposes to Lucy at 5.52
Lucy accepts a proposal of Grumpy at 5.52
Linus proposes to SnowWhite at 8.03
SnowWhite breaks her engagement to Kermit at 8.03 !
SnowWhite accepts a proposal of Linus at 8.03
SnowWhite panics and settles on Linus at 8.03 !

***** Linus has successfully courted SnowWhite !

Kermit proposes to Henrietta at 9.31
Henrietta rejects Kermit's proposal at 9.31
Henrietta panics and settles on Skooter at 9.31 !

***** Skooter has successfully courted Henrietta !

***** MissPiggy agrees to marry BigBird !

***** Henrietta agrees to marry Skooter !

Lucy rejects Grumpy's proposal at 20.00
Lucy panics and settles on Grumpy at 20.00 !

***** SnowWhite agrees to marry Linus !
```

```
INPUT
MissPiggy
8 1 3 4 2 5 6 7
Henrietta
8 3 2 4 5 6 7 2
Lucy
8 1 2 4 7 6 5 2
SnowWhite
8 1 3 4 5 2 6 7
Kermit
4 2 3 1
Skooter
2 3 4 1
Linus
4 2 1 3
Animal
4 2 1 3
Beauregard
4 3 2 1
BigBird
1 2 3 4
Grumpy
3 1 2 4
SuperMan
4 2 3 1
```

```
********************************** Congratulations ! ***********************************
```

Figure 3.10 Stochastic stable marriages: The results

then the *postlude*. This can be forced by using the compiler directive **inner**. Now the description of the life cycle of the subclass will be spliced into the body of the superclass at whatever point **inner** occurs:

> **process class** *Person* (. . .);. . .
>
> . . .
>
> **comment**: *Body*;
> *prelude*;
> **inner**;
> *postlude*;
>
> **end** *of Person class*;

```
begin

External class simulation;

comment: A SIMULA Model of Morven's Marriage Tournament (modified rules)
         The BASIC TIME UNIT is 1 minute !;

   ref (infile)  Data;
   ref (outfile) Results;
   integer       NoOfFemales, NoOfMales;
   integer       Seed;
   real          LengthOfSim;

NoOfFemales := inint;
NoOfMales   := inint;
Seed        := inint;
LengthOfSim := inreal;

comment: Set up Files and dimension the problem;

Data    :- new infile  ("CUPIDDATA");
Results :- new outfile ("CUPIDRESULTS");
Data    .open (blanks (80));
Results .open (blanks (80));

simulation begin

comment: Model Block;

   ref (Female) array TableOfFemales (1: NoOfFemales);
   ref (Male  ) array TableOfMales   (1: NoOfMales  );
   ref (head  ) QofEligibleFemales;

   procedure ExchangeRings (bride, groom); ref (Female) bride;
                                           ref (Male  ) groom;
   begin
   inspect Results do
     begin
     outtext (bride.aName) outtext (" accepts a proposal of ");
     outtext (groom.aName); outtext (" at "); outfix (time, 2, 7);
     outimage;
     end of inspection;
   bride.out;
   bride.currentFiancee  :- groom;
   bride.engaged         := true;
   groom.currentFiancee  :- bride;
   groom.engaged         := true;
   end of ExchangeRings procedure;

process class Person (noOfCandidates, anIndex); integer noOfCandidates;
                                                 integer anIndex;
```

Program 3.4 *Stochastic stable marriages (a SIMULATION model)*

```
  begin
    text          aName;
    integer array preferences (1 : noOfCandidates);
    ref (Person) currentFiancee;
    boolean       stillUndecided, engaged;

    procedure WriteNameOfFiancee (aFile); ref (outfile) aFile;
      if currentFiancee =/= none then
        aFile.outtext (currentFiancee.aName)
      else
        aFile.outtext ("nobody (sigh !)");

comment: A Person's Body;

comment: initialize fiancee and preferences;

comment: *** prelude ***;
currentFiancee :- none;
stillUndecided := true;
engaged        := false;
Data.inimage;
aName :- copy (Data.image.strip);
Data.inimage;
  begin
    integer partnerIndex;
  for partnerIndex := 1 step 1 until NoOfCandidates do
    preferences (partnerIndex) := Data.inint;
  end;

inner;

comment: *** postlude ***;
inspect Results do
  inspect this Person
    when Male do
      begin
      outimage;
      outtext ("***** ");
      outtext (aName);
      outtext (" has successfully courted "); WriteNameOfFiancee (Results);
      outtext (" !"); outimage; outimage;
      end
    when Female do
      begin
      outimage;
      outtext ("***** ");
      outtext (aName);
      outtext (" agrees to marry "); WriteNameOfFiancee (Results);
      outtext (" !"); outimage; outimage;
      end;

  end ***** of Person class *****;
```

(cont.)

```
Person class Male;
begin
  ref (Female) currentProposee;
  real          meanLengthOfCourtship, sigmaLengthOfCourtship;

  procedure ProposeToNextBestFemale (nextBestChoice);
                                        ref (Female) nextBestChoice;
  begin
  this Male.into (nextBestChoice.qOfSuitors);
  passivate;
  if time <= LengthOfSim then
     begin
     this Male.out;
     currentProposee :- nextBestChoice;
     DelayProposalForLengthOfCourtship:
     hold (normal (meanLengthOfCourtship, sigmaLengthOfCourtship, Seed));
     if currentProposee.stillUndecided and this Male.stillUndecided then
        begin
        Results.outtext (aName); Results.outtext  (" proposes to ");
        Results.outtext (currentProposee.aName); Results.outtext (" at ");
        Results.outfix  (time, 2, 7);                Results.outimage;
        currentProposee.currentProposer :- this Male;
        activate currentProposee;
        end;
     end;
  end of ProposeToNextBestFemale procedure;

comment: A Male Person's Body;

meanLengthOfCourtship := 5;
sigmaLengthOfCourtship := 2;

comment: His lifecycle;

begin
  integer index;
for index := 1 step 1 until NoOfFemales do
  begin
  if stillUndecided and time <= LengthOfSim then
     ProposeToNextBestFemale (TableOfFemales (preferences (index)));
  end;
end;

end ***** of class Male *****;

Person class Female;
begin
  ref (head) qOfSuitors;
  ref (Male) currentProposer;
  real        timeToOldMaidPanic;

  procedure TentativelyAcceptOrRejectProposal;
  begin
```

Program 3.4 (cont.)

```
if qOfSuitors.first =/= none then
    begin
    activate qOfSuitors.first;
    passivate;
    comment: wait to hear proposal;
    if preferences (currentProposer.anIndex) <
      preferences (currentFiancee .anIndex) then
        begin
        comment: this guy is better than the one I am engaged to, so accept
                 his proposal;
          ref (Male) yesterdaysHero;
        yesterdaysHero :- currentFiancee;
        Results.outtext(aName); Results.outtext(" breaks her engagement to ");
        Results.outtext(yesterdaysHero.aName); Results.outtext(" !");
        Results.outtext (" at "); Results.outfix (time, 2, 7);
        Results.outimage;
        this female      .engaged := false;
        yesterdaysHero.engaged := false;
        if currentProposer =/= none then
          ExchangeRings (this Female, currentProposer);
        activate currentFiancee;
        end
    else
      inspect Results do
        begin
        comment: blissfully happy with the current engagement, so send
                 proposer away, disappointed;
        outtext (aName); outtext (" rejects "); outtext (currentProposer.aName);
        outtext ("'s proposal at "); outfix (time, 2, 7); outimage;
        end of inspection;
    end outer if;
  if time <= timeToOldMaidPanic then
    begin
    comment: too afraid to stay unwed to stand the strain any longer !
             Settle on current fiancee !;
    inspect Results do
      begin
      outtext (aName); outtext (" panics and settles on ");
      if currentFiancee =/= none then
        outtext (currentFiancee.aName)
      else
        outtext (" Nobody (sigh)");
      outtext (" at "); outfix (time, 2, 7);
      outtext (" !"); outimage;
      end of inspection;
    stillUndecided := false;
    if currentFiancee =/= none then
      currentFiancee.stillUndecided := false;
    passivate;
    end
  else
    passivate;
end of TentativelyAcceptOrRejectProposal procedure;
```

(cont.)

```
comment: A Female Person's Body;

qOfSuitors :- new head;
timeToOldMaidPanic := uniform (30, 120, Seed);

passivate;
while (qOfSuitors.first == none) and (time < timeToOldMaidPanic) do
  passivate;
if time <= timeToOldMaidPanic then
  begin
  activate qOfSuitors.first;
  passivate;
  comment: Wait to hear his proposal;
  ExchangeRings (this Female, currentProposer);
  end;

comment: her life cycle;

while stillUndecided and time < timeToOldMaidPanic do
  TentativelyAcceptOrRejectProposal;

end ***** of class Female *****;

comment: ********** main body **********;

comment: Set up the females !;

  begin
    integer personIndex;
    ref (Female) hilda;
    ref (Male   ) stan;
  for personIndex := 1 step 1 until NoOfFemales do
    begin
    hilda :- new Female (NoOfMales, personIndex);
    TableOfFemales (personIndex) :- hilda;
    activate hilda;
    end;
```

Program 3.4 (cont.)

```
Person class Female;
...
comment: a Female Body;
while stillUndecided do PonderProposals;
end of Female class;
```

will be executed as:

⟨*prelude*; **while** *stillUndecided* **do** *PonderProposals*; *postlude*;⟩

Program 3.4 is now prefixed by **simulation**. It is good practice to include a comment about the length of the basic unit in which all of a model's

```
  for personIndex := 1 step 1 until NoOfMales do
    begin
    stan :- new Male (NoOfFemales, personIndex);
    TableOfMales (personIndex) :- stan;
    activate stan;
    end;

comment: Now let the females send the males hunting !;

  for personIndex := 1 step 1 until NoOfFemales do
    activate TableOfFemales (personIndex);

end of Initialization block;

hold (LengthOfSim);

comment: all females should have panicked by now !;
begin
comment: Flush out the shy ones !;
  integer personIndex;
for personIndex := 1 step 1 until NoOfFemales do
  activate TableOfFemales (personIndex);
for personIndex := 1 step 1 until NoOfMales do
  activate TableOfMales (personIndex);
end;

Results.outimage; Results.outimage;
Results.outtext (" ********** Congratulations ! ********** ");
Results.outimage;

end of Model;

Data  .close;
Results.close;

end of Program;
```

timings are expressed; here it is one minute. A reference to a list

ref (head) *QofEligibleFemales*;

replaces the *TableOfFemales* array in our previous model. *stillUndecided* is a new *Person* attribute, which remembers whether a final partner decision has already been made. It will be initialized to **true**. Note that if we relied on default initialization it would be set to **false**, causing an error in the model's logic. Note also that the program has to provide a seed value for the random-number generators:

integer *Seed*;

. . .

DelayProposalForLengthOfCourtship;

hold (**normal** (*5*, *2*, *Seed*));

Unlike PASCAL, SIMULA permits statements to be labelled. Labels are identifiers terminated by a colon. In this program labels are used as comments, but SIMULA also has a **go to** statement, which may be used to hop around in a program. If good programming style is observed, this should, however, rarely be necessary.

The rest of the program should largely be self-documenting. Note that instances of *Person* classes are processes, which must be scheduled by statements like:

activate *TableOfMales* (*personIndex*);
passivate;
hold (*LengthOfCourtship*);

Note also that females can move themselves into

into (*QofEligibleFemales*);

and out of lists

out;

This can also be done for them by other processes. Dot notation is used in these cases:

SnowWhite.**into** (*QofEligibleFemales*);
MissPiggy.**out**;

The lists themselves must of course be instantiated and bound to a reference variable. This is typically done in a model's outer shell:

ref (**head**) *QofEligibleFemales*;

. . .

QofEligibleFemales:– **new head**;

The preceding discussion has shown that SIMULA is an extremely flexible and versatile modelling and programming tool, and that its class concept makes it particularly well suited for easy extension. Birtwistle *et al.* (1985) contains a sequence of tutorials on the use of these features, looking at the way DEMOS was written.

In the following sections we will examine some SIMULA-based programming systems (classes *MiniQnContext*, *MCcontext*, *CScontext*, DEMOS and DISCO) that are implemented in this way.

3.4.2 Scenarios for Monte Carlo simulation

Figure 3.11 shows a scenario of the potion effectiveness study described in Chapter 2. State-transition graphs such as this one have long been used for network analysis. GERTS [Moore and Clayton (1976)] is a software system for simulating stochastic networks from which the symbols in Figure 3.11 have been adopted. Nodes denote states and state transitions are shown by arrows. Circles stand for sources (which generate samples) and sinks (which

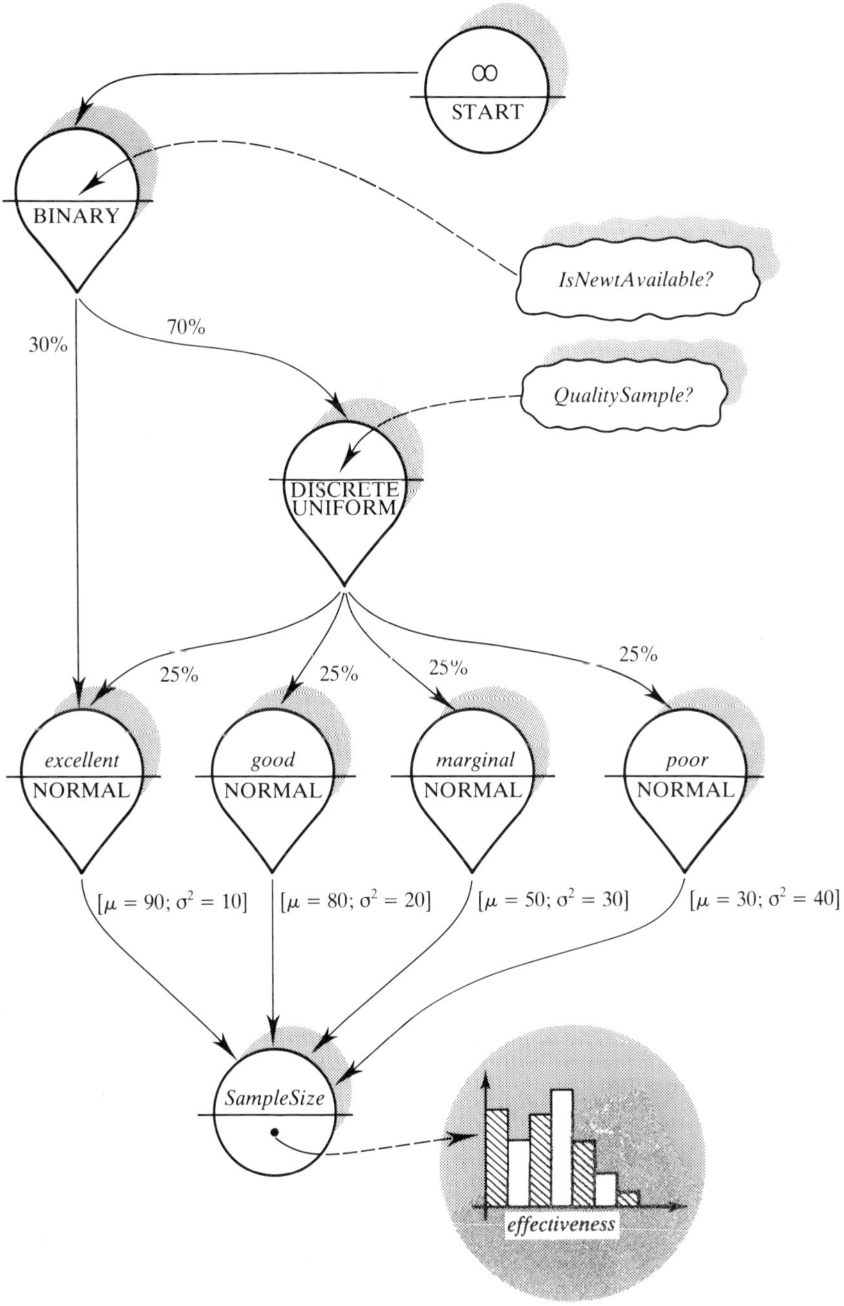

Figure 3.11 Lovepotion effectiveness: A state-transition graph

```
begin
external class MCcontext;
MCcontext begin
  ref (bdist)       IsNewtAvailable;
  ref (idist)       QualitySampled;
  ref (rdist) array distributions (1:4);
  ref (histogram)  results;
  integer sample, SampleSize;
  integer QualityIndex; comment: 1 is "excellent"    , 2 is "good",
                                 3 is "satisfactory", 4 is "poor";
  real    effectiveness;
  integer HistLb, HistUb, HistClasses;

comment: ***** Initializations *****;

outf :- new outfile ("ANALYSISOUTPUT");
outf.open (blanks (80));

box ("An Effectiveness Study of Manuel's LovePotions");

SampleSize := inint;
readdist (IsNewtAvailable, "Newt Avail. ");
readdist (QualitySampled , "Potion Qual.");
for QualityIndex := 1 step 1 until 4 do
  readdist (distributions (QualityIndex), "Qual. Dist.");
HistLb := inint; HistUb := inint; HistClasses := inint;
results  :- new histogram ("Potion Eff.", HistLb, HistUb, HistClasses);

comment: ********* Sampling Cycle *********;

for sample := 1 step 1 until SampleSize do
  begin
  if IsNewtAvailable.sample then
    QualityIndex := 1
  else
    QualityIndex := QualitySampled.sample;
  effectiveness := distributions (QualityIndex).sample;
  if effectiveness <   0 then effectiveness := 0;
  if effectiveness > 100 then effectiveness := 100;
  results.update (effectiveness);
  end of sampling loop;

end of MCcontext prefix;
end of model;
```

Program 3.5 *An effectiveness study of Manuel's Lovepotions (SIMULA class MCcontext)*

record observations), while raindrop-like symbols are used to define stochastic processes. A graphical representation such as this one can be mapped almost directly into the primitives provided by appropriate scenario languages.

Class *MCcontext* is a SIMULA context for Monte Carlo methods adopted from DEMOS [Birtwistle (1979)]. It is prefixed by class **simset** and

```
*******************************************************************************
*                                                                             *
*                 An effectiveness study of Manuel's lovepotions              *
*                                                                             *
*******************************************************************************

*******************************************************************************
*                                                                             *
*                                   REPORT                                    *
*                                                                             *
*******************************************************************************

                                DISTRIBUTIONS
                                *************

     TITLE         (RE)SET        OBS/TYPE          A         B          SEED
  Newt Avail.       0.000      1000 DRAW          0.300                33427485
  Potion Qual.      0.000       703 RANDINT       1         4          22276755
   Qual. Dist.      0.000       491 NORMAL       90.000    10.000      46847980
   Qual. Dist.      0.000       178 NORMAL       80.000    20.000      43859043
   Qual. Dist.      0.000       165 NORMAL       50.000    30.000      64042082
   Qual. Dist.      0.000       166 NORMAL       30.000    40.000      44366385

                                 HISTOGRAMS
                                 **********

                                  SUMMARY

    TITLE        (RE)SET      OBS       AVERAGE       EST.ST.DV      MIN      MAX
  Potion Eff.     0.000       1000       72.023         29.132       0.000   100.000

CELL  LOWER LIM    N    FREQ    CUM
                                    | ---------------------------
  0   -INFINITY    0    0.00    0.00 |
  1     0.000     72    0.07    7.20 |**********
  2    10.000     26    0.03    9.80 |***
  3    20.000     30    0.03   12.80 |****
  4    30.000     35    0.03   16.30 |*****
  5    40.000     41    0.04   20.40 |*****
  6    50.000     40    0.04   24.40 |*****
  7    60.000     59    0.06   30.30 |********
  8    70.000    149    0.15   45.20 |********************
  9    80.000    227    0.23   67.90 | *******************************************
 10    90.000    197    0.20   87.60 |**************************
 11   100.000    124    0.12  100.00 |****************
                                    | ---------------------------
```

Figure 3.12 Effectiveness of Manuel's lovepotions: A histogram

offers predefined objects to represent a variety of distributions and data-collection devices. It also provides methods for their initialization, sampling and updating.

Program 3.5 shows the lovepotion analysis in this context. It is compact and easy to follow. **bdist**, **idist** and **rdist** references are bound to Boolean-, integer- and real-valued distributions. Histograms record and display frequency information. Samples can be taken from all distribution objects. Procedure **readdist** permits information about class and parameters of distributions to be read from a file:

ref (**idist**) *QualitySampled*;

. . .

readdist (*QualitySampled*, *"Potion Qual."*);
with input line:
"PotionQual. RANDINT *1 4 "*
will create an integer valued uniform distribution ranging from 1 to 4 and bind it to reference variable *QualitySampled*.

QualitySampled.**sample**; can then be used to sample this distribution

Message **update** is used for statistical instrumentation. Such information is collected, formatted and printed automatically. **outf** is a predefined reference to file objects (80 character lines) into which a report will be written. **box** can be used to print boxed headings.

Figure 3.12 shows the resulting report. Automatically generated reports like this one are typical for scenario languages, since the programmer should be relieved of all unnecessary drudgery. The standard report can be overridden (procedure **noreport**) or supplemented with additional detail. In this example we get information about the distributions, their seeds (predefined) and how many samples were taken from each of them. The histogram shows a profile of expected potion effectiveness for a sample of 1000 observations.

3.4.3 Scenarios for simulating continuous systems

A. System dynamics

System dynamics has become a popular modelling style in a multitude of diverse disciplines. The systems-dynamics approach focuses on aggregate levels and rates of flow, posing questions about the stability of a system and its response to external perturbations or parameter changes. Randers (1980) is a good survey of the method and its application, while Lebel (1982) offers an extensive bibliography.

System-dynamics scenarios use boxes to symbolize levels and valve-like symbols for rates. Initial states are indicated by crossed circles. We show state transitions as solid and influences as dashed or dotted lines. Sometimes additional symbols are employed to denote flows of particular kinds of entities (people, products, goods, money, information, etc.).

Figure 3.13 shows a system-dynamics scenario for modelling DreamWorld's lovesickness epidemics. The use of system-dynamics diagrams for lumped-parameter systems with feedback was first proposed by J. W. Forrester in 1968. Such models describe a miniworld in terms of states, called **levels**, and their quasi-continuous changes (**rates**). Levels and rates can be repre-

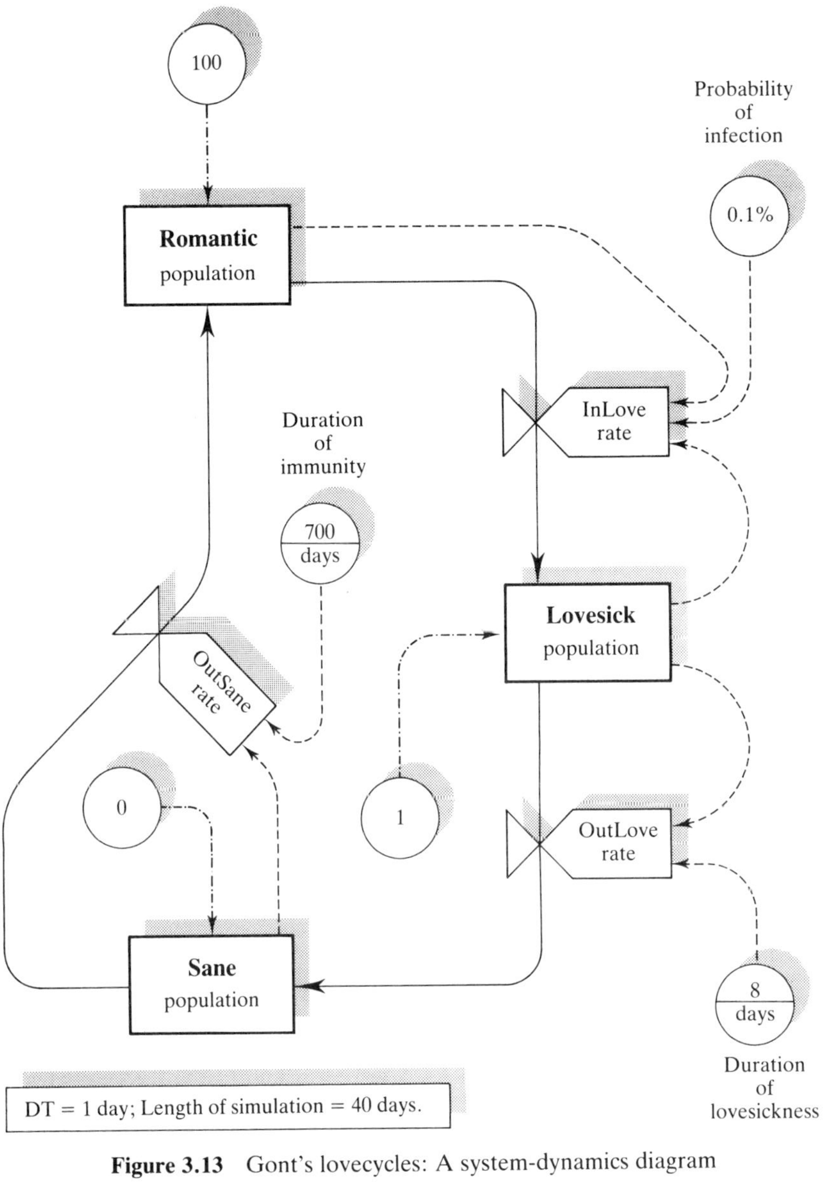

Figure 3.13 Gont's lovecycles: A system-dynamics diagram

sented by variables and expressions, both of which are bound into a set of difference equations defining a model's behaviour.

DYNAMO [Pugh (1973)] is a FORTRAN-based programming language, which can be used to animate models of the system-dynamics type. Unfortunately it inherits FORTRAN's rigid formatting and naming conventions. All identifiers must be in capitals. DYNAMO models are line-oriented, with column 1 serving as a label field. An asterisk flags the rest of the line as a comment. Models consist of a set of equations, which start in column 7. These define initial values, constants, levels or rates. Lines tagged with 'L' describe levels, those with 'R' rates. 'C' stands for constant and 'N' for initial-value definitions. DYNAMO's scheme for time indexing can be visualized as a window sliding across a linear time axis. 'J', 'K' and 'L' refer to past, present and future. Each level variable carries a single index value, because it defines a state persisting at some time instant. Each rate variable, which symbolizes a change over some time interval, requires two indices.

A DYNAMO model is moved through simulated time by a monitor, which evaluates *all* equations at each time step. A SPEC statement specifies the grain of time (DT) and the intervals in which model states should be reported: plotted (PLTPER) or printed (PRTPER). At the end of a simulation the model's behaviour is automatically summarized. A PLOT statement can be used to define the plotting symbols if a plot of levels against time, a chronological graph, is requested.

```
*
*  LOVE CYCLES IN GONT
*
*     LEVELS
*      ++++
*
L      ROMPOP.K=ROMPOP.J+(DT)(OUTSNE.JK-INLOVE.JK)
N      ROMPOP=1000        INITIAL SIZE OF ROMANTIC POPULATION
L      LSKPOP.K=LSKPOP.J+(DT)(INLOVE.JK-OUTLVE.JK)
N      LSKPOP=1           INITIAL SIZE OF LOVESICK  POPULATION
L      SNEPOP.K=SNEPOP.J+(DT)(OUTLVE.JK-OUTSNE.JK)
N      SNEPOP=0           INITIAL SIZE OF "SANE"     POPULATION
*                         (IMMUNE AGAINST LOVESICKNESS)
*
*     RATES
*      +++
*
R      INLOVE.KL=INFPRB*ROMPOP.K*LSKPOP.K
*                         (RATE AT WHICH PEOPLE FALL IN LOVE)
C      INFPRB=0.001       (PROBABILITY OF INFECTION)
R      OUTLVE.KL=LSKPOP.K/LVETME
*                         (RATE AT WHICH PEOPLE FALL OUT OF
*                         LOVE AND BECOME IMMUNE)
C      LVETME=8           (DURATION OF LOVESICKNESS IN DAYS)
R      OUTSNE.KL=SNEPOP.K/SANITY
*                         (RATE AT WHICH PEOPLE LOOSE THEIR
*                         IMMUNITY AND BECOME LOVESICK AGAIN)
C      SANITY=700         (DURATION OF IMMUNITY IN DAYS)
*
*
PLOT SNEPOP=O/ROMPOP=+/LSKPOP=*
SPEC DT=1/LENGTH=40/PLTPER=1/PRTPER=0
```

Program 3.6 *The dynamics of lovesickness (in DYNAMO)*

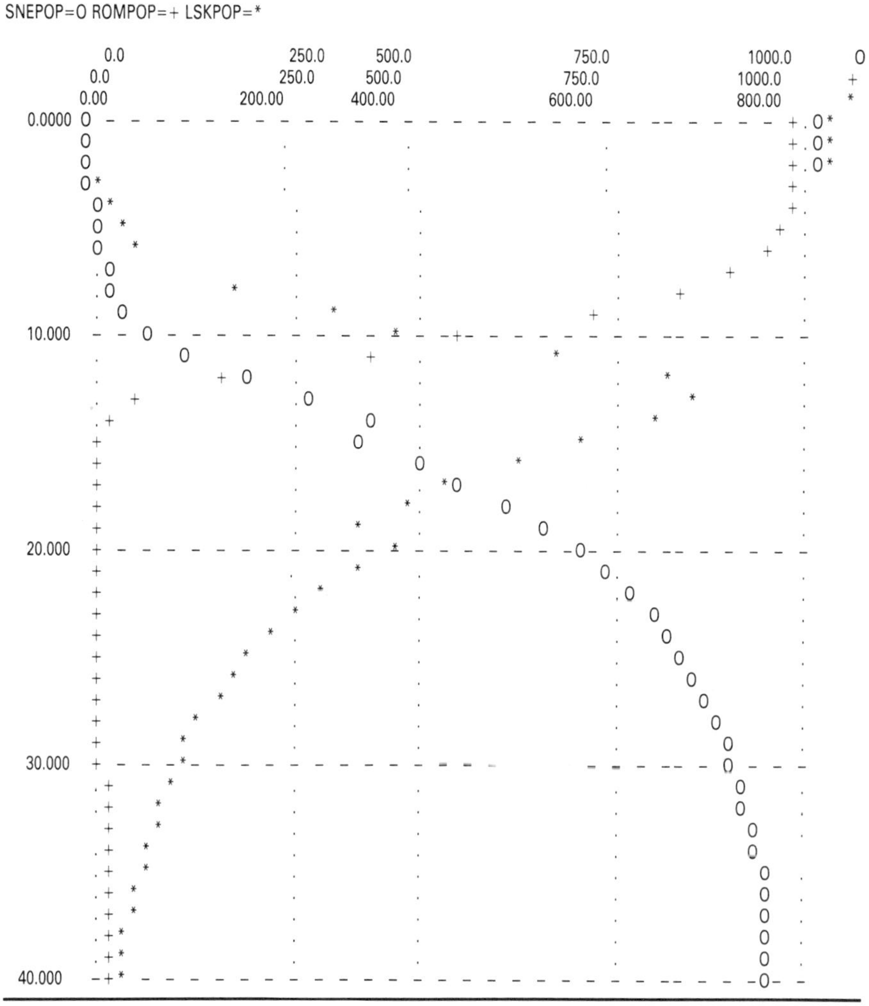

Figure 3.14 Gont's lovecycles: Chronological graph of an epidemic

Program 3.6 and Figure 3.14 show a DYNAMO model of DreamWorld's lovecycles and the output produced by the model.

The system dynamics paradigm has proved to be popular among, for example, engineers, biologists, economists and sociologists, and has successfully been applied to a wide variety of problems (such as control systems, studies of long-term trends in processes of growth and decay). Quite a number of bells and whistles can be added to a simple model like the one shown, although system dynamics and quasi-continuous simulation models typically tend to have a higher degree of aggregation and a lower degree of complexity than discrete-event simulations.

```
begin
external class CScontext;
CScontext begin
  integer         SimLength;
  real            InfectProb;
  integer         InLoveTime, SanityTime;
  ref (variable) RomanticPop, LovesickPop, SanePop;

  continuous class world;
  begin
  RomanticPop .rate := ((SanePop.state / SanityTime)  -
                        (InfectProb * RomanticPop.state * LovesickPop.state));
  LovesickPop .rate := ((InfectProb * RomanticPop.state * LovesickPop.state) -
                        (LovesickPop.state / InLoveTime));
  SanePop     .rate := ((LovesickPop.state / InLoveTime) -
                        (SanePop     .state / SanityTime));
  end of class world;

  reporter class ChronologicalGraph;
  begin
  plot ('+', RomanticPop .state, time);
  plot ('*', LovesickPop .state, time);
  plot ('o', SanePop     .state, time);
  end of class ChronologicalGraph;

comment: ********** Body of main program **********;

InfectProb  := inreal;
InLoveTime := inint;  SanityTime:= inint;
SimLength  := inint;
```

Program 3.7 *The dynamics of lovesickness on Gont (SIMULA class CScontext)*

B. Systems of differential equations

Languages like CSMP [Speckhart and Green (1976)], CSSL [Strauss *et al.* (1967)] and ESL [Crosbie *et al.* (1985)] offer alternative views of continuous models. While DYNAMO uses differential equations to evaluate change in a discrete grid of time, these systems typically provide a variety of procedures to numerically integrate flows of a quantity across some space–time continuum.

Program 3.7 uses a SIMULA context called *CScontext* to model DreamWorld's lovecycles. *CScontext* provides predefined classes to represent quasi-continuously fluctuating state variables (class **variable**) and continuous processes (class **continuous**).

There is a **reporter** entity to gather and plot the data for chronological graphs, using symbols (here '+', '*', 'o') defined in **plot** statements. The sampling frequency can be set by the user, with **setfrequency** (0) recording every change in state.

rate and **state** are predefined properties of continuous variables. After they have been instantiated and initialized, processes associated with **variable** and **continuous** objects can be started and stopped. The user can request

```
dtmin := 0.01; dtmax := 1;
maxrelerror:= 1&-2;
euler        := true;

inspect new ChronologicalGraph do
  begin
  setfrequency (0);
  start;
  end inspection;

RomanticPop :- new variable (inint);
LovesickPop :- new variable (inint);
SanePop     :- new variable (inint);
RomanticPop .start;
LovesickPop .start;
SanePop     .start;

new world.start;
hold (SimLength);

outtext ("Love Cycles in Avalon"); outimage;
outtext ("  +++++++++++ "); outimage; outimage;
outtext (" + indicates the size of the romantic population "); outimage;
outtext (" *  indicates the size of the lovesick  population "); outimage;
outtext (" o  indicates the size of the sane      population "); outimage;

end of CScontext prefix;
end of model;
```

minimum and maximum step sizes (**dtmin** and **dtmax**) for integration and may also specify an absolute or relative bound on the integration error. A spectrum of integration methods (**euler**, **simpson**, **improved_heun**), can be chosen from.

CScontext expects the user to write equations contained within **continuous** process definitions in proper order for evaluation.

Classes COMBINEDSIMULATION and DISCO [Helsgaun (1980)] are other powerful and flexible SIMULA contexts designed to support system models defined by a differential-equation scenario. They can also handle models of the system-dynamics type.

3.4.4 Queueing-network scenarios

A. Material- and machine-oriented world views

Although simple job-shop models can be studied analytically [Coffman (1976)], the exploration of more complex ones was one of the earliest

```
begin

external class MiniQNcontext;

comment: Manuel's Shop from Manuel's point of view
         ***********************************.

MiniQNcontext begin
  ref (aDist        ) arrivalDist, serviceDist;
  ref (ourSource    ) arrivals;
  ref (aFIFOqueue) lovesickCustomers, happyCustomers;
  ref (ourServer    ) manuel;
  ref (ourSink      ) departures;
  ref (aTally       ) flowTimes;
  integer presimulationSamples, simulationSamples;

aSource class ourSource;
begin
  ref (aCustomer) romeo;

  ref (aCustomer) procedure createaCustomer;
  begin
  createaCustomer :- new aCustomer ("romeo", time);
  recordCreation;
  end of createaCustomer procedure;

tasks:
  while true do
    begin
    romeo :- createaCustomer;
    anOutputQueue.insertaCustomer (romeo);
    if manuel.isIdle then activate manuel;
    hold (sampleTimeToNextArrival);
    end while;
end of ourSource;

aSingleServer class ourServer;
begin
    ref (aCustomer) romeo;
tasks:
  while true do
    begin
    if anInputQueue.isEmpty then
      passivate
    else
      begin
      romeo :- anInputQueue.isFirstItem;
      anInputQueue.removeaCustomer (romeo);
      makeBusy;
      hold (sampleServiceTime);
      makeIdle;
      anOutputQueue.insertaCustomer (romeo);
      activate departures;
      end if;
    end while;
end of ourServer;
```

Program 3.8 *Manuel's shop from Manuel's point of view*

```
aSink class ourSink;
begin

  procedure eataCustomer;
  begin
    ref (aCustomer) fred;
  if not (anInputQueue.isEmpty) then
    begin
    fred :- anInputQueue.isFirstItem;
    anInputQueue.removeaCustomer (fred);
    recordDeparture;
    flowTimes.update (time - fred.hasEntryTime);
    end if;
  end of eataCustomer procedure;

tasks:
  while hasEaten < hasSampleSize do
    begin
    eataCustomer;
    passivate;
    end of while;
  activate main;
end of ourSink;

procedure ReadAndInitialize;
begin
arrivalDist       :- new anExpoDist  ("arrival process", inint, inreal);
serviceDist       :- new anExpoDist  ("service process", inint, inreal);
lovesickCustomers :- new aFIFOqueue  ("Lovesick ones");
happyCustomers    :  new aFIFOqueue  ("Happy ones");
manuel            :- new ourServer   ("Manuel", serviceDist,
                                      lovesickCustomers, happyCustomers);
flowTimes         :- new aTally      ("Flowtimes");
arrivals          :- new ourSource   ("arrivals", arrivalDist,
                                      lovesickCustomers);
activate manuel;
activate arrivals;
end of ReadAndInitialize procedure;

procedure ResetStatistics;
begin
outtext ("Statistics reset after "); outint (departures.hasEaten, 5);
outtext (" samples at time ");   outfix (time, 2, 7);
outimage;
outtext ("===========");
outimage; outimage;
arrivalDist       .reset;
serviceDist       .reset;
arrivals          .reset;
lovesickCustomers .reset;
happyCustomers    .reset;
manuel            .reset;
flowTimes         .reset;
departures        .reset;
end of ResetStatistics procedure;
```

(cont.)

```
procedure Summarize;
begin
outtext ("Manuel's Love Potion Shop");
outtext (" ************************"); outimage; outimage;
outtext ("Statistics after  "); outint (departures.hasEaten, 5);
outtext (" samples at time "); outfix (time, 2, 7);
outimage; outimage;
arrivalDist        .report;
serviceDist        .report;
arrivals           .report;
lovesickCustomers  .report;
happyCustomers     .report;
manuel             .report;
flowTimes          .report;
departures         .report;
end of Summarize procedure;

comment: ************** main program body **************;

ReadAndInitialize;
departures :- new ourSink ("departures", inint, happyCustomers);
activate departures;
passivate; comment: for presimulation samples;

ResetStatistics;
departures :- new ourSink ("departures", inint, happyCustomers);
activate departures;
passivate; comment: for sampling experiment;

Summarize;

end of model;
end of program block;
```

Program 3.8 (cont.)

applications of simulation as a decision-making tool. The purpose of job-shop simulations is to obtain statistics about machine utilizations, queue lengths, processing and waiting times, and the throughput of the system as a whole. These can then be used to tune configurations and procedures (for example, to handle routing of material, priority rules for queues, interrupt schemes for server units) in order to optimize some aspect of the system's operation (such as machine loadings, throughputs, flow times).

Job-shop models are typical representatives of **queueing networks**, which can formally be described in a number of ways. Two extreme viewpoints are referred to as machine and material orientation, and the basic associated world views have been characterized in Chapter 2.

Machine-oriented models describe a system's operation from the machines' point of view: they treat materials as purely passive objects that are created, deposited into some input queue, selected for processing, operated upon, and deposited into some output queue, from which they are eventually removed. A machine-oriented model therefore consists of declarations of arrival- and service-time distributions, queues, transaction counters or records (material), and data-collection devices. Model behaviour is defined via life-cycle descriptions for all active entities: sources, servers (machines) and sinks. The transient entities (transactions or tasks) are passive data structures, routed through a network of active servers. Task selection and routing strategies are encoded in the servers' life cycles.

Program 3.8 shows a SIMULA model for machine-oriented simulations of Manuel's shop. It uses the *MiniQNcontext* described in Section 3.4.1. We use class *ourSource* to prefix our source object. It creates customers (*tasks*), filing them into Manuel's input queue and notifying him of arrivals. Successive customer creations are separated by time intervals of random duration. These are determined by sampling from the inter-arrival-time distribution (for example, **hold**). The message *createaCustomer* (declared as a **virtual** method in *MiniQNcontext*) is implemented: creating and time-stamping a customer record, and recording its arrival. A SIMULA module with references to virtual procedures can be compiled without having to supply their actual definitions. These can be provided later, in a block or class prefixed by it. This feature can, for example, be used to clearly signal that a context expects a user's program to provide its own implementations for a number of methods before it can execute. Some SIMULA compilers produce warnings to remind a programmer of this.

manuel is modelled as a subclass of class *SingleServer*. The server (*machine*) waits idly while its input queue is empty. As long as there are still customers to serve, he takes them from the head of the input queue (FIFO), performs a service according to the service-time distribution, files them at the tail of the output queue, and notifies the sink that a customer is ready to leave. If it is ever passivated, it will be activated again by a new customer arrival.

The *Sink* implements message *eataCustomer* by taking the first customer out of its input queue, recording its departure, and measuring the time it took to obtain and endure Manuel's service (entry to exit). A sink's life cycle is simple: it eats a customer and falls asleep, to be activated again after Manuel has finished with yet another customer.

A *Tally* object called *FlowTimes* is used to record the time intervals customers spent in the shop, and later compute and report their mean value.

Procedure *ReadAndInitialize* creates instances of all of these entities. It also starts *arrivals* and *manuel* on their life cycles.

Note how the parameter lists of classes and subclasses are concatenated. Class *arrivalDist* is *anExpoDist*, which in turn is prefixed by *aDist* and *aStructureObject*. It therefore inherits a text-valued name from *aStructureDist*, a seed from *aDist*, and adds a rate as a parameter of its own. Seed and rate are read from an input stream at the time a new instance of *anExpoDist* is created.

In this model again a presimulation period is used for model initialization. Procedure *ResetStatistics* resets all entities' statistics, which procedure

Summarize reports at the end of a simulation. The program body starts and terminates a simulation experiment.

Figure 3.15 shows the report generated by a simulation with parameters bound to the values shown in Figure 2.4.

Material-oriented models describe a system's operation from the viewpoint of transactions. After being created, transactions step through a cycle of queueing for service, acquiring server facilities, being delayed in service activities, releasing server facilities, gathering statistics, and departing from the model.

Machines (servers) are treated as passive objects that can be grabbed or released, if they are available. A material-oriented model declares distributions for sampling arrival and service time, queues, server objects (Boolean flags, counters or records) and data collection devices. In open queueing models facilities for simulating sources and sinks are also needed. Model behaviour is defined through life cycles associated with transient entities

```
Statistics reset after 1000 samples at TIME 1292.59

Manuel's lovepotion shop
**********************

Statistics after 1000 samples at TIME 2504.38
Seed value for this arrival process object = 12345
Seed value for this service process object = 67891

1001 customers were created by these arrivals

Reporting on queue: Lovesick ones

Current   queuelength is  : 2
The max.  queuelength was: 15
The mean  queuelength was: 3.00

Reporting on queue: Happy ones

Current   queuelength is  : 1
The max.  queuelength was: 1
The mean  queuelength was: 1.00

Manuel's utilization was: 0.83

Reporting on tally: Flowtimes

1000 customers were observed leaving the shop

The distribution of times the customers spent in the shop:

Mean = 4.37   Standard deviation = 3.44

departures ate 1000 customers.
```

Figure 3.15 Queueing delays in Manuel's shop

```
begin

external class miniQNcontext;

comment: Manuel's Shop from a Customer's point of view ;
         ****************************************

miniQNcontext begin
  ref (aDist      ) arrivalDist, serviceDist;
  ref (ourSource  ) arrivals;
  ref (aFIFOqueue) lovesickCustomers, happyCustomers;
  ref (ourServer  ) manuel;
  ref (ourSink    ) departures;
  ref (aTally     ) flowTimes;
  integer presimulationSamples, simulationSamples;

aSource class ourSource;
begin
  ref (ourCustomer) romeo;

  ref (aCustomer) procedure createaCustomer;
  begin
  createaCustomer :- new ourCustomer ("romeo", time);
  recordCreation;
  end of createaCustomer procedure;

tasks:
  while true do
    begin
    romeo :- createaCustomer;
    anOutputQueue.insertaCustomer (romeo);
    activate romeo;
    hold (sampleTimeToNextArrival);
    end while;
end of ourSource;

  aCustomer class ourCustomer;
  begin
    ref (ourCustomer) self;
  tasks:
    self :- this ourCustomer;
    manuel.seize;
    lovesickCustomers.removeaCustomer (self);
    hold (manuel.sampleServiceTime);
    manuel.release;
    happyCustomers.insertaCustomer (self);
    activate departures;
  end of ourCustomer;

  aSingleServer class ourServer;
  begin
```

Program 3.9 *Manuel's shop from a customer's point of view* (cont.)

```
   procedure seize;
   begin
   while isBusy do passivate;
   makeBusy;
   end of seize procedure;

   procedure release;
   begin
   makeIdle;
   lovesickCustomers.wakeupNextInLine;
   end of release procedure;

end of ourServer;

aSink class ourSink;
begin

   procedure eataCustomer;
   begin
      ref (ourCustomer) fred;
   if not (anInputQueue.isEmpty) then
        begin
        fred :- anInputQueue.isFirstItem;
        anInputQueue.removeaCustomer (fred);
        recordDeparture;
        flowTimes.update (time - fred.hasEntryTime);
        end if;
   end of eataCustomer procedure;

tasks:
   while hasEaten < hasSampleSize do
      begin
      eataCustomer;
      passivate;
      end of while;
   activate main;
end of ourSink;

procedure ReadAndInitialize;
begin
arrivalDist        :- new anExpoDist  ("arrival process", inint, inreal);
serviceDist        :- new anExpoDist  ("service process", inint, inreal);
lovesickCustomers  :- new aFIFOqueue  ("Lovesick ones");
happyCustomers     :- new aFIFOqueue  ("Happy ones");
manuel             :- new ourServer   ("Manuel", serviceDist,
                                       lovesickCustomers, happyCustomers);
flowTimes          :- new aTally      ("Flowtimes");
arrivals           :- new ourSource   ("arrivals", arrivalDist,
                                       lovesickCustomers);
activate manuel;
activate arrivals;
end of ReadAndInitialize procedure;
```

Program 3.9 (cont.)

```
procedure ResetStatistics;
begin
outtext ("Statistics reset after "); outint (departures.hasEaten, 5);
outtext (" samples at time ");    outfix (time, 2, 7);
outimage;
outtext ("=========  ");
outimage; outimage;
arrivals          .reset;
lovesickCustomers .reset;
happyCustomers    .reset;
manuel            .reset;
flowTimes         .reset;
departures        .reset;
end of ResetStatistics procedure;

procedure Summarize;
begin
outtext ("Manuel's Love Potion Shop"); outimage;
outtext (" ************************"); outimage; outimage;
outtext ("Statistics after  "); outint (departures.hasEaten, 5);
outtext (" samples at time "); outfix (time, 2, 7);
outimage; outimage;
arrivalDist       .report;
serviceDist       .report;
arrivals          .report;
lovesickCustomers .report;
happyCustomers    .report;
manuel            .report;
flowTimes         .report;
departures        .report;
end of Summarize procedure;

comment: ************** main program body **************;

ReadAndInitialize;
departures :- new ourSink ("departures", inint, happyCustomers);
activate departures;
passivate; comment: for presimulation samples;

ResetStatistics;
departures :- new ourSink ("departures", inint, happyCustomers);
activate departures;
passivate; comment: for sampling experiment;

Summarize;

end of model;
end of program block;
```

(transactions). Permanent server entities are purely passive. Server selection and routing decisions are therefore woven into transactions' behaviour patterns.

Program 3.9 shows a model of Manuel's shop written in such a material-oriented style. The source now creates customer instances and files them into a queue called *lovesickCustomers*. Each customer knows his true name (*self*). He is therefore able to remove himself from this queue once Manuel has been successfully seized. The service-time distribution is sampled and produces a value for the time interval by which the service activity delays the customer's progress. Manuel is then released and the customer inserts himself at the tail of queue *happyCustomers*, ready to leave the shop. *departures* is signalled to measure delays (and 'open the door').

The server now responds to two new messages. *seize* passivates a waiting customer while Manuel is busy. It then grabs the server and inverts its status (to *busy*). *release* makes Manuel idle and then sends a wake-up message to the next waiting customer. This customer will be activated (by *wakeupNextInLine*) to test the server's *busy* status again. The rest of the code is almost identical to that of Program 3.8.

Figure 3.16 gives a bird's eye view of the model's structure. It shows the objects' properties and methods and the pattern of message passing taking place among them. Dashed lines denote the creation of class instances (objects), with solid lines showing scheduling and other messages. Bold, italicized attributes are implemented by procedures. Underlined attributes are **virtual**, which means they are defined or redefined at lower levels of a prefix chain. To avoid cluttering the picture even further, messages spawned by the *ResetStatistics* and *Summarize* procedures are not shown.

Observe that **main** and the Run-Time Control System (RTCS) drive the simulation. The body of **main** describes the protocol of a simulation experiment. Because it is prefixed by **simulation**, it becomes an instance of class *aSimulation*, inheriting an RTCS and scheduling procedures. The RTCS owns a clock (called **time**) and agenda (called **sqs**). It performs all actions associated with event sequencing, time and memory management.

arrivals, *departures*, *romeo* and *manuel* are instances of classes ultimately prefixed by class *aProcessObject*. They are therefore **process** objects, which can be scheduled (**activate**, **passivate**, **hold**). All other objects (prefixed by class *aStructureObject*) inherit linkages only, allowing us to move them into and out of list structures.

Figure 3.16 shows all the objects' property inheritance. Only Manuel's will now be considered.

manuel is an instance of class *ourServer*. He therefore responds to requests to seize and release him. Since this model is material oriented, he has no life cycle of his own. The customers' life cycle drives the simulation. Multiple customers may be in the shop: waiting in the queue *lovesickCustomers*, being served by *manuel*, or leaving the shop (in the queue *happyCustomers*). The server is a subclass of *aSingleServer*, which means that *manuel* inherits methods to set his state to *idle* or *busy* and answer queries about it. Since

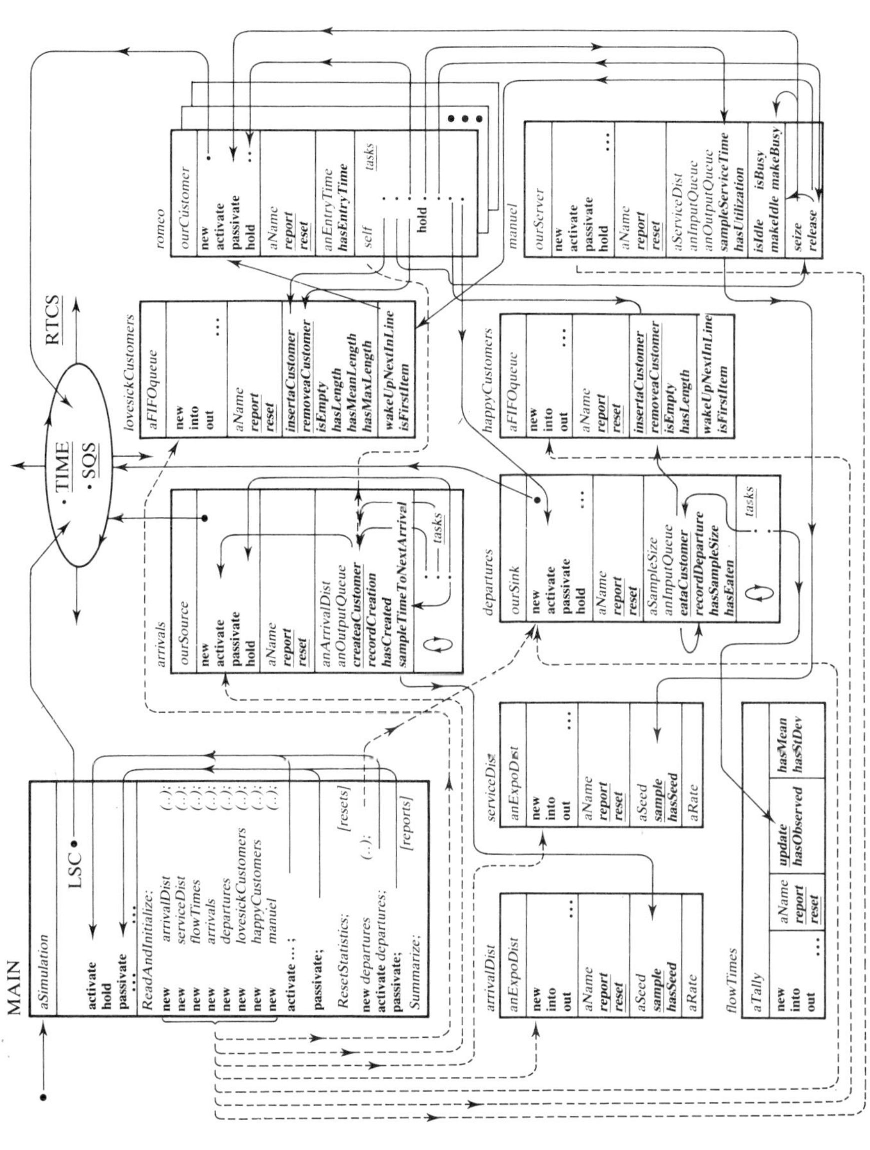

Figure 3.16 Manuel's shop as a set of communicating objects (material-oriented view)

aSingleServer is a specialization of *aServer*, *manuel* also owns a number of references (to his service-time distribution, his input and output queue). All servers know how to sample a service time and how to compute their utilization. Class *aServer* is a subclass of *aProcessObject*, which supplies a *name* attribute and default *report* and *reset* procedures. Class *aProcessObject* is prefixed by **process**, a system-defined class that is itself prefixed by system class **link**. Among other things this makes scheduling procedures **activate**, **passivate**, and **hold** available.

Class **link** provides responses to messages (**into**, **out**) for list manipulation. Finally, a number of basic messages (such as **new**) are understood by all SIMULA class objects.

The message patterns associated with *manuel* can be interpreted as follows: The dashed line indicates that class *ourServer* responds to an instantiation message (**new**) sent by **main**. This causes it to create a new instance of itself, called *manuel. romeo* triggers a service by sending a *sampleServiceTime* message to *manuel*, suspending (**hold**) himself for the answered time period. *manuel* in turn executes this message by sending a *sample* message to his service-time distribution, obtaining a value that can be returned to *romeo.romeo* also sends messages to seize *manuel* prior to a service activity and to release him afterwards. In response to this, *manuel* makes himself *busy* and *idle*, gathering statistics as a side effect. If *manuel* is *busy*, the method associated with the *seize* message prompts him to **passivate** *romeo*, in which case an appropriate message is sent (via *romeo*) to the RTCS. *release* results in a wake-up message sent to the next waiting *romeo*, so that he can proceed on his life cycle. In order to find the right object to wake up, *manuel* inquires about the identity of the first customer linked into his input queue (message to *lovesickCustomers*). *manuel* sends an **activate** if there is anyone waiting. With the exception of *report*, *hasUtilization* and *reset* (not shown here), no other methods in *manuel's* protocol are activated.

The same kind of interpretation can be performed for all objects in Figure 3.16. Since this is a material-oriented model, *arrivals*, *departures* and all customer objects have life cycles (tasks) that drive a simulation. They spawn chains of messages to generate customers, move through sequences of tasks, remove customers from the shop and gather statistics about any delays they encounter. Customers move themselves into and out of queues, request, occupy and release the server. The requested sample size causes *departures* to trigger *reset* messages and, finally, to report on the model's behaviour.

Although it is quite common to use a purely machine or material-oriented modelling style, it is often the best strategy to opt for a mixed approach. Here all objects are represented by both structures and behaviour patterns, preferably described within the same module. This is the basic idea behind object-oriented programs.

Class *MiniQNcontext* and the two models discussed in this section are written in an object-oriented style. This is not always efficient, but it leads to exceptionally clean, extensible and portable implementations. These models could, for instance, quite easily be reimplemented for a SMALLTALK-based simulation environment.

B. Activity-cycle scenarios

The use of life- or activity-cycle diagrams for model conceptualization and structuring has a long history. Tocher (1964) was probably the first to employ them for this purpose. Although they are an excellent tool for design and documentation of any discrete model, their use is particularly encouraged by systems like ECSL [Clementson (1966)], HOCUS [Poole and Szymankiewicz (1977)], CAPS [Hutchinson (1975b)], DRAFT [Mathewson (1974)] and DEMOS [Birtwistle (1979)].

Figure 3.17 shows a life-cycle description of Manuel's shop. Such a diagram captures the main entities' behaviour. Each life cycle is a sequence of states and state-changing activities. Objects synchronize their actions by taking part in common activities, triggered if at least one incarnation of all participating classes of objects can proceed (that is, a customer is waiting for a service and Manuel is idle).

Life-cycle diagrams can serve as blueprints for later encoding in a programming language. In a process-oriented framework each life cycle is mapped into a process class. For activity-oriented models the diagram shows all states, activity sequences and their interactions. Event-oriented modelling styles can use the diagram to graphically highlight event routines (start and finish of activities) and all relevant sequencing relationships between them.

We will use the DEMOS system to demonstrate how the task of modelling quite complicated patterns of synchronization between entities can be very much simplified by a well-designed SIMULA context.

DEMOS (Discrete-Event Modelling On SIMULA) is a class of about 3000 lines of SIMULA code, written by Graham Birtwistle (1979). SIMULA programs can use its concepts to simulate queueing network scenarios when prefixed accordingly (DEMOS **begin**... **end**;).

Additional to all standard concepts of SIMULA, a number of special building blocks (data types and operations) become available. DEMOS supports the same random drawing features as class *MCcontext*. Many statistics are automatically collected, formatted and written into designated files. Debugging support is offered through excellent dumping and tracing facilities. Predefined templates for different kinds of resources and queues are provided, as well as some other utilities.

DEMOS encourages the use of activity-cycle diagrams as a graphical structuring device. It is quite straightforward to code life-cycle descriptions into a DEMOS model.

Queueing-network scenarios consist of sources, sinks, servers and workload items. All these can be thought of as active entities, whose actions are independent of one another. Since it is also typically desired to model these entities' interactions, their life cycles must be synchronized via common states or activities.

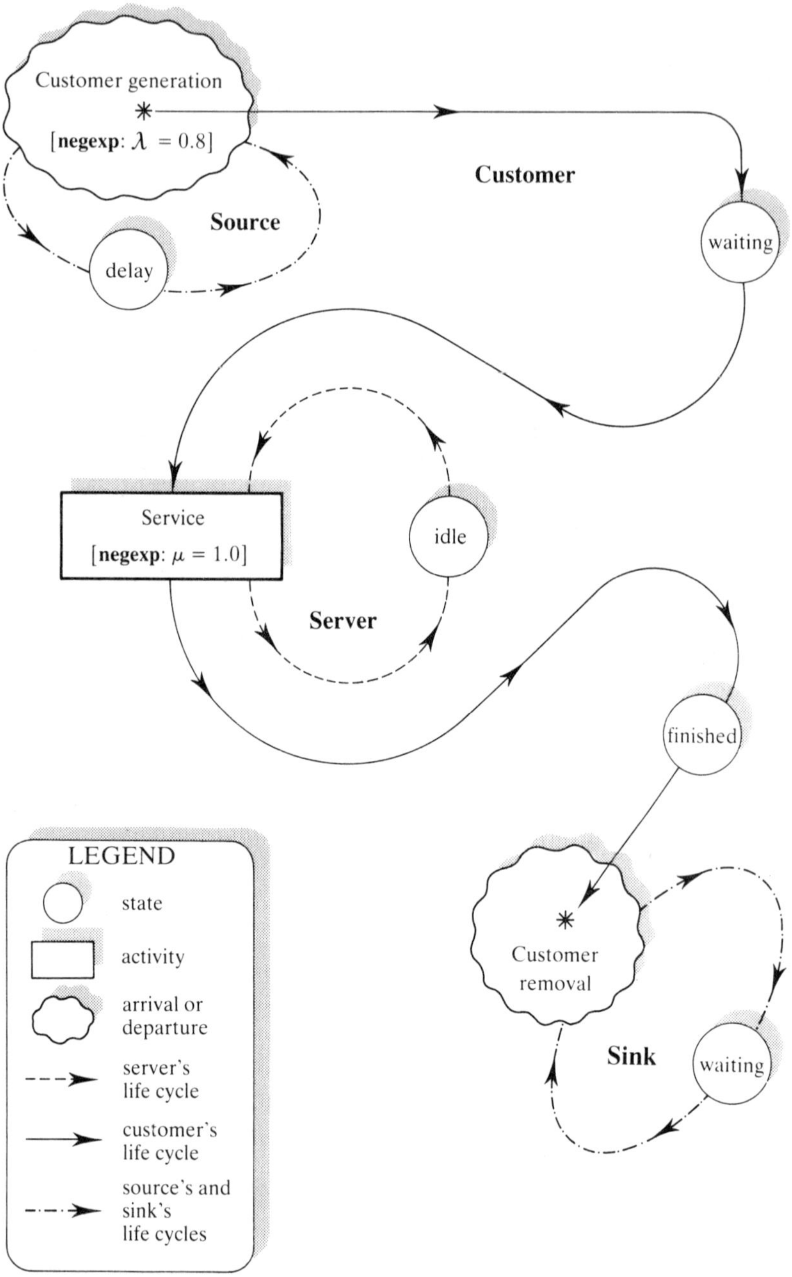

Figure 3.17 A life-cycle diagram of Manuel's shop as an open queueing model ($M/M/1$)

There is a large body of literature on parallelism in hardware [Lorin (1972)] and software [Andrews and Schneider (1983)]. Most research focuses on different schemes for process synchronization in genuinely parallel systems. Although in simulation programming we are primarily interested in pseudo-parallelism (coroutines), similar principles still apply.

Using DEMOS-style life-cycle diagrams of Manuel's shop, Figures 3.18, 3.20, 3.22 and 3.24 illustrate four typical patterns of process synchronization. Cases A and C are machine-oriented scenarios, whereas B and D view customers as the predominant class of entities (material orientation).

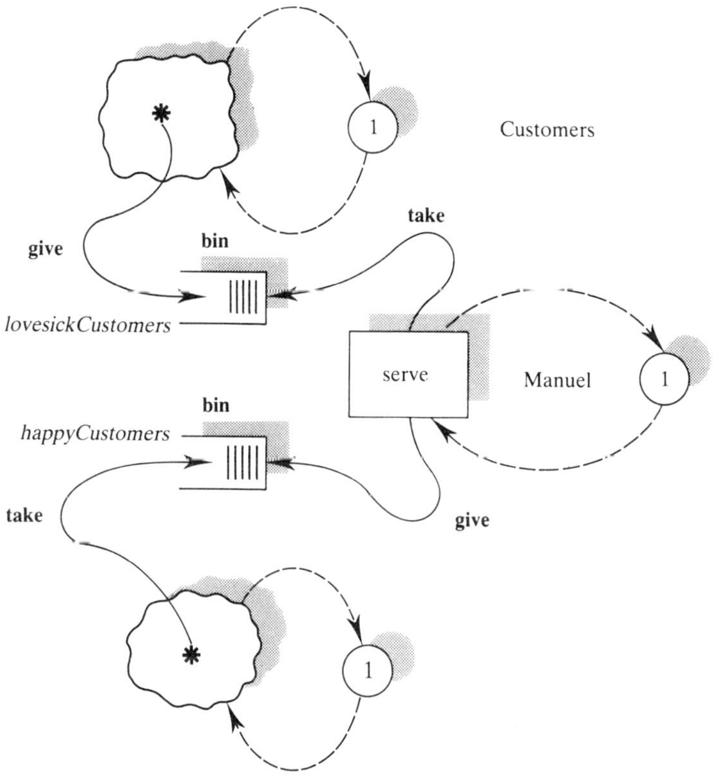

Figure 3.18 Case A: Producer–consumer synchronization

Case A is an example of the **producer–consumer paradigm** (Figure 3.18). Buffers are used for synchronization between processes. Lovesick customers are produced by a source (procedure **give**) and consumed by *manuel* (procedure **take**). The service activity produces happy customers, who are then consumed by a sink.

This pattern can be implemented in DEMOS by using a data type called

```
begin
external class demos;
demos begin
  ref (aSource) Arrivals;
  ref (rdist   ) ArrivalProcess, ServiceProcess;
  ref (bin     ) LovesickCustomers, HappyCustomers;
  ref (aServer ) Manuel;
  ref (aSink   ) Departures;
  ref (count   ) LostCustomers;

  entity class aSource (lambda); value lambda; real lambda;
    while true do
      begin
      if LovesickCustomers.avail > 10 then
        LostCustomers.update (1)
      else
        begin
        LovesickCustomers.give (1);
        if Manuel.idle then
          Manuel.schedule (now);
        end if;
      hold (ArrivalProcess.sample);
      end while & source;

  entity class aServer (mu); value mu; real mu;
    while true do
      begin
      LovesickCustomers.take (1);
      hold (ServiceProcess.sample);
      HappyCustomers.give (1);
      end while and Server;
```

Program 3.10 *Producer–consumer synchronization with* **bins**

bin. Bins will represent those customers waiting for service and those who are ready to leave the shop:

```
ref (bin) Lovesicks, Happies;
. . .
Lovesicks:- new bin ("Lovesicks", 0);
Happies:- new bin ("Happies", 0);
```

In this model customers are represented implicitly by the contents of the two bins. Case B will consider them as individual objects.

To justify our choice of data type, we further assume that there are no constraints on the number of customers Manuel's shop may hold. Bins have no capacity limit. A producer therefore never has to wait. If a bin is empty, consumers will be delayed until a producer deposits another item. If the shop were to hold only a finite maximum number of customers we would have to model it as a capacity-constrained resource.

Program 3.10 shows a corresponding DEMOS model. **rdist** represents a

```
entity class aSink;
  while true do
    HappyCustomers.take (1);

comment: ********** Body of main program **********;

outf :- new outfile ("Man1report");
outf.open (blanks (80));

Arrivals          :- new aSource("source",  inreal        );
Manuel            :- new aServer ("manuel", inreal        );
ArrivalProcess    :- new negexp ("arrivals", Arrivals.lambda);
ServiceProcess    :- new negexp ("service ", Manuel .mu    );
LovesickCustomers :- new bin    ("lovesicks", 0           );
HappyCustomers    :- new bin    ("happies ", 0            );
Departures        :- new aSink  ("departures"             );
LostCustomers     :- new count  ("lost cust."             );

Arrivals   .schedule (0.0);
Manuel     .schedule (0.0);
Departures.schedule (0.0);

box ("Manuel's Shop: Statistics.");
hold (inreal);

end of prefixed block;
end of model;
```

real-valued distribution object; here negative exponential distributions are used for arrival and service processes.

Let us assume that Manuel loses some of his business if customers must wait for too long. A **count** can be used to gather occurrence statistics; here it counts the number of those who, we assume, are discouraged from entering the shop whenever the queue exceeds ten waiting customers.

Source, server and sink are active objects. **schedule** messages cause an entity to proceed on its life cycle (similar to SIMULA's **activate**), after some specified delay.

```
ref (aServer) Manuel;
...
Manuel:- new aServer ("Manuel");
Manuel.schedule (0.0);
```

give (*n*) puts *n* items into a **bin**, from which they can be removed by **take** (*n*).

```
******************************************************************************
*                                                                            *
*                         Manuel's shop: Statistics                          *
*                                                                            *
******************************************************************************

                           CLOCK TIME = 200.000
******************************************************************************
*                                                                            *
*                                  REPORT                                    *
*                                                                            *
******************************************************************************

                               DISTRIBUTIONS
                               *************
```

TITLE	(RE)SET	OBS	TYPE	A	B	SEED
arrivals	0.000	147	NEGEXP	0.800		33427485
service	0.000	147	NEGEXP	1.000		22276755

COUNTS
****

TITLE	(RE)SET	OBS
lost Cust.	0.000	3

BINS
***

TITLE	(RE)SET	OBS	INIT	MAX	NOW	AV. FREE	AV. WAIT	QMAX
lovesicks	0.000	163	0	11	10	2.600	0.369	1
happies	0.000	152	0	1	0	0.000	1.309	1

Figure 3.19 Statistics produced by Program 3.10

{*in aSource's life cycle*}
Lovesicks.**give** (*1*);

{*in aServer's life cycle*}
Lovesicks.**take** (*1*);

If *LostCustomers* refers to a count, the message *LostCustomers*.**update** (x) records lost customers, incrementing the count by x items. **avail** returns the number of items currently deposited in a bin:

ref (**count**) *LostCustomers*;

LostCustomers:− **new count** ("*Lost Cust.*");

. . .

if *Lovesicks*.**avail** > *10* **then**
LostCustomers.**update** (*1*);

DEMOS automatically reports a model's performance by printing a number of standard statistics for each entity (Figure 3.19). Note the use of

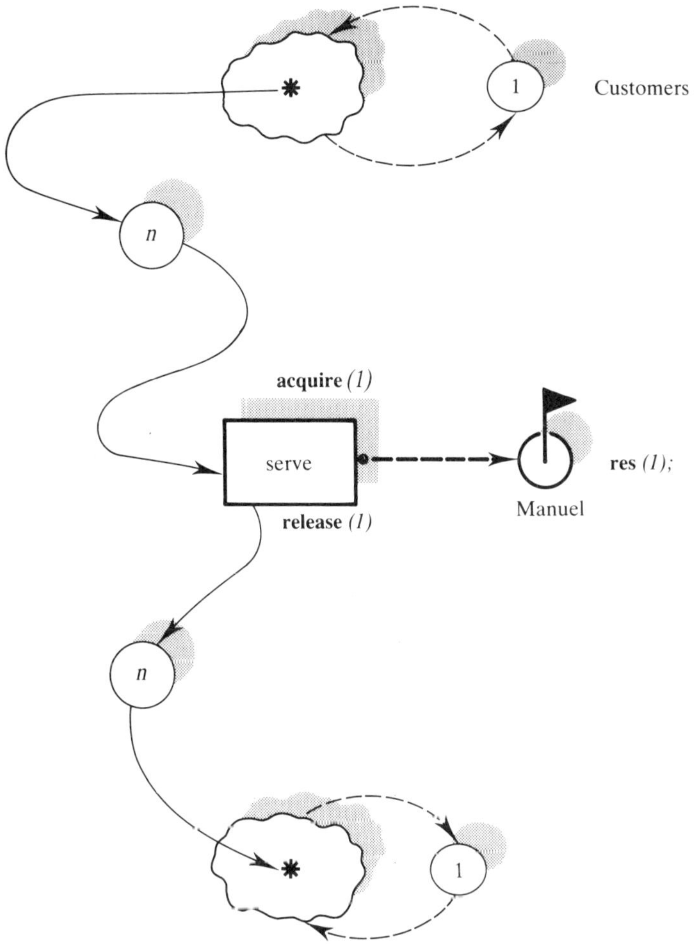

Figure 3.20 Case B: Mutual exclusion through capacity-constrained resources

outf, a reference to a file object. It is normally associated with the standard output device (the terminal screen), but it can easily be redirected. Here it is set to a file called *Man1report*, into which statistics will now be written. The **open** message initializes the file buffer to 80 blank characters. Prior to a model's termination, **close** messages are automatically sent to any file objects still open:

outf:– **new outfile** (*"Man1report"*);
outf.open (*80*);

Note also that all **entity** objects inherit a text-valued parameter, used to label reports. The initial contents of a bin must be stated when it is instantiated. Procedure **box** prints a (boxed) title at the head of a report.

Figure 3.19 shows the result of simulating this model for 200 time units.

Case B demonstrates how DEMOS models synchronization through **capacity-constrained resources** (Figure 3.20). The basic idea is similar to that of semaphores. A resource could be implemented by a Boolean flag, integer counter or some structured object. There must be a pair of operations for its acquisition and release, implemented by set and reset, increment and decrement (P and V), or some other procedures.

Since a finite capacity is the defining property of a resource, an item requesting its use may have to wait until there is enough room to accommodate it. Queues of items are therefore typically associated with resources. DEMOS models resources as instances of class **res**, which have name and capacity parameters and own an implicit queue.

In our case we need to represent Manuel as a single server:

ref (**res**) *Manuel*;

. . .

Manuel:− **new res** (*"Manuel"*, *1*);

Customers can **acquire** him, delay themselves for as long as a service is going to take, and then issue a **release** message, freeing Manuel for other customers:

Manuel.**acquire** (*1*);
hold (*ServiceTime*);
Manuel.**release** (*1*);

Statistics about utilization, queue lengths and waiting times are automatically recorded and printed at request (message **report**) or at the end of a simulation.

The DEMOS program for this scenario is very simple (Program 3.11). There is only one class of active entities. Upon entering the shop, each customer records his entry time and schedules the next customer's arrival (what the source did in Program 3.10). The customer proceeds to acquire Manuel's services, being delayed in a queue, if unsuccessful. Once the **acquire** request is granted, he or she waits for the time it takes to be served, thereafter releasing Manuel for the next customer. Before leaving the model the time spent in the shop is recorded in a **tally** object called *FlowTimes*. All instances of **tally** respond to **update** messages and DEMOS automatically computes and prints statistics about them.

Note that resources must be initialized to some maximal capacity, but this may be changed during model execution. A simulation is started by the arrival of the first customer, triggered by the **main** program. Figure 3.21 shows the statistics after a presimulation period of 50 time units, followed by a **reset** of statistics and a proper simulation of another 100 time units.

Resources are similar to semaphores, as they are used in computer operating systems. They can control access to data or processes by decrementation and incrementation of availability statistics.

While **res** objects provide a simple and straightforward way to model many common situations, this scheme is unfortunately not flexible enough to

```
begin
external class demos;
demos begin
  ref (rdist ) ArrivalProcess, ServiceProcess;
  ref (res   ) Manuel;
  ref (aSink) Departures;
  ref (tally ) FlowTimes;

  entity class aCustomer;
  begin
    real anEntryTime;
  anEntryTime := time;
  new aCustomer ("romeo").schedule (ArrivalProcess.sample);
  Manuel.acquire (1);
  hold (ServiceProcess.sample);
  Manuel.release (1);
  FlowTimes.update (time - anEntryTime);
  end of Customer;

comment: ********** Body of main program **********;

outf :- new outfile ("Man2report");
outf.open (blanks (80));

Munuel          :- new res     ("Manuel", 1);
ArrivalProcess :- new negexp ("arrivals", inreal);
ServiceProcess :- new negexp ("service ", inreal);
FlowTimes       :- new tally   ("FlowTimes");

new aCustomer ("romeo").schedule (now);

hold (inreal); comment: Presimulation period;
reset;
hold (inreal); comment: Length of Simulation Experiment;

end of prefixed block;
end of model;
```

Program 3.11 *Mutual exclusion with* ***resources***

capture more complex synchronization patterns. All items in the queues associated with the **res** objects of DEMOS can only wait for a single event. They will proceed on their individual life cycles if this event occurs. Sometimes, however, processes may have to be delayed until a combination of conditions become true simultaneously. Because such situations cannot adequately be represented by simply queueing for resources, DEMOS offers **condition queues (condq)** as an alternative. Condition queues are queues of entities waiting on arbitrarily complex conditions. Entities may file themselves into a **condq** by sending a **waituntil** (condition) message to it:

```
ref (condq) Lovesicks;
...
```

```
                              CLOCK TIME = 150.000
*******************************************************************************
*                                                                             *
*                                   REPORT                                    *
*                                                                             *
*******************************************************************************

                                DISTRIBUTIONS
                                *************

   TITLE      (RE)SET      OBS       TYPE        A          B          SEED
  arrivals     50.000       88      NEGEXP      0.800                33427485
  service      50.000       89      NEGEXP      1.000                22276755

                                  TALLIES
                                  *******

   TITLE      (RE)SET      OBS      AVERAGE    EST.ST.DV     MIN        MAX
 Flowtimes     50.000       77       3.420       2.712    3.711&-02   11.080

                                 RESOURCES
                                 *********

   TITLE      (RE)SET    OBS  LIM   MIN  NOW   %USAGE    AV. WAIT     QMAX
  Manuel       50.000     77    1     0    1   74.407      2.454        8
```

Figure 3.21 Statistics from Program 3.11

```
Lovesicks:- new condq ("Lovesicks");
...
Lovesicks.waituntil ((x and y) or z);
...
z:= true;
Lovesicks.signal;
...
x:= true;
Lovesicks.signal;
...
```

Entities in the same queue may wait on different conditions. A **signal** message must be sent to ensure that the conditions of waiting entities are re-evaluated each time an object associated with any of these conditions changes its state. Each evaluation will look at the first, or, if the **condq** attribute **all** is set, every item waiting in the queue. If the condition it is waiting for becomes true, an item can proceed on its life cycle, grabbing any resources it requires.

To demonstrate this feature we will extend our example slightly. Let us assume that during periods of slack demand for lovepotions (when no lovesickness epidemic is raging) Manuel lets half of his shop to Samantha, a young aspiring sorceress who specializes in rain-dance recipes and common household spells for purposes such as turning your neighbour into a toad. When in doubt about anything magical, wizards on DreamWorld use a secret book, the Necronomicon, whose pages abound with recipes, spells, curses, incantations, and other generally good advice. Samantha, having not yet passed all of her professional examinations, does not own a copy herself. She has to time share Manuel's. Manuel is quite an experienced old wizard and needs the book only to diagnose genuinely difficult cases; say for about 10% of all his customers. Samantha has less practise and must therefore look up about half of her spells and recipes.

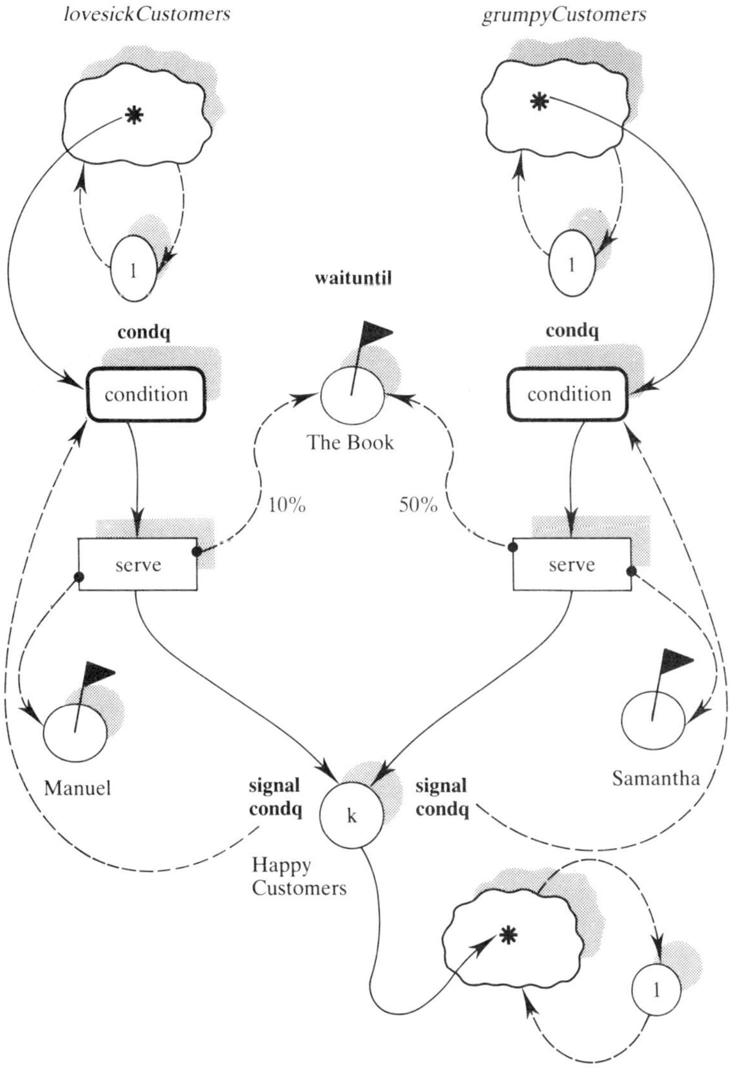

Figure 3.22 Case C: Waiting on semaphores

```
begin
external class demos;
demos begin
  ref (aManuelSource  ) LovesickSource;
  ref (aSamanthaSource) GrumpySource;
  ref (rdist          ) LovesickArrivals, GrumpyArrivals,
                        PotionService    , SpellService;
  ref (bdist          ) ManuelMustLookUpRecipe,
                        SamanthaMustLookUpSpell;
  ref (condq          ) LovesickCustomers, GrumpyCustomers;
  ref (res            ) Manuel, Samantha, TheBook;
  ref (accumulate     ) MeanWaiting;

  entity class aManuelSource;
    while true do
      begin
      new aLovesickCustomer ("romeo").schedule (now);
      hold (LovesickArrivals.sample);
      end while and ManuelSource;

  entity class aSamanthaSource;
    while true do
      begin
      new aGrumpyCustomer ("hilda").schedule (now);
      hold (GrumpyArrivals.sample);
      end while and SamanthaSource;

  entity class aLovesickCustomer;
  begin
    boolean manuelNeedsBook;
  MeanWaiting.update (LovesickCustomers.length + GrumpyCustomers.length + 1);
  manuelNeedsBook := ManuelMustLookUpRecipe.sample;
  if manuelNeedsBook then
    LovesickCustomers.waituntil (Manuel.avail > 0 and TheBook.avail > 0)
  else
    LovesickCustomers.waituntil (Manuel.avail > 0);
  MeanWaiting.update (LovesickCustomers.length + GrumpyCustomers.length);
  Manuel.acquire (1);
  if manuelNeedsBook then TheBook.acquire (1);
  hold (PotionService.sample);
  if manuelNeedsBook then TheBook.release (1);
  Manuel.release (1);
  LovesickCustomers.signal;
  GrumpyCustomers .signal;
  end LovesickCustomer;
```

Program 3.12 *Conditional waits in* **condq**

```
  entity class aGrumpyCustomer;
  begin
    boolean samanthaNeedsBook;
  MeanWaiting.update (LovesickCustomers.length + GrumpyCustomers.length + 1);
  samanthaNeedsBook := SamanthaMustLookUpSpell.sample;
  if samanthaNeedsBook then
    GrumpyCustomers.waituntil (Samantha.avail > 0 and TheBook.avail > 0)
  else
    GrumpyCustomers.waituntil (Samantha.avail > 0);
  MeanWaiting.update (LovesickCustomers.length + GrumpyCustomers.length);
  Samantha.acquire (1);
  if samanthaNeedsBook then TheBook.acquire (1);
  hold (SpellService.sample);
  if samanthaNeedsBook then TheBook.release (1);
  Samantha.release (1);
  LovesickCustomers.signal;
  GrumpyCustomers .signal;
  end LovesickCustomer;

comment: ********** Body of main program **********;

outf :- new outfile ("Man3report");
outf.open (blanks (80));

LovesickSource            :- new aManuelSource    ("arr. lovesicks");
GrumpySource              :- new aSamanthaSource("arr. grumpies");
LovesickArrivals          :- new negexp ("ls. arrivals", inreal        );
GrumpyArrivals            :- new negexp ("gr. arrivals", inreal        );
PotionService             :- new negexp ("potion service", inreal     );
SpellService              :- new normal ("spell   service"),inreal, inreal);
SamanthaMustLookUpSpell   :- new draw    ("SamanthaLooksUp", inreal );
ManuelMustLookUpRecipe    :- new draw    ("ManuelLooksUp    ", inreal );
LovesickCustomers         :- new condq ("Q of lovesicks");
LovesickCustomers.all     := true;
GrumpyCustomers           :- new condq ("Q of grumpies ");
GrumpyCustomers .all      := true;
Manuel                    :- new res     ("manuel    ", 1);
Samantha                  :- new res     ("samantha", 1);
TheBook                   :- new res     ("book      ", 1);
MeanWaiting               :- new accumulate ("Mean Number Waiting in Shop");

LovesickSource.schedule (now);
GrumpySource .schedule (now);
trace;
hold (inreal);      comment: Trace Time;
notrace;
hold (inreal);      comment: Length of Simulation Experiment;

end of prefixed block;
end of model;
```

Case C shows this constellation (Figure 3.22). Lovesick customers arrive to queue for one of Manuel's famed potions, whereas generally grumpy customers queue in front of Samantha to acquire an odd spectrum of simple

```
                         CLOCK TIME = 0.000
*************************************************************************************
*                                                                                   *
*                         TRACING COMMENCES                                         *
*                                                                                   *
*************************************************************************************

TIME   CURRENT        AND ITS ACTION(S)

0.000  DEMOS          HOLDS FOR 10.000, UNTIL 10.000
0.159  arr. loves 1   SCHEDULES romeo 2 NOW
       romeo 2        W'UNTIL IN Q of lovesic
       arr. loves 1   HOLDS FOR 0.654, UNTIL 0.813
0.415  romeo 1        RELEASES 1 TO manuel
                      SIGNALS Q of lovesic
                      SIGNALS Q of grumpie
                      ***TERMINATES
       romeo 2        LEAVES Q of lovesic
                      SEIZES 1 OF manuel
                      HOLDS FOR 0.306, UNTIL 0.721
0.721                 RELEASES 1 TO manuel
                      SIGNALS Q of lovesic
                      SIGNALS Q of grumpie
                      ***TERMINATES
0.813  arr. loves 1   SCHEDULES romeo 3 NOW
       romeo 3        SEIZES 1 OF manuel
                      HOLDS FOR 1.511, UNTIL 2.324
       arr. loves 1   HOLDS FOR 0.247, UNTIL 1.060
1.051  arr. grump 1   SCHEDULES hilda 2 NOW
       hilda 2        W'UNTIL IN Q of grumpie
       arr. grump 1   HOLDS FOR 4.741, UNTIL 5.792
1.060  arr. loves 1   SCHEDULES romeo 4 NOW
       romeo 4        W'UNTIL IN Q of lovesic
       arr. loves 1   HOLDS FOR 2.017, UNTIL 3.077
2.146  hilda 1        RELEASES 1 TO book
                      RELEASES 1 TO samantha
                      SIGNALS Q of lovesic
                      SIGNALS Q of grumpie
                      ***TERMINATES
       hilda 2        LEAVES Q of grumpie
                      SEIZES 1 OF samantha
                      SEIZES 1 OF book
                      HOLDS FOR 1.511, UNTIL 3.657
2.324  romeo 3        RELEASES 1 TO manuel
                      SIGNALS Q of lovesic
                      SIGNALS Q of grumpie
                      ***TERMINATES
```

Figure 3.23 Statistics produced by Program 3.12

spells and recipes. In 10% of all cases Manuel has to refer to the book, but he will not start serving any such customer if he sees that the book is currently used by Samantha. Samantha also makes sure that the book is

```
     romeo 4        LEAVES Q of lovesic
                    SEIZES 1 OF manuel
                    HOLDS FOR 1.180, UNTIL 3.505

                              CLOCK TIME = 10.000
*******************************************************************************
*                                                                             *
*                           TRACING SWITCHED OFF                              *
*                                                                             *
*******************************************************************************

                              CLOCK TIME = 200.000
*******************************************************************************
*                                                                             *
*                                  REPORT                                     *
*                                                                             *
*******************************************************************************

                                DISTRIBUTIONS
                                *************

    TITLE       (RE)SET     OBS      TYPE        A         B        SEED
  ls.arrivals    0.000      147     NEGEXP     0.800               33427485
  gr.arrivals    0.000      169     NEGEXP     0.800               22276755
  potion servi   0.000      147     NEGEXP     1.000               46847980
  spell servic   0.000      163     NORMAL     1.000     0.500     43859043
  samanthalook   0.000      169      DRAW      0.500               64042082
  manuellooksu   0.000      147      DRAW      0.100               44366305

                                ACCUMULATES
                                *************

    TITLE       (RE)SET     OBS     AVERAGE   EST.ST.DV    MIN        MAX
  Mean Number    0.000      519      6.060     3.430      0.000     18.000

                                 RESOURCES
                                 **********

    TITLE       (RE)SET   OBS  LIM  MIN  NOW   %USAGE      AV. WAIT    QMAX
    manuel       0.000    149   1    0    1    74.290       0.000        1
    samantha     0.000    107   1    0    0    93.834       0.000        1
     book        0.000     74   1    0    1    63.080       0.000        1

                              CONDITION QUEUES
                              *****************

    TITLE       (RE)SET    OBS    QMAX   QNOW   Q AVERAGE  ZEROS   AV. WAIT
  Q of lovesic*  0.000     149      8      0      1.541      43     2.068
  Q of grumpie*  0.000     108     11      5      4.520       6     8.112
```

available before she starts serving any of the 50% of her customers with less than straightforward desires. If the book is needed by any of the two, it is required for the whole length of a service activity. Once it is released this is signalled to the customers of both Manuel and Samantha.

Program 3.12 shows a corresponding DEMOS program. Two classes of sources are needed, one generating lovesick customers for Manuel and the other generating generally grumpy customers for Samantha. Accordingly, four **rdist** objects are now needed, defining Manuel's and Samantha's arrival- and service-rate distributions. Whether the book is required for serving a customer is determined by sampling from a Boolean distribution, separate instances of which are associated with Manuel and Samantha. Two **condq** are also needed – one for lovesick and one for grumpy customers. Samantha and Manuel themselves are instances of class **res**.

MeanWaiting, a predefined data collector of class **accumulate**, is used to determine the (time-weighted) mean of all (Manuel's and Samantha's) customers waiting in the shop.

DEMOS's **accumulate** is similar to the *measure* type used in Chapter 2. This statistic is updated after lovesick customers enter the shop. It is then determined whether or not the book is required. Customers queue until Manuel and possibly the book become available, whereupon they acquire and release these resources, subjecting themselves to Manuel's service. Prior to their departure they signal Manuel's and the book's availability.

The life cycle of grumpy customers is almost identical, with Manuel's place taken by Samantha.

The model is initialized in the main body. Negative exponential and normal distributions are created and associated with **rdist** references. **bdist** is instantiated to binary **draw** distributions. Note that **all** is set for both **condq**, prompting DEMOS to evaluate conditions for all waiting entities each time a **signal** is received.

Figure 3.23 shows the results of simulating this model. The tracing facility of DEMOS is demonstrated here. It is switched on by a **trace** message to the monitor and can be switched off again by **notrace**.

When drawing activity-cycle diagrams one often finds activity activations to depend on the availability of instances of two different classes of entities. Sometimes their life cycles are then coupled together throughout a number of related activities. Consider a model of a car ferry, where cars stay on the ferry for the whole duration of a crossing.

If one class of entities is singled out as the dominant one, we have the case of a **master-slave synchronization**. A sequence of activities can be started by one instance of the master and a fixed number of instances of slave objects. The master is active: co-opting a specified number of slaves, moving through a sequence of activities, and, finally, scheduling her slaves to proceed on life cycles of their own. DEMOS supports such a scheme through a device called a **waitq**.

Case D shows our original example, with customers treated as slaves and Manuel as the master (a machine-oriented model). Each **waitq** has two queues associated with it: one for waiting slaves (labelled by '*' in reports), and one for waiting masters (Figure 3.24). Lovesick customers file themselves into the slave queue by issuing **wait** messages. If there is at least one customer waiting, Manuel co-opts her from the head of the queue and starts a service activity; alternatively he has to wait in the master queue. The customer is scheduled to proceed and Manuel tries to co-opt the next one after the service is completed:

{*the slave's life cycle*}
LovesickCustomers. **wait**;

...

{*the master's life cycle*}
romeo:– *LovesickCustomers.* **coopt**;

...

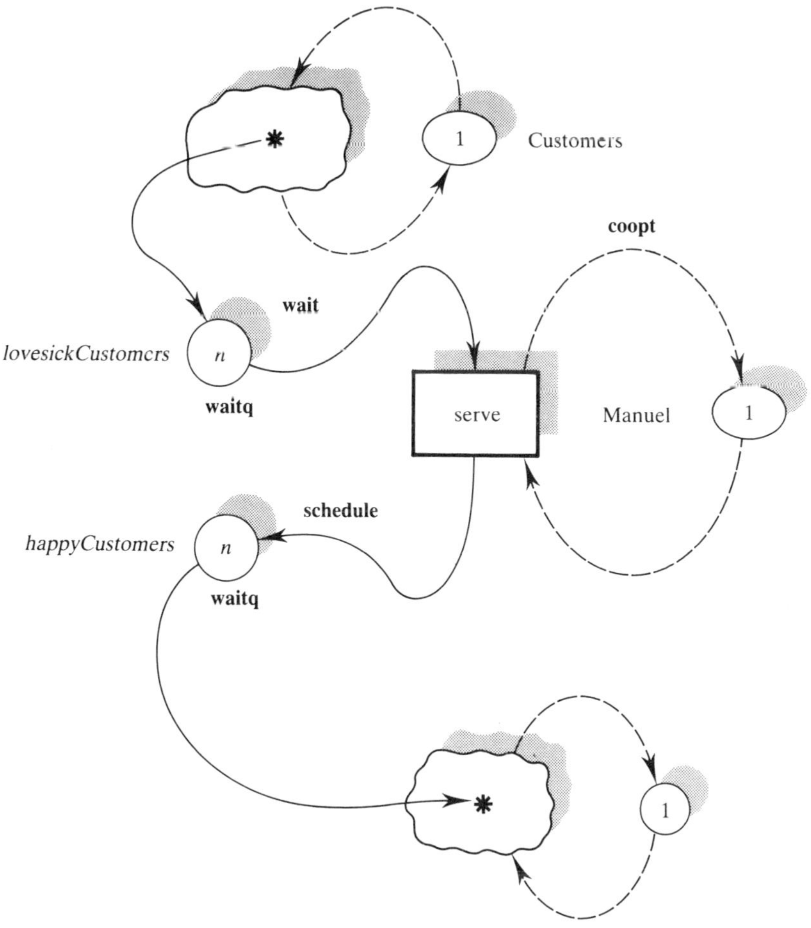

Figure 3.24 Case D: Master–slave synchronization

```
{a string of common activities}
  ...
  romeo. schedule (now);
  ...
```

Program 3.13 shows the corresponding DEMOS model. At one end *aSource* generates customers and activates them, at the other *aSink* takes instances from the queue of happy customers and determines the time they spent in the shop. This information is recorded in a **histogram**. DEMOS provides histograms as another class of predefined data collection devices.

```
begin
external class demos;
demos begin
  ref (aSource  ) Arrivals;
  ref (rdist    ) ArrivalProcess     , ServiceProcess   ;
  ref (waitq    ) LovesickCustomers, HappyCustomers;
  ref (aServer  ) Manuel;
  ref (aSink    ) Departures;
  ref (histogram) FlowTimes;

  entity class aSource (lambda); value lambda ; real lambda;
    while true do
      begin
      new aCustomer ("romeo").schedule (now);
      hold (ArrivalProcess.sample);
      end while and source;

  entity class aCustomer;
  begin
    real anEntryTime;
  anEntryTime := time;
  LovesickCustomers.wait;
  HappyCustomers   .wait;
  end of Customer;

  entity class aServer (mu); value mu; real mu;
  begin
    ref (aCustomer) romeo;
  while true do
    begin
    romeo :- LovesickCustomers.coopt;
    hold (ServiceProcess.sample);
    romeo.schedule (now);
    end while;
  end of Server;
```

Program 3.13 *Master–slave synchronization in* ***waitq***

Note that histograms need to be initialized with respect to lower bound, upper bound, and the number of frequency classes they can handle.

Customers have simple life cycles. They record their entry time, slavishly wait for Manuel to take them out of the queue of lovesick customers, and, after being served, wait for *departures* to remove them from the shop. Manuel performs service activities, repeatedly co-opting waiting customers, who are then activated to wait for their departure.

Instead of specifying a time interval we want to simulate this model for a given number of customer departures. We therefore associate a sample size with the sink, program it to count the number of departing customers, and

```
  entity class aSink (aSampleSize); integer aSampleSize;
  begin
    ref (aCustomer) romeo;
    integer         samples;
  samples := 0;
  while aSampleSize > samples do
    begin
    samples := samples + 1;
    romeo  :- HappyCustomers.coopt;
    FlowTimes.update (time - romeo.anEntryTime);
    end while;
  demos.schedule (now);
  end of Sink;

comment: ********** Body of main program **********;

outf :- new outfile ("Man4rcport");
outf.open (blanks (80));

Arrivals           :- new aSource   ("source ", inreal);
Manuel             :- new aServer   ("manuel", inreal);
ArrivalProcess     :- new negexp    ("arrivals", Arrivals.lambda);
ServiceProcess     :- new negexp    ("service ", Manuel .mu      );
LovesickCustomers  :- new waitq     ("lovesicks");
HappyCustomers     :- new waitq     ("happies" );
FlowTimes          :- new histogram ("FlowTimes", 0, 10, 10);

Arrivals  .schedule (0.0);
Manuel    .schedule (0.0);
departures :- new aSink ("departures", inint);
departures.schedule (now);
passivate;

snapsqs;
snapqueues;

end of prefixed block;
end of model;
```

```
                              CLOCK TIME = 136.697
*********************************************************************************
*                                                                               *
*                                  EVENT LIST                                   *
*                                                                               *
*********************************************************************************

N      EV. TIME     OBJ. TITLE      LL             BL              RL
1      136.697      DEMOS                          departures 1
2      136.697      departures 1    DEMOS          source 1
3      137.362      manuel 1                       source 1
4      138.936      source 1        departures 1                   manuel 1


                              CLOCK TIME = 136.697
*********************************************************************************
*                                                                               *
*                            LIST OF PASSIVE OBJECTS                            *
*                                                                               *
*                                                                               *
*********************************************************************************

                         ENTITIES WAITING IN lovesicks
                        ******************************

                        ***** THE QUEUE IS EMPTY. *****

                         ENTITIES WAITING IN lovesicks
                        ******************************

                        NO OBJECT    PRIORITY ENTRY IN Q
                        1 romeo 2           0    130.001
                        2 romeo 3           0    131.052
                        3 romeo 4           0    136.641

                          ENTITIES WAITING IN happies
                          ***************************

                          *** THE QUEUE IS EMPTY. ***

                          ENTITIES WAITING IN happies
                          ***************************

                          *** THE QUEUE IS EMPTY. ***
```

Figure 3.25 Statistics for Program 3.13

reactivate the model's body (called **demos**) as soon as this sample size has been reached.

Figure 3.25 shows this model's output. **snapqueues** and **snapsqs** messages are included in the model to demonstrate these debugging features. Notice

```
                                CLOCK TIME = 136.697
*********************************************************************************
*                                                                               *
*                                    REPORT                                     *
*                                                                               *
*********************************************************************************

                                   DISTRIBUTIONS
                                   *************

   TITLE      (RE)SET      OBS       TYPE        A          B          SEED
  arrivals     0.000       106      NEGEXP     0.800                 33427485
  service      0.000       101      NEGEXP     1.000                 22276755

                                    HISTOGRAMS
                                    ***********

                                     SUMMARY

   TITLE      (RE)SET      OBS      AVERAGE   EST.ST.DV      MIN         MAX
 FlowTimes     0.000       100       2.255      2.351     3.711&-02     9.143

 CELL  LOWER LIM   N   FREQ    CUM
                                  | -------------------------------
   0   -INFINITY   0   0.00    0.00 |
   1     0.000    37   0.37   37.00 | ***********************************************
   2     1.000    30   0.30   67.00 |************************
   3     2.000     9   0.09   76.00 |*******
   4     3.000     4   0.04   80.00 |***
   5     4.000     2   0.02   82.00 |**
   6     5.000     6   0.06   88.00 |*****
   7     6.000     4   0.04   92.00 |***
   8     7.000     6   0.06   98.00 |*****
   9     8.000     1   0.01   99.00 |*
  10     9.000     1   0.01  100.00 |*
  11    10.000     0   0.00  100.00 |
                                  | -------------------------------

                                   WAIT QUEUES
                                   ************

   TITLE      (RE)SET      OBS    QMAX  QNOW   Q AVERAGE   ZEROS    AV. WAIT
 lovesicks     0.000       101      1     0      0.376       64       0.509
 lovesicks *   0.000       101      8     3      1.168       38       1.458

  happies      0.000       100      1     0      1.000        0       1.367
  happies *    0.000       100      1     0      0.000      100       0.000
```

that as the simulation shuts down there are still three customers waiting. Manuel is busy and both the master and slave queue associated with departures (*HappyCustomers*) are empty. A printout of the event list is shown, with each scheduled object, its time aspect, *LL*, *BL* and *RL* links.

DEMOS organizes the event set (*SQS*) as a tree structure. Entities with equal event times form a linear sublist connected by *BL* and *LL*, with the *RL* of all but the first set to **none**.

DEMOS is a very flexible and powerful modelling framework, which deserves to be more widely known than it is. In the author's opinion it is currently the best tool available for any kind of discrete-event simulation. It can also be customized to give more support for specialized scenarios. OASIS [Lomov *et al.* (1984)] is an example, catering for computer-system simulation and software prototyping.

C. Transaction-flow scenarios

GPSS [Bobillier *et al.* (1976)] is both the oldest and the most widely used scenario language. GPSS programs can be viewed as one-to-one translations of a so-called GPSS block diagram. Figure 3.26 uses this graphical notation to model Manuel's shop.

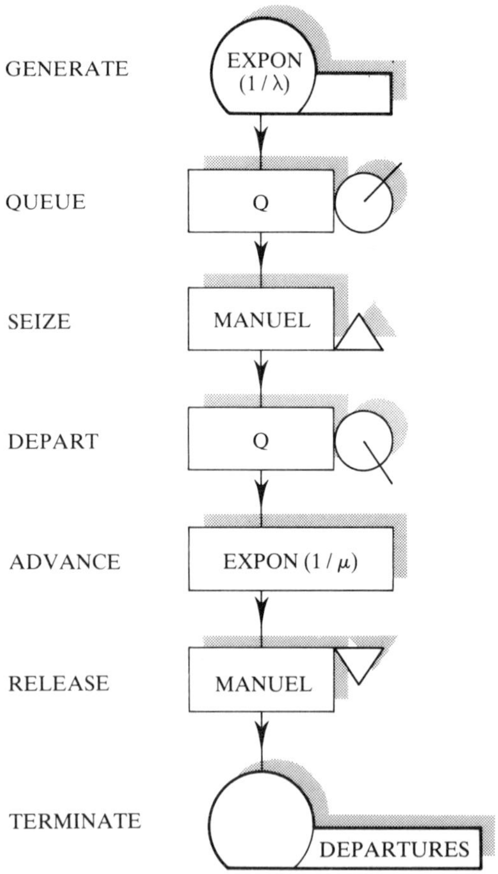

Figure 3.26 Queueing analysis of Manuel's shop: A GPSS block diagram

GPSS block diagrams are not often used by more experienced programmers, since they suffer from being bound at too low a level (one block maps into one statement) and there are too many different symbols cluttering the model structure.

GPSS models are driven by dynamic items called **transactions**, whose life cycles describe the model's behaviour. There can be any number of those within one model. GPSS can therefore be viewed as a process-oriented language, with each transaction pattern corresponding to a process class. Transactions are generated, queue for resources and are delayed (**advance**) for the time they take to be serviced. This is the typical structure of material-oriented modelling styles.

GPSS provides predefined queue, server, and list objects (called **sets**, **groups** and **chains**). Servers can be single servers (**FACILITY**) or multiple server stations (**STORAGE**). Parameterized **blocks** (similar to procedure calls) are used to operate on these entities from within a transaction's life

```
JOB
Arrve   FUNCTION RN1, EXPON (1/λ)
Serve   FUNCTION RN2, EXPON (1/μ)

        GENERATE  FN$Arrve
        QUEUE     q
        SEIZE     manuel
        DEPART    q
        ADVANCE   FN$Serve
        RELEASE   manuel
        TERMINATE 1

START 500
END
```

SIMULATION RESULTS

START SIMULATION 500 TERMINATIONS

FACILITY	NUMBER ENTRIES	AVERAGE TIME/TRANS	---AVERAGE UTILIZATION DURING--- TOTAL TIME	AVAILABLE TIME	UNAVAILABLE TIME
manuel	500	0.865	0.76	0.76	N.A.

QUEUE	MAXIMUM CONTENTS	AVERAGE CONTENTS	TOTAL ENTRIES	ZERO ENTRIES	PERCENT ZEROS	AVERAGE TIME/TRANS
q	12	2.5	500	116	0.23	3.15

Program 3.14 *A queueing analysis of Manuel's shop (in GPSS)*

cycle. Present GPSS versions offer approximately 50 different block types.

Program 3.14 shows a GPSS model of Manuel's shop. Most data structures are implicit rather than declared explicitly. This is just one of the various remnants of GPSS's FORTRAN heritage.

In our model Manuel is a **FACILITY** and *Q* a **QUEUE** object. Distributions are represented by **FUNCTION** entities (*FN*). The **GENERATE** block acts as a source, repetitively spawning new transactions according to a number of parameters. The number of predefined and user-defined parameters associated with each transaction is fixed (usually 10). **FN$** (fn-name) returns a function value, drawn from a distribution with the function argument (here a reference to a random-number generator) bound to a specified value. Sinks are represented by **TERMINATE** blocks. **TERMINATE** *1* increments a count by one each time a customer leaves the shop. **START** *500* specifies that the model is to be simulated until this termination count reaches 500.

The conciseness of the GPSS model reflects the fact that the language was explicitly designed for models such as this one. Its clean structure tends to rapidly disappear as soon as GPSS is applied to tasks outside a relatively narrow class of problems.

The language also copes well with single-level queueing networks. It lacks the necessary encapsulation facilities to be usable for multi-level models, where server or other nodes at one level must be expanded into individual scenarios at lower levels.

Computer-system simulations are an example of the advantages of such multi-level scenarios. Here programs can simultaneously be viewed as capacity-constrained resources and resource-consuming processes. In their resource aspect they can be acquired and released by a task, with queues forming in front of them. In their process aspect they themselves request and queue for processors, data paths and devices. GPSS is not well equipped to deal with such complicated scenarios. Models get hopelessly cluttered with irrelevant detail and their overall structure is very difficult to retrieve.

Some specialized extensions for computer-system simulation have therefore been proposed. CSS [IBM (1972)] is one of the more widely known, offering predefined building blocks for representation of programs, processors, devices, etc. This approach does not really solve the problem, since these GPSS offsprings still lack concepts for decent data and control flow structuring, which makes it very difficult to implement complex process interactions correctly. Provisions for linking FORTRAN, PL/1 or assembly-language procedures to a GPSS model via **HELP** blocks are clumsy and difficult to use. They also require a detailed knowledge of the interpreter's internal table organization.

Note that GPSS is assumed to support sampling from predefined exponential functions directly. Many older systems define all distributions by a set of data points, between which linear interpolations have to be made. The reasons for this 'stone-axe' approach are hard to understand. GPSS also allows only integer values. It automatically collects, computes, formats, and prints statistical measures of system congestion, server utilization and delay times. This is one of the most convincing reasons for its continuing popularity.

GPSS programs consist of declaratives (**JOB**, **END**), control (**START**) and block statements. Almost all versions, with the exception of GPSS/H [Henriksen (1975, 1983)], are interpreted. There is a preinterpretation phase, where blocks are translated into a set of tables. A run-time control system then performs the model interpretation, using an integer-valued clock and a number of event lists (*CEC*: current-events chain; *FEC*: future events chain; *IR*: interrupt chain).

GPSS is an old language. Most of the popular implementations suffer from a number of defects. 'Normal' input/output is not part of the original design. Interpretative execution and fixed sized tables make GPSS models slow and bulky. Arithmetic is awkward (only integers and clumsy **VARIABLE** entities); control structures are poor (basically **GO TO** branches only); procedures and separate compilation of modules are not supported. Much of this criticism does not hold for more modern systems like GPSS/H.

Apart from the convenience of automatic report generation, GPSS offers good error tracing and debugging aids. Although it is still primarily a batch-oriented language, interactive versions with limited graphics capabilities have been developed [see Henriksen (1983a)].

The continuing popularity of the language has spawned dialects implemented in FORTRAN [Schmidt (1980)], PASCAL [Uyeno (1980)], and SIMULA [Vaucher (1974)]. An integration of GPSS and GASP to provide both discrete and continuous modelling primitives has also been attempted [Hooper and Reilly (1983)]. INS [Roberts and Sadlowski (1975)], SIMON75 [Hills (1975)] and BORIS [Wendt (1978)] are other scenario languages with similar world views.

3.4.5 Putting it all together: Combined systems

For a long time the user communities of continuous and discrete simulation were almost completely isolated from one another, both by a separation of interests in different kinds of systems and by different traditions and terminologies.

Continuous-system simulation was primarily used in engineering, physics, chemistry, biology, ecology, economics and sociology, where the behaviour of dynamic systems is traditionally captured by differential equations, either because this is considered a 'natural' representation for some process (engineering, physics, chemistry) or because of the high level of aggregation at which phenomena are studied (biology, ecology, economics, sociology). Systems are typically represented at a uniform, fairly low level of detail. Irregularities, expressed by complex interactions between individual events, do not occur or are abstracted away.

Discrete-event simulation, on the other hand, originated in studies of inventory and manufacturing systems. It has been popular with operations researchers, mathematicians and computer scientists. Here, a phenomenon can be studied at any desired level of detail. Arbitrarily complex process interactions and any relevant irregularities may be explicitly modelled. Unified theories (although always desirable) are not required.

Many modelling studies can greatly benefit from an integration of both techniques, a fact that is recently becoming more widely recognized. It is hoped that an increased awareness of characteristics, strengths and limitations of different simulation frameworks will help to finally overcome any remaining barriers.

Combined models can be driven by a discrete-event simulator, with some components represented as continuous processes. A predominantly continuous model may alternatively be augmented by incorporating a number of special events describing irregularities in its behaviour.

Figure 3.27 returns us to DreamWorld to a situation that maps quite naturally into a continuous model, but in which certain exceptional events may occur.

The highlands of Avalon are populated by wizards and unicorns. Both populations live and die according to some constant relationships (unicorns' birth rate, unicorns' death rate, wizards' birth rate, wizards' death rate). Wizards use unicorn horns in some of their ancient rituals, and for this purpose a number of unicorns are 'cropped' each year, the exact number depending only on the level of the wizard population. Such a two-species ecology can be studied by using a simple system-dynamics model, similar to the one discussed in Section 3.4.3. To complicate this scenario slightly, we will introduce two irregularities.

For the year 21 476 (unicorn calendar) a **time event** is unconditionally scheduled. A big flood will occur at this date, wiping all unicorns from the face of DreamWorld.

A **state event** also becomes necessary whenever the wizard population reaches a state of overcrowding – say more than 1000 individuals. Strife between rival tribes will then reduce it to half this number.

These two irregularities in the otherwise continuous growth of unicorn and wizard populations are reflected in the system-dynamics diagram (Figure 3.27) and the associated *DiscoContext* program.

DiscoContext [Helsgaun (1980)] is a SIMULA context that adds continuous-simulation capabilities (various integration methods, time plots, etc.) to the discrete-event framework carried by **class simulation**. It is an extension of class *CScontext* used in Section 3.4.3. *DiscoContext* can cater for discrete, distributed-parameter and lumped-parameter systems of both the differential- and difference-equation (system dynamics) varieties.

unicorns and *wizards* are declared as subclasses of *beings*, whose numbers fluctuate according to a continuous process. **state** values of a **variable** object grow quasi-continuously at a specified **rate**. The message **start** is used to initiate a variable's dynamics. Growth rates and initial populations can easily be deduced from Program 3.15. The rate equations in class *growth* define the interaction of the unicorn (prey) and wizard (predator) populations.

A **reporter** object gathers annual (**setfrequency**) statistics about the model's time-varying behaviour. These statistics are stored in a standard output file called **sysplot**, for which appropriate labels and scales are also defined.

The main program instantiates and initializes all data structures (called *wizpop*, *ucornpop* and *popdynamics*). **dtmax** and **maxrelerror** specify the

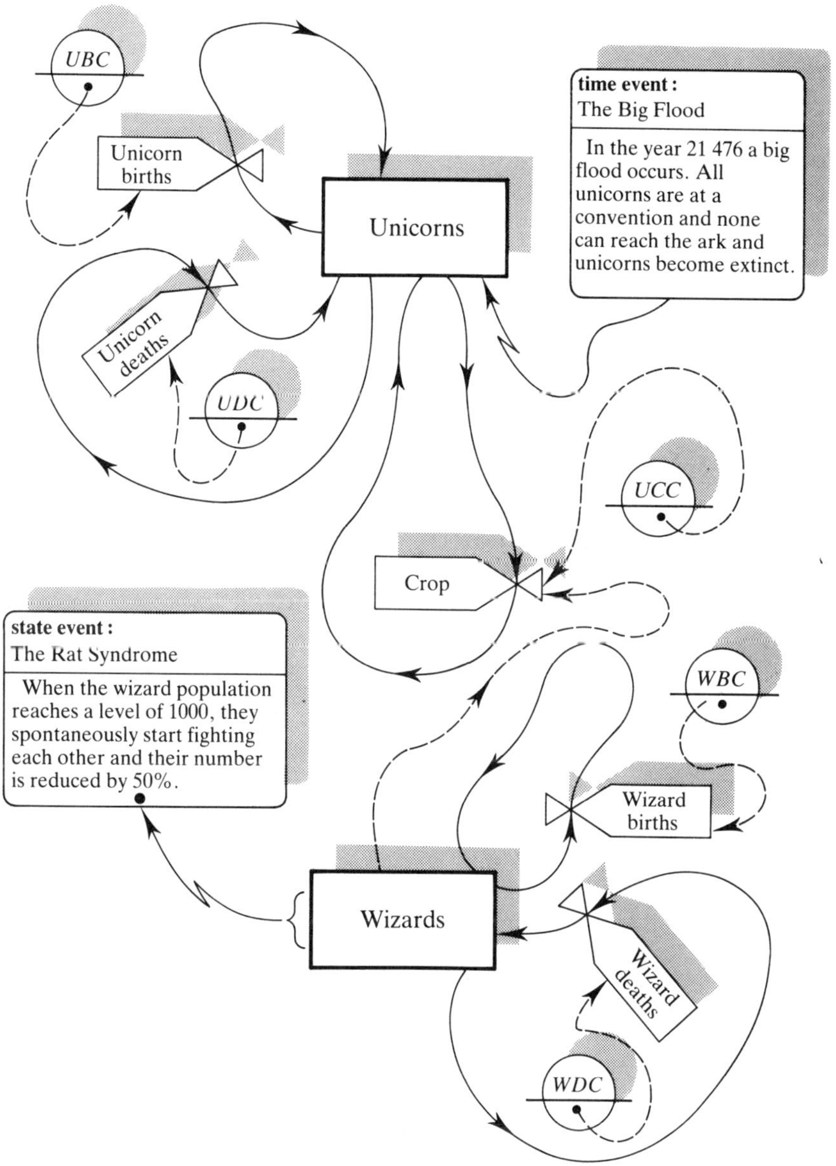

Figure 3.27 Wizards and unicorns: A scenario for combined discrete/continuous system simulation

```
begin

comment: A combined discrete/continuous model of the
         unicorns/wizards ecology;

external class DISCOcontext;

  ref (printfile) out;

DISCOcontext begin

  ref (wizards)        wizpop;
  ref (unicorns)       ucornpop;
  ref (growth)         popdynamics;
  ref (popmonitor)     war;
  ref (floodscheduler) doomsday;
  real                 end_of_time;

process class beings;
begin
comment: This class prefixes both species. It defines properties
         common to unicorns and wizards;
  ref (variable) numbers;
  real percent_births, percent_deaths;
end of beings;

beings class unicorns;
begin
comment: Initializes unicorn properties;
percent_births  := 0.002;          comment: UBC: each unicorn has a calf
                                            every 500 years;
percent_deaths  := 0.001;          comment: UDC: each unicorn lives for
                                            1000 years;
numbers:- new variable (1000);     comment: Initial unicorn population
                                            in Avalon;
numbers.start;
end of unicorn description;

beings class wizards;
begin
comment: Defines additional wizard properties and initializes all
         of them;
  real ucorn_need;
percent_births  := 0.03;           comment: WBC: every witch has 3
                                            children in her lifetime;
percent_deaths  := 0.02;           comment: WDC: every wizard can expect
                                            to live for 200 years;
ucorn_need  := 0.02;               comment: UCC: Every 50 years each
                                            wizard needs a unicorn horn
                                            for sinister purposes;
numbers:- new variable (200);      comment: Initial number of wizards
                                            in Avalon;
numbers.start;
end of wizard description;
```

Program 3.15 *A combined discrete/continuous model of a unicorn–wizard ecology*

```
continuous class growth (predator, prey); ref(wizards) predator;
                                          ref(unicorns) prey;
begin
  comment: describes the growth processes of both populations;
inspect predator do
  numbers.state:= numbers.laststate + dt*((percent_births -
                   percent_deaths)* numbers.state);
prey.numbers.state:= prey.numbers.laststate + dt*((prey.percent_births
                   - prey.percent_deaths) * prey.numbers.state
                   - (predator.ucorn_need * predator.numbers.state));
end of growth process;

process class popmonitor (population); integer population;
begin
  comment: When state of a population reaches a certain level,
           it is reduced by half;
while true do
  begin
  waituntil (wizpop.numbers.state >= population);
  outtext ("War"); outimage;
  wizpop.numbers.stop;
  wizpop.numbers.state:= wizpop.numbers.state / 2.0;
  wizpop.numbers.start;
  end;
end of population censor;

process class floodscheduler (date); integer date;
begin
  comment: Floods occur at the specified date and wipe out the
           whole unicorn population;
  integer currentdate;

currentdate:= inint;
hold (date - currentdate);
ucornpop.numbers.stop;
ucornpop.numbers.state:= 0.0;
ucornpop.numbers.start;
end of flood scheduler;

reporter class plotter;
begin
plot ('*', time, ucornpop.numbers.state);
plot ('+', time, wizpop.numbers.state);
end of plotter;

comment:  body of main program ;
          +++++++++++

dtmax:= 1.0;
euler:= true;
end_of_time:= inreal;
out:- new printfile ("UCORNOUT");
```

(cont.)

```
wizpop      :- new wizards; activate wizpop;
ucornpop    :- new unicorns; activate ucornpop;
war         :- new popmonitor (1000); activate war;
doomsday    :- new floodscheduler (21476); activate doomsday;
popdynamics :- new growth (wizpop, ucornpop); popdynamics.start;

inspect new plotter do
  begin
  setfrequency (1);
  start;
  end;
inspect sysplot do
  begin
  linesperplot (40);
  setkey ('*', "Ucornpop"); setkey ('+', "Wizpop");
  setxlabel ("Time");
  setymin ('*', 0); setymax ('*', 2000);
  setymin ('+', 0); setymax ('+', 2000);
  end of plotfile definition;

hold (end_of_time);
outtext ("Simulation stops at "); outfix (time,2,10); outimage;
end ***** of DISCOcontext model *****;

end of program block;
```

Program 3.15 (cont.)

time grain and desired accuracy (1%), according to which *DiscoContext* selects an appropriate time step and integration technique. The simulation then proceeds until **end-of-time**, whose value is read from the terminal.

A **waituntil** command implements the state event, waiting for overcrowding to occur. **waituntil** is used by process class *popmonitor*, the only instance of which (called *strife*) is created and activated by the main program. *strife* will monitor the unicorn population and detect any numbers exceeding *popthreshold*, a parameter set to 1000 by the main program. It will perform its sad duties at that time. In artificial intelligence circles this kind of pattern-driven procedure is often referred to as a **demon**.

Triggering the time event (flooding) is delegated to a process of class *floodscheduler*. *doomsday* is created, instantiated with the **date** of the flood as a parameter, and activated at the start of the simulation. It reads the *currentdate* (21 400) and then falls asleep until it is time to cause the flood [**hold** (*date* − *currentdate*)].

At the **end-of-time** the simulation stops, when model control passes through the last **end** statement. The code associated with **inspect new plotter** causes a plot of the two populations' temporal evolution.

Note that process classes execute their bodies in parallel after being activated, until either passivated, delayed or having exhausted all actions of their life cycles. Continuous processes, once started, loop through their life cycles until stopped or until the end of a simulation run. Figure 3.28 shows a graph of both populations' time paths.

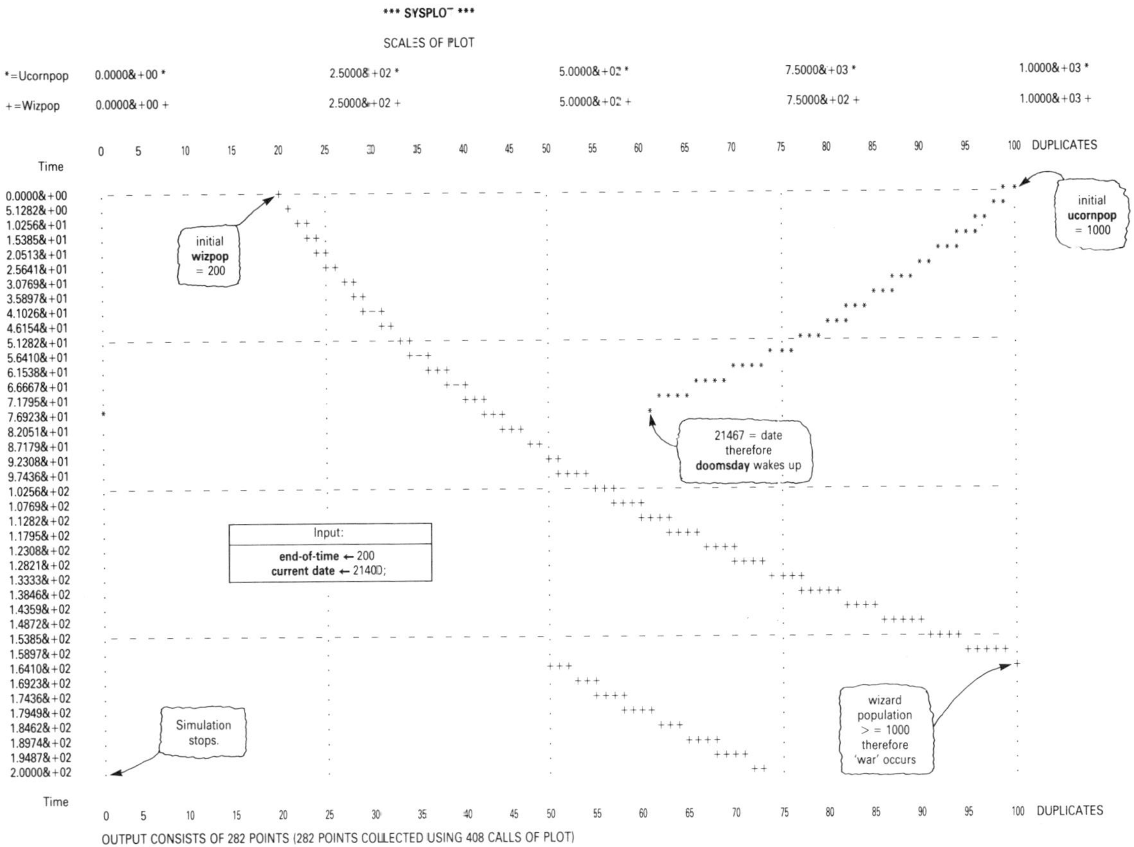

Figure 3.28 Wizards and unicorns: A chronological graph

Note that in Chapter 2 Manuel's shop was used to demonstrate a different kind of continuous simulation. There the model was basically a discrete-event one, with only one component characterized by a quasi-continuous process.

Flexible combined discrete–continuous modelling tools are only slowly becoming more widely available. GASP IV [Pritsker (1974, 1979)] was the first language allowing for this kind of event/process interaction. SLAM [Pritsker and Pegden (1979), Pritsker (1983)] is another more recently developed package of FORTRAN procedures for combined modelling. There is an extension to SIMSCRIPT II.5 [Delfosse (1976)] that makes this language suitable for combined scenarios. DISCO and CADSIM are SIMULA classes catering for this modelling style.

To implement combined simulation tools, extensions for representing continuous processes were typically made to discrete simulation programming languages. It is obviously much easier to provide solution procedures for differential equations in a discrete-event framework than to build such a framework from scratch and impose it on a continuous-system simulator. Fahrland (1970) and Oren (1977) are good, although now somewhat dated surveys of the state of the field.

3.5 Model generators

Model generators offer very high-level user environments. They are, however, restricted to relatively narrow areas of application. There are special-purpose program packages for computer-system simulation, such as SCERT [COMTEN (1976)] and generator systems [Mathewson (1974, 1983), Subrahmanian and Cannon (1981), Davies (1979)] that aid inexperienced users in building models based on activity cycles [Hutchinson (1975a)].

Program 3.16 shows how the Manuel model can be defined with an interactive model generator called CAPS (Computer Aided Programming of Simulations) [Hutchinson (1975b), Clementson]. CAPS is based on life-cycle diagrams and assembles a model from a dialogue with its users, mapping it into an activity-oriented ECSL program [Clementson (1966)]. This program is then compiled and executed.

Note that a 'NO' reply can be given by typing 'LINEFEED', 'NO' or 'N'. The input format of CAPS has been slightly altered to make the session protocol more readable, although all essential features are maintained. The user's responses are italicized.

3.6 Developments

In the last few years the development of new simulation software has slowed considerably. Languages like GASP, SIMSCRIPT, SIMULA and GPSS are well established and have not changed dramatically since the 1968 IFIP conference on simulation programming languages [Buxton (1968)]. Only a few, mainly cosmetic changes, have been made and some bells and whistles have been added.

PASCAL's success, particularly in university environments, has prompted the design and implementation of a few PASCAL-based simulation systems, but not as many as expected, attracting user communities, which are not as large as anticipated. One of these, SIMPAS, was discussed in Section 3.3.3.

SIMONE [Perrot *et al.* (1980)], PASCAL PLUS [Welsh (1979)] and PASSIM [Uyeno (1980)]; Smith and Smith (1984), Frantz and Trott (1984), Vaucher (1984) and Marsden (1984) are other systems that have not achieved wider popularity for a number of reasons.

It is difficult to convince people of the merits of switching to a different notation, regardless of its intellectual or technological superiority. Once user communities have entrenched themselves, they remain very loyal to their chosen programming languages. Proponents (inmates) of a particular language culture hardly ever switch, but they will eventually retire. This may explain why most simulation projects are still carried out in FORTRAN.

Kleine (1970, 1971) reports that in the early 1970s in the USA FORTRAN, GASP, GPSS and SIMSCRIPT were almost exclusively used for discrete-event simulations. It is likely that this is still true today. SIMULA is quite popular in other parts of the world.

The main continuous-system simulation languages are probably DYNAMO, CSMP, CSSL IV, SIMNON, DARE and ACSL. GASP and SLAM are the best-known combined systems.

Some interesting developments, however, are clearly visible. With the dramatic drop in hardware prices the use of powerful personal work stations to host modelling toolboxes has now become economically viable. The trend is towards integrated and interactive programming environments and graphically supported user interfaces.

ADA [Haberman and Dewayne (1983)] is one of the languages for which such an environment (ADAPSE) has been designed and which promises to become an attractive host language for simulation software. A number of ADA-based modelling systems have been suggested [Adelsberger (1982, 1984); Bruno (1982, 1984); Lomov *et al.* (1982, 1984); Sheppard *et al.* (1984)].

Program 3.17 shows an ADA model of Manuel's shop, patterned after Bryant's (1981b) proposal. *Queue*, *Simulation*, and *Random* are separately compiled packages for queueing network simulations. Our model is called *ManuelsShop* and contains definitions for customer and queue objects engaged in arrival and service tasks. Variables and procedures for statistics gathering are not shown here. To simplify the program further we also assume that all parameters are compile-time constants.

Both arrival and service tasks have an entry point called *start*, which initializes a simulation. Entry *wakeUp* restarts the service process at the end of an idle period. All customers are dynamic, transient entities. Their *ShopEntryTime* is recorded when they enter the model.

Do you wish to have instructional comments ? *NO*
Do you wish to start a new problem ? *YES*
Problem Name: *MANUELSSHOP*

Type name of entity: *ARRIVALPROCESS*
How many ? *1*
Type a list of states through which these entities pass
(precede queues by "Q" and activities by "A"):

QWAITLAMBDA
AARRIVAL
QWAITLAMBDA

Type name of entity: *CUSTOMER*
How many ? *100*
Type a list of states as above:

QOUTSIDE
AARRIVAL
QLOVESICKCUSTOMERS
ASERVICE
QOUTSIDE

Type name of entity: *MANUEL*
How many ? *1*
Type a list of states as above:

QIDLE
ASERVICE
QIDLE

Type name of entity: * ; *no further entities*

Are there any activities which use more than one entity of a
particular type ? *NO*

Do you wish to see a summary of life cycles ? *YES*

ARRIVALPROCESS	*1*	,*QWAITLAMBDA, AARRIVAL, QWAITLAMBDA.*
CUSTOMER	*100*	,*QOUTSIDE, AARRIVAL, QLOVESICKCUSTOMERS,*
		ASERVICE, QOUTSIDE.
MANUEL	*1*	,*QIDLE, ASERVICE, QIDLE.*

Do you want to make any changes to the logic section ? *NO*
Are there any queues whose discipline is not FIFO ? *NO*

I propose putting the activities in the following order:
ARRIVAL, SERVICE

Do you wish to raise the priority of any activity ? *NO*

ARITHMETIC

After each activity name type the formula for its duration:
ARRIVAL = *NEGEXP (1.25, 1)* ; *rn−generator #1*
SERVICE = *NEGEXP (1.0 , 2)*

Program 3.16 *Using CAPS to model Manuel's lovepotion shop*

Do you wish to define any other attribute ? *NO*
Do you wish to make any changes to the arithmetic section ? *NO*

RECORDING

WAITLAMBDA = *0* ; *no recording*
OUTSIDE = *0*
LOVESICKCUSTOMERS = *3* ; *record length & waiting time*
IDLE = *0*

Delay histogram ranges = ; *no delay statistics*

Do you wish to record any (more) attributes ? *NO*
What length of run–in period is required ? *0* ; *no presimulation phase* !
Do you wish to make changes to the recording section ? *NO*

INITIAL CONDITIONS

Are there any activities in progress ? *YES*

activity – *ARRIVAL*
Termination time: *0*

activity – ; *no further active activities*

How many entries should there be in each queue (listed after name) ?

ARRIVALPROCESS – 1 entities
1 used by activity in progress
WAITLAMBDA – *0*

CUSTOMER – 100 entities
1 used by activity in progress
OUTSIDE – *99*
LOVESICKCUSTOMERS – *0*

MANUEL – 1 entities
IDLE – *1*

Please give the simulation duration – *200*

Do you wish to make any changes to the initial condition section ? *NO*
Do you wish to have the program typed ? *NO*

Program stored in file "MANUELSSHOP".
Do you wish the details of the problem to be kept ? *NO*

COMPILE MANUELSSHOP ; *file, compile, execute the model.*
*
EXECUTE MANUELSSHOP
* ; *system will now print reports on length and*
; *waitingtime for lovesick customers on the screen*
STOP
*

```
with Queue, Simulation, Random; -- bring in the required packages.

procedure ManuelsShop is

use Simulation, Queue;
SimLength : constant := 200.0;-- run simulation for 200 time units
Lambda    : constant :=   0.8;-- arrivals per second
Mu        : constant :=   1.0;-- services per second

task Arrival is
  entry start;
end Arrival;

task Service is
  entry start;
  entry wakeUp;
end Service;

type Customer;
type CustomerPtr is access Customer;
type Customer is
  record
    anEntryTime : float;
  end record;

CustomerQueue is new Queue (Customer, CustomerPtr);
LovesickCustomers : CustomerQueue.queue;

-- Variables used for statistics collection would be declared here
    ...

task body Arrival is
  romeo   : CustomerPtr;
  samples : integer := 0;
  use Random;
begin
accept start;                     -- Wait for a simulation to start.
loop                              -- Forever
                                  -- Wait for next arrival.
  hold (Exponential (lambda));    -- Exponential is defined in package "Random".
  romeo             := new Customer; -- Create a new customer.
  romeo.anEntryTime := SimTime;
  Samples           := Samples + 1;
                                  -- Place romeo in waiting line.
                                  -- If Manuel is idle, wake up the service task.
  if LovesickCustomers.empty then
    CustomerQueue.put (romeo, LovesickCustomers);
    Service.wakeUp;
  else
    CustomerQueue.put (romeo, LovesickCustomers);
  end if;
```

Program 3.17 *Manuel's shop in ADA*

```
end loop;
end Arrival;


task body Service is
  romeo : CustomerPtr;
  flowTime : float;
  use Random;
begin
accept Start;       -- Wait for a simulation to start.
loop                -- Forever
  if LovesickCustomers.empty then
                    -- No customers to service. Sleep until customer arrives.
    accept wakeUp;
  else
                    -- Take the first Customer out of the Waitingline.
    romeo := CustomerQueue.take (LovesickCustomers);
                    -- Delay for duration of service activity.
    hold (Exponential (Mu));
                    -- Observe the time a customer spent in the shop and
                    -- record this statistic.
    flowTime := SimTime - romeo.anEntryTime;
    ...
                    -- Let storage allocator reclaim customer.
    romeo := null;
  end if;
end loop;
end Service;


begin               -- main procedure
                    -- Initialize the Waitingline.
CustomerQueue.Init (LovesickCustomers);
                    -- Scheduler, arrival, service are all blocked now.
                    -- Start them and let the simulation begin.
Scheduler.start;
Arrival.start;
Service.start;
                    -- Wait until time for simulation run has expired.
Hold (SimLength);
                    -- Terminate the simulation.
abort Arrival;
abort Service;
abort Scheduler;


                    -- Print statistics.
                       ...


end of program;
```

A generic instantiation of package *Queue* is used to represent queues of customers. Note that the data structure actually linking queueing customers together is just one of each queue's attributes (for example, *CustomerQueue. queue*).

ADA models must be described at separate specification (**task is**) and implementation (**task body is**) levels. The *Arrival* task acts as a source, creating customer records, time stamping them, counting their number and filing them into the queue of lovesick customers. If he is idle, Manuel must be sent a *wakeUp* message. Customer creations must occur in regular intervals, sampling the arrival process.

Random provides an exponential generator and package *Simulation* furnishes a clock (*SimTime*), an agenda and sequencing procedures (*hold*, etc.).

The generic instantiation of package *Queue* (called *CustomerQueue*) accepts messages *empty*, *put* and *take*.

A service task is triggered by *start* or *wakeUp* signals. The task is passivated if there are no waiting customers. A service activity removes the first waiting customer from the queue, delays both server (implied by the service task) and customer for however long the service is expected to take, and measures the time the customer spent in the shop. Finally, the customer leaves and his frame is reclaimed by the storage allocator.

The main procedure instantiates a queue of lovesick customers, starts the tasks, and then blocks itself until the simulation time has expired.

Scheduler is the name of the predefined event monitor, which is part of the *Simulation* package.

At the end of a run all tasks are terminated and statistics printouts are produced before the simulation shuts down.

Note that this particular model uses a machine-oriented viewpoint; material-oriented models can, of course, be written just as easily.

While both languages can support process-oriented simulation styles, ADA and SIMULA view models from quite different angles. Apart from some minor syntactic differences there are a number of more significant ones. The following discussion draws on arguments first put forward by Davis (1980) and Loyer (1983).

The major contributor to **data abstraction** in ADA is the package. A package provides the means of grouping components of objects together. It hides all details of the abstraction's implementation, which can also be compiled separately.

ADA does not support subclassing: its discriminant records and derived type concepts come closest to this notion, without offering its full power. For instance, SIMULA encourages top-down design through **virtual** procedures. In ADA all variants of records must be known when a record is declared. Packages can therefore not be used to build property inheritance hierarchies, although they can of course be nested within one another.

SIMULA's class concept is more powerful than ADA's package. It groups all relevant aspects of some world model into trees of class definitions.

Since it focuses attention on commonalities and differences between real-world entities, its style of describing new objects as subclasses of already known ones has recently been referred to as **differential programming**. The basic limitation of SIMULA's class–subclass concept stems from its strictly hierarchical nature: a server cannot be a subclass of two different superclasses without also making one a subclass of the other. This was the reason for introducing separate process (prefixed by **process**) and structure (prefixed by **link**) objects in class *MiniQNcontext*.

The motivation behind these differences in approach to data abstraction can be found in their different histories and objectives. ADA is very much committed to software reliability and efficient execution, calling for stack-based memory allocation. SIMULA's philosophy asks for a close correspondence between world models and their programmed representations. It requires more expensive heap allocation and considerable run-time support. ADA's packages are therefore expanded at compile time (**new** is a compile-time command, similar to the one in MODULA!), while SIMULA's classes are instantiated at program execution time (**new** is a run-time command!).

Since the multi-tasking primitives were developed to suit each language's main purpose, ADA's and SIMULA's models of parallelism also differ widely. ADA was designed for applications with real-time parallelism (multiple processors), while SIMULA caters only for interleaved execution of pseudo-concurrent processes (coroutines) on a uniprocessor system. ADA lacks any mechanism to enable tasks to explicitly schedule other tasks, whereas SIMULA's model of parallelism is specifically geared to the implementation of process-oriented simulation models. There is also no common ancestor of task types (such as class **process**) in ADA, so that it is impossible to define a set of scheduling entries accessible to all of them. Scheduling decisions must therefore be distributed throughout all the tasks comprising a model.

For purposes of simulation, SIMULA's class–subclass idea is definitely superior to ADA's package. It combines several ideas (such as the template for objects' properties and behaviour, reference variables for accessing instances, the distinction between visibility and existence of objects, prefix hierarchies, quasi-concurrent processes) into one uniform concept. This has inspired many mechanisms found in newer languages like CLU, ALPHARD, EUCLID, etc., and has led to a further generalization in the SMALLTALK context of viewing computations as messages passing between coexisting objects.

An ADA package is not an object template. There is no special allocation machinery. Objects inside a package are allocated memory as if they were not embedded. The package just introduces a syntactic barrier for filtering visibility. This idea is related to similar mechanisms found in other modern systems-programming languages like MESA and MODULA-2.

Packages and tasks can, however, still be used to write process-oriented simulations. These features, its programming support environment and the

```
class name                  LovepotionShop
superclass                  Simulation
instance variable names     flowTimes  Samples
instance methods

initialization

initialize ||
  super initialize
  samples ← 0
  flowTimes ← Histogram from: 0 to: 20 by: 2

defineArrivalSchedule ||
  self scheduleArrivalOf (LoveSickCustomer new)
       accordingTo: (Exponential parameter: 0.8)

defineResources ||
  self produce: 1 of: 'wiżard'

accessing

exit: aSimulationObject ||
  super exit: aSimulationObject
  Samples ← samples + 1
  flowTimes store: currentTime – aSimulationObject shopEntryTime

printFlowStatistics: aStream ||
  flowTimes printStatisticsOn: aStream
```

Program 3.18 *Manuel's shop in SMALLTALK*

anticipated widespread availability of ADA compilers will make it an important base for simulation tools.

SMALLTALK-80 [Goldberg and Robson (1983)] already offers a highly sophisticated programming environment. Its object-oriented flavour makes it particularly attractive for simulations in the style of the SIMULA systems illustrated above.

SMALLTALK was influenced by SIMULA's model of computation. Alan Kay, its designer, added the message-passing paradigm, creating a programming style, which has since become known as **object-oriented programming**.

SMALLTALK offers a powerful, mouse-driven and window-based programming environment for modeless, 'exploratory' programming. In this sense it has similarities with INTERLISP [Sandewall (1978)]. Its excellent tools for staging graphical animations can be very profitably employed by simulation models [see, for example, Goldberg and Robson (1979)].

Program 3.18 shows how Manuel's shop could be modelled in SMALL-

class name	**LovesickCustomer**
superclass	SimulationObject
instance variable names	*shopEntryTime*
instance methods	

initialization

```
initialize ||
   super initialize
   shopEntryTime ← ActiveSimulation currentTime
```

accessing

```
shopEntryTime ||
   ↑ shopEntryTime
```

```
tasks
   | manuel |
   manuel ← self acquire: 1 ofResource: 'wizard'
   self holdFor: ServiceTimeSample
   self release: manuel
```

private

```
serviceTimeSample ||
   ↑ (Exponential parameter: 1.0) next
```

TALK. We assume that a simulation driver, such as the one described in Goldberg and Robson (1983; pp. 417–537) is available.

SMALLTALK needs single-user work stations with sufficient processing power and memory (at least a megabyte) to support its model of computation at a satisfactory level of performance. In addition, a memory-mapped graphics screen and a pointing device is required for its user interface. SMALLTALK's user interface is probably the most advanced integrated computing environment that has been successfully implemented [see Guedji *et al.* (1980)]. A program is defined through a collection of interacting objects by a combination of mouse-controlled selection commands and keyboard activity, using a syntax-driven editor.

Each SMALLTALK program consists of a number of class definitions. A class is named, bound to a superclass, and may have a number of class and instance variables and methods. Class variables and methods are shared among all objects instantiated to be of that class, while instance variables and methods are private to individual objects. All messages an object can respond to form its **protocol**. Class definitions associate methods (implementations) with these messages. Method descriptions can be structured into chapters (such as initialization methods, accessing methods, private methods), but this is transparent to the interpreter.

Note that SMALLTALK, like LISP, is a typeless language. The types of objects should only be deducible from their names. This is not really a shortcoming, since in interpreted languages like LISP, PROLOG or SMALLTALK the notion of 'strong typing' does not buy any advantages for the user. Since programs in these languages are developed in a highly interactive fashion, type (and other) errors can easily be detected and corrected at the keyboard; powerful tools for doing this are part of the programming environment. This is different from compiled languages, where the increases in program reliability and efficiency achievable through strong typing are invaluable.

Note that for the purpose of the following discussion many syntactic considerations will be ignored.

SMALLTALK assumes that all global variables start with an upper-case letter. Arguments to methods are enclosed by vertical bars. Square brackets delimit blocks, whose evaluation is deferred. As in LISP, all SMALLTALK messages return some value, possibly themselves. An up-arrow can be used to explicitly denote what should be returned.

Our model consists of two classes: *LovepotionShop*, a subclass of class *Simulation* [see Goldberg and Robson (1983; pp. 417–537)] and *LovesickCustomer*, a *SimulationObject*. We are using a material-oriented modelling style, describing events in Manuel's shop from a customer's viewpoint. Class *LovepotionShop* can respond to messages requesting a shop simulation to be initialized, asking to define an arrival process and resources, triggering the observation of customer departures and the reporting of statistics. The corresponding methods are shown.

self and *super* are **pseudo-variables**. They allow references to the current object and its superclass. In this case we assume that all the necessary actions for initializing a simulation are provided by class *Simulation*, the superclass of our *LovepotionShop*. Sending the *initialize* message to *super* should therefore trigger these actions.

flowTimes is instantiated to a histogram and used for measuring customer delays. As in our previous models, we want to create lovesick customers according to an exponential distribution with a mean of 0.8 customers per minute. Again we assume that *Exponential* and the *scheduleArrivalOf* method are inherited from class *Simulation*.

new is one of a number of class messages understood by all SMALLTALK objects. Class *LovesickCustomer* will create a new lovesick customer in response.

Our simulation assumes one permanent resource, called *wizard*. *exit* is a message known to *Simulation* objects, *store* is known to histograms, *currentTime* returns the current clock time and *shopEntryTime* the time a customer entered the shop. We can treat *Streams* as file objects into which statistics will be written.

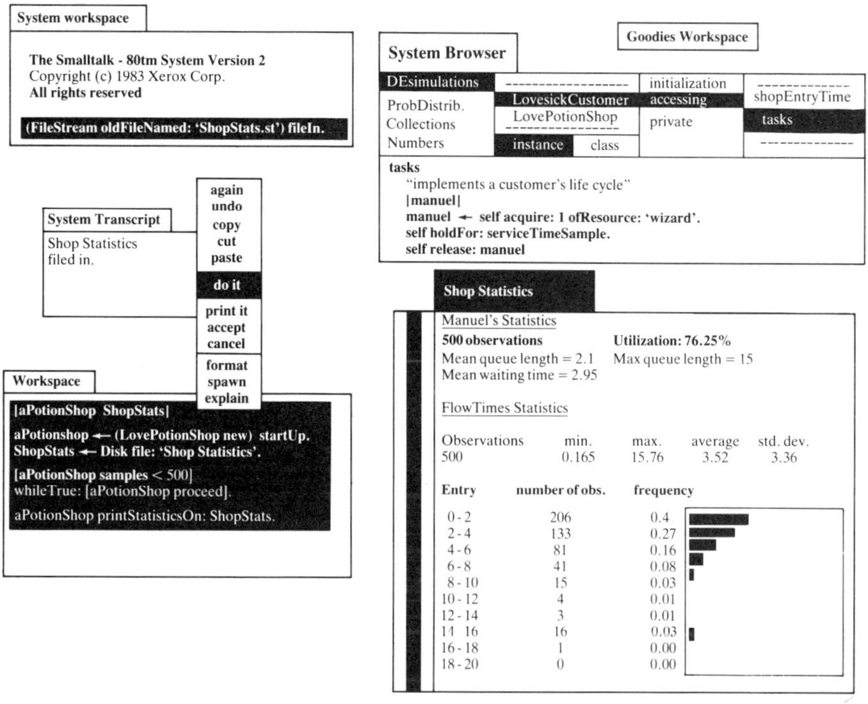

Figure 3.29 Simulating Manuel's shop: A SMALLTALK screen

The definition of *LovesickCustomers* follows the same pattern. We assume that some instance of *LovepotionShop* is the currently active *simulation*. Manuel is a wizard who can be acquired, held for a specified time period and released again. Messages *acquire*, *holdFor*, and *release* are part of the predefined protocol for resource objects.

Figure 3.29 demonstrates how this model can be executed and shows a snapshot of a typical SMALLTALK screen just after such a simulation (called *manuel*) has terminated, serving 500 customers. Notice that a number of overlapping windows are visible on the screen. One shows a **browser** on the class *LovesickCustomer*. Browsers can be used to view methods. Here we selected the *tasks* method within the *accessing* category of the instance methods for class *LovesickCustomer*. This class was selected from a number of alternatives within class category *DEsimulations*. One window, called *ShopStatistics*, shows the results of our simulation.

The expression that was last evaluated is visible in the **Workspace** window in the lower left-hand corner of the screen. It created a new *LovepotionShop* and sent it a *startUp* message, thereby spawning a *LovepotionShop* simulation. It then ran the simulation until 500 samples were taken, whereupon statistics were written into the *ShopStats* file.

SMALLTALK allows selection of objects and command execution

through mouse action. The mouse is moved into a menu field after one of its buttons is pushed. Selected commands and windows are shown in reverse video (white on black). The thick black lines at the left rim of active windows are **scroll bars**. They can be used to move a view up or down. At the time this snapshot was taken the *ShopStats* file had just been selected. The **System Workspace** window shows the expression responsible for filing it in. This was the last action executed (*doIt* menu item) before the snapshot was taken. The contents of that file is visible in the **Shop Statistics** window. It shows the *FlowTimes* histogram and some other statistical measures for the simulation. The **Goodies** window has been collapsed, showing only its title. It may at any time be opened again. **System Transcript** records the system's current state.

One way to classify programming tools is by their relative levels. The general idea is that a language is bound at a higher level if we can express the same problem with less detail, which means that the language processor performs more tasks automatically. An alternative way of stating this is that a higher-level language is less 'procedural' than a lower-level language.

PROLOG [Clocksin and Mellish (1981), MacLennan (1983)] can be called a very high-level language in this sense. It uses the Horn clause form of symbolic logic to describe non-numerical problems and their solutions. Programs are expressed as rules and assertions. The deductive machinery of PROLOG implements a theorem prover which then constructs the desired solution as a side effect of proving its existence.

Since the mid-1970s PROLOG has been rising in popularity as a language for artificial intelligence applications, particularly in Europe. Its proponents have suggested it as a successor to LISP for symbol processing, compiler writing, deductive information retrieval, symbolic algebra, computer-aided design, expert system development and complex problem solving in general.

PROLOG's deductive machinery may also be used to perform sequences of simulation experiments automatically, searching for some predefined goal. T-Prolog [Futo and Szeredi (1982)] adds a temporal frame to PROLOG's world view. It extends standard PROLOG by some built-in primitives to handle a time dimension, processes, resources, messages between processes and conditions processes may be waiting on.

Program 3.19 presents a particularly simple-minded kind of deterministic scheduling problem. This time it is assumed that there are two serial servers, Manuel and a pixie. Manuel now specializes in diagnosing a customer, who then has to wait for the pixie to hand out an appropriate amount of potion and collect his donation.

It is necessary to look at three customers arriving on the same broomstick (simultaneously!) and find a sequence that allows all of them to be served in 15 minutes or less. The three customers (Romeo, Kermit and Skooter) all place different time demands on the two servers; it is assumed that these are deterministic and known to Manuel.

The program therefore sets up two resources, each with a capacity of one customer. A customer's life cycle is defined as a goal, expressed in

```
resource(manuel, 1).
resource(pixie   , 1).

customerLifeCycle(DiagnosisTime, DrinkAndPayTime) :-
      seize(manuel),
      hold (DiagnosisTime),
      release(manuel),
      seize(pixie),
      hold (DrinkAndPayTime),
      release(pixie).

   :- new( customerLifeCycle(2,4),  romeo, 0, 15),
      new( customerLifeCycle(1,7),  kermit, 0, 15),
      new( customerLifeCycle(6,3), skooter, 0, 15).
```

Program 3.19 *Manuel's shop in PROLOG*

terms of a sequence of subgoals. All subgoals have to be fulfilled to realize the main goal. During a deduction process, variables can be bound to different values. Note that PROLOG expects all variables to start with an upper-case letter, whereas all facts and rules have to start with a lower-case symbol. Boldface keywords (**resource**, **seize**, **hold**, **release**, **new**) are primitives of T-PROLOG. The symbol **:–** is used to indicate that a goal on the left can be realized by fulfilling the subgoals listed on its right-hand side. Commas separate conjunctions of subgoals. The left-hand side of a goal sequence may be empty.

A material-oriented viewpoint has been chosen. To fulfill his lifetime's aspirations each customer must therefore successfully seize Manuel, hold him for diagnosis, release him, seize the pixie, hold her for drinking and paying, and release her. A simulation of this model is conducted by asking PROLOG to find a solution (sequence of state transitions). The interpreter then tries to realize the goals of three newly created customer processes (*romeo*, *kermit*, *skooter*). Each of these processes is patterned after our declaration of a *customerLifeCycle*. It has an identifying name, starts (*arrives*) at time 0 and must be finished (*depart*) at time 15.

Executing this model causes T-PROLOG to explore possible sequences of model states, back-tracking automatically if a particular subgoal cannot be met (depth-first search). In our example it would reject schedules (*romeo*, *kermit*, *skooter*) and (*romeo*, *skooter*, *kermit*) because they both finish only at time 16. Eventually the schedule (*kermit*, *romeo*, *skooter*) would be found.

PROLOG has been called a declarative language because one can specify how a solution should look without actually describing the process by which it is reached. Getting PROLOG to satisfy goals (prove theorems) reasonably efficiently currently requires some compromise on this idea and is a challenging research problem.

T-PROLOG needs a significant amount of procedural information (see goal *customerLifeCycle*) and is far too unsophisticated for complex models. It can only deal with deterministic simulations of simple resource-constrained sequencing strategies. The problem of generalizing it to incorporate stochastic processes is a difficult one and certainly warrants further investigation.

Chapter 4

The Simulation Programming Landscape: Analysis and Trends

System simulation is always a spiral process, as shown in Figure 4.1. Its methodology provides tools for all phases of a simulation project. This book has concentrated on model design and program implementation, but this should not be taken as an indication that other activities, such as system identification, model validation and experimentation, are less important. It rather reflects the fact that modelling and simulation is a multi-faceted enterprise [see Zeigler (1984)] with many important viewpoints, which cannot all be covered completely in a single book.

There are various good textbooks on statistical and experimental aspects of simulation methodology. Most of these are listed in Appendix 3. Different styles of model structuring and simulation programming have so far received much less satisfactory coverage.

In Chapter 1 two different objectives of simulation studies were identified. One view is reflected in the use of models predominantly for decision making. This is the classical style of simulation in operations research, where models are employed to numerically predict a system's likely responses. Its successful application is critically dependent on acceptable model validation and skilful variations of policy parameters. This modelling style therefore suffers from all the difficulties associated with model calibration and validation. It is also very expensive to experiment with. A simulation run generates only a single point on a model's response surface and, if it contains stochastic components, has to be replicated a considerable number of times to establish statistical credibility. While this kind of simulation remains an important method of last resort for a wide variety of problems that cannot easily be solved by analytical means, it is the second view of system simulation that will reap the full benefits of new technological developments.

Modelling for insight is greatly simplified by object-oriented structuring techniques and interactive programming environments with integrated

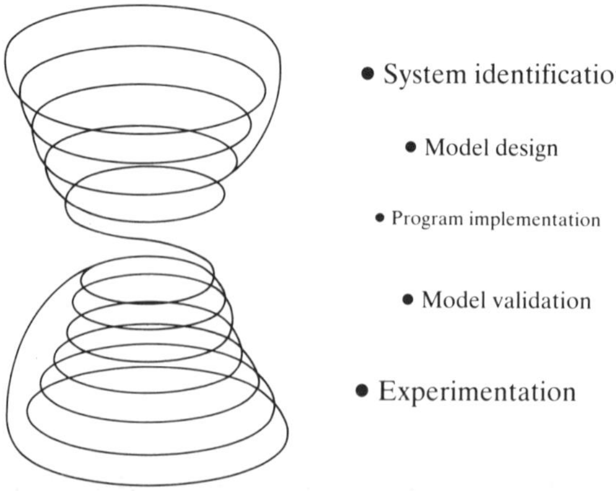

Figure 4.1 A spiral view of simulation

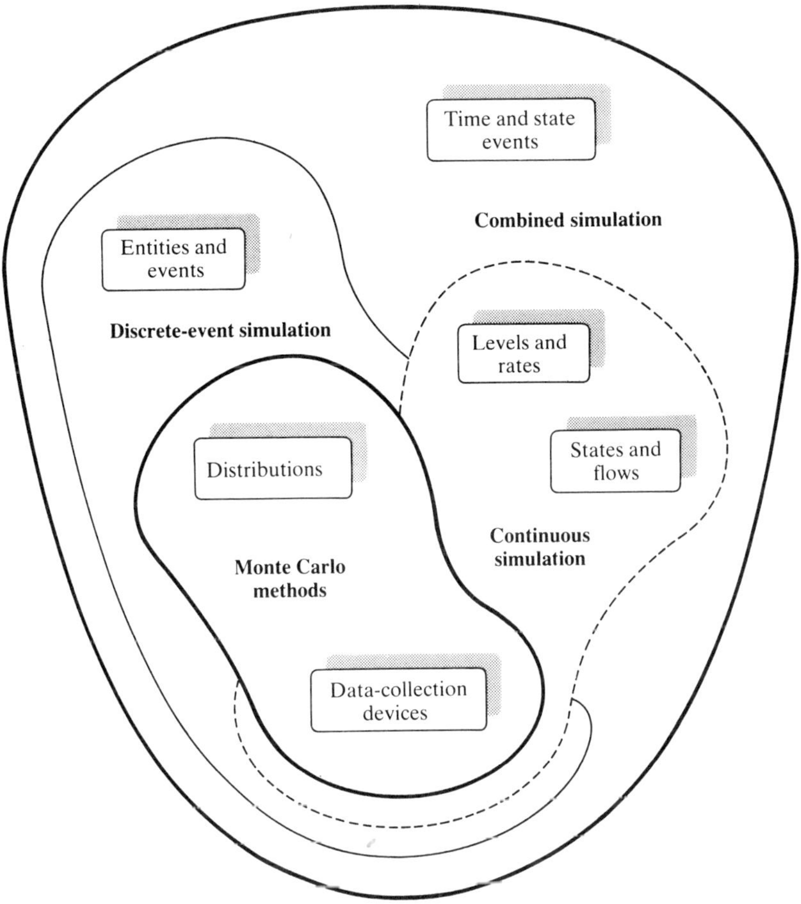

Figure 4.2 Some paradigms of system simulation

graphics. Sloman (1978) refers to this modelling style as **experimental epistemology**, a name reflecting its suitability for explorations of complex theories across a wide range of sciences. Such simulations have traditionally been employed to animate artificial intelligence miniworlds, a domain where tools as well as flexible and habitable modelling environments for using personal work stations were originally developed.

Figure 4.2 characterizes the four main modelling paradigms discussed in this book.

Data-collection devices will always be needed for **simulation**, since it is by definition an experimental technique where data generated by each experiment has to be collected, summarized and represented in a meaningful way. We have demonstrated the use of objects for time-independent (count, tally, histogram) and time-weighted (accumulate, chronograph) statistics through messages to initialize, update, reset and report their values.

In addition, a number of distribution objects were defined to represent stochastic model components, as were methods for their initialization and sampling. These utilities were first required in the context of **Monte Carlo simulation**, a method that is often applied in quality control and project scheduling. The motivation behind this technique is to approximate an aggregated, long-term distribution of a system's likely states by sampling from sets of stochastically specified influences. This can be done under control of a simple monitor routine guarded by a specified sample size.

Although questions of causality are largely irrelevant to the analysis of Monte Carlo models, this notion is at the heart of all dynamic-modelling techniques.

In the case of **continuous-system simulation** we are interested in summaries of some system's time-dependent response. This may be modelled by sets of interacting processes describing state fluctuations. A number of subparadigms were identified, with system dynamics and differential-equation scenarios among them. Motivations for simulations of this type are often the study of cyclic processes in economics (trade cycles), business (spreadsheets for financial analysis), biology (population dynamics), engineering (control systems) and other sciences. These models are typically composed of some levels or state variables bound into sets of difference or differential equations reflecting their rates of change. Objects respond to requests to initialize, step or integrate their values. A simulation analysis is performed through simple time-slicing techniques, controlled by some clock-driven monitor procedure. All of a model's equations are repeatedly evaluated at a fixed level of resolution over some specified time interval.

Discrete-event models can be employed in more detailed studies of a system's dynamic behaviour. Sequences of activities are directly represented through event procedures and scheduling commands. This paradigm permits representation of systems at an essentially unlimited level of detail. A number of subparadigms (inventory models, queueing networks) can again be identified.

The queueing-network approach is the most popular one, particularly for computer scientists. This framework uses a world view where different categories of entities (sources, sinks, transactions, servers, queues) interact through messages, triggering initializations and sequences of state transformations (*generate*; *terminate*; *arrive*, *route*, *depart*; *enter*, *leave*; *seize*, *release*; etc.).

Simulation experiments are performed under control of a monitor entity with either an event, activity or process orientation. All monitors must keep a clock (advanced in a next event fashion) and an agenda of pending events. Model executions are again guarded by some termination criterion. Discrete-event methods form a popular technique for studying a wide range of problems involving assignment (such as stable marriages) and resource allocation (such as Manuel's shop). Modelling service contentions in shops or other institutions is one of the most important of these.

Finally it was demonstrated how all of these paradigms can be integrated into a uniform framework for **combined simulation**, modelling systems in all of their stochastic, continuous and discrete aspects. Object-oriented modelling seems to offer natural representations for systems of this kind. Here entity descriptions encapsulate a name, structure and behaviour into one organizational unit – an object class. Objects' behaviour can then equally well be implemented by sequences of discrete states or continuous processes. Either the discrete or the continuous part of a model can be the dominant component. In order to integrate these parts, the monitor routine must continually check for both state and time events.

Almost any programming language can and has been used for simulation model building, although prepackaged objects and procedures can ease the modelling effort considerably. It has been shown how basic utilities (random-number and random-variate generators), model instrumentation and reporting facilities are relatively easy to implement in general-purpose languages such as PASCAL. The primitives necessary for representing a time dimension, concurrency, the creation of transient objects and the handling of dynamic relationships are not adequately supported by any of the more widely used programming systems.

The system dynamics approach, differential-equation scenarios and the event-oriented framework are the easiest world views to implement. Event orientation unfortunately does not offer very user-friendly modelling environments. The flexibility and convenience of process- and object-oriented frameworks is bought through more sophisticated schemes for storage allocation and control, which can no longer be so easily implemented. Use of a general-purpose language is therefore not an option to be recommended for complex models. It is very much a case of reinventing the wheel, and the necessary redesigning, coding and debugging is always time consuming, tedious and, above all, very error prone.

Whereas preprocessors can often provide simulation-specific programming contexts through user-friendly environments without extraordinary amounts of implementation effort, languages implemented by self-contained translators are to be preferred for reasons of execution efficiency. Interpretative systems are typically expensive to use, a disadvantage that can be offset by a high degree of user convenience, good diagnostics and symbolic tracing facilities. One must, however, be alert to grossly inefficient systems, 'good' examples of which are some GPSS implementations.

Well-designed scenario languages can directly carry concepts and terminology of specific application areas, providing very effective and user-friendly modelling frameworks. If based on an extensible general-purpose language they can be easy to build; we have used the SIMULA classes *MCcontext*, *CScontext*, *MiniQNcontext*, DEMOS and DISCO to demonstrate this. A reasonable amount of flexibility is important, although any specialized instrument suffers a loss of effectiveness if it is misapplied.

GPSS is again a good example. It is a pleasant tool for building small models that can naturally be defined as material-oriented queueing networks. Unfortunately, it has been extensively misused for tasks it was not designed to cope with.

Scenario languages lose much of their appeal if the user cannot be shielded from knowing about idiosyncrasies of implementation structures, particularly if one then sometimes has to program 'around' rather than 'in' a language. This is not to say that this cannot be done (it certainly has been!), but rather that it is ineffective, clumsy and error prone to use any programming tool this way.

Model and program generators can be useful for rapid prototyping of stereotyped applications. Unfortunately they lack the flexibility to cope with more complicated problems.

The choice of world view, programming style and language to model a particular project is never an easy one. The preceding chapters give some guidelines by demonstrating a few widely used tools and techniques. Other considerations like availability, personal fluency, costs and time constraints of course also influence such decisions – apart from less tangible personal preferences and dislikes. This is as it should be, as long as an awareness of the full spectrum of tools and techniques, their characteristics, strengths and weaknesses exists. This book has tried to lay a foundation for this. Tables 4.1, 4.2 and 4.3 summarize the characteristics of some of the most important simulation languages. Of these, ADA will surely become a popular host to simulation tools, but a SMALLTALK toolkit catering for the variety of modelling styles and scenarios discussed in our survey seems a much more exciting prospect. Very high-level languages like PROLOG have yet to show whether they can make any significant impact. To the author, object-oriented programming styles seem to offer the best, most powerful and flexible modelling frameworks.

Model implementation is just one activity amongst others to be performed during a simulation project. The others include system identification, model specification and documentation, data analysis, model validation and experimentation. Currently different notations are used for all of these. However, it would be much more convenient to have a single unified framework [see Oren and Zeigler (1979)] to work in.

SMALLTALK's programming environment can provide an excellent base for such a system, since it allows us to interact in rich and flexible ways with several tasks at once. This language and the programming paradigm it carries ('any computation is ultimately a simulation') are based on the fundamental principle that it must suit humans rather than computers. It therefore tries to reflect the way we think and supports the categories we use to perceive and organize reality through the concepts of objects, classes of objects, reference, inheritance and message passing.

An object-oriented program is just like the representation of some part of the world in a human's mind, with all its facts, associations and beliefs closely corresponding to our intuitions. Simulation is then akin to thought experiments: an animation of this symbolic structure according to some rules of interaction.

Object orientation has additional advantages for managing large and complicated software systems. Experiences with design and implementation of complex programs, such as SMALLTALK's user interface, animation laboratories [THINGLAB: Borning (1979)] and a Norwegian project to provide a flexible design context for engineering work stations, have shown that its strong localization of logically related information, property inheritance and the capability of objects to defend themselves against invalid requests make it a very safe programming tool. The technique is certainly not restricted to simulation tasks.

For these reasons a number of object-oriented languages (such as OBJECTIVE C, CLASCAL, OBJECT PASCAL) have appeared. This is particularly obvious in the field of artificial intelligence programming. There they are called **actor languages** and are usually based on LISP (ZetaLisp's Flavours, GLISP, ObjTalk, ACT). It is also interesting to note that almost all programming tools for knowledge representation in expert systems (KEE, LOOPS, BABYLON) have chosen an object-oriented viewpoint as their main structuring paradigm.

The **class** concept of SIMULA and SMALLTALK also offers an ideal base for the definition of picture primitives and graphic data structures patterned after the GKS standard [Enderle *et al.* (1984)]. Because of poor hardware efficiencies, implementations of such systems used to be prohibitively expensive. The last few years have brought a rapid and unprecedented development of cheap and powerful hardware components, built as integrated one-chip units. Price–performance ratios of processor and memory resources have changed tremendously, thereby justifying much more emphasis on effective problem-solving environments at the expense of hardware requirements. The Apple MacIntosh and similar systems promise to become excellent hosts for sophisticated modelling toolboxes at very reasonable costs.

As in most other areas of computing, there is as yet no generally agreed standardization of symbols, tools and techniques. Many people believe that there should be a standard for simulation: currently two committees are working on standards for continuous simulation languages (CSSL committee of Simulations Council (SCS); TC3 of IMACS) and one SCS committee is working on discrete-event simulation language standardization.

The author believes that system simulation and simulation programming are not quite settled fields yet. Many fundamental issues of data and process encapsulation and important aspects of control-flow structuring are still not completely understood. Research is taking place and more is required. This holds particularly for discrete-event simulation, whereas there is a much better agreement on basic concepts and techniques for simulating continuous systems.

It is, however, true that very few significantly new concepts have been proposed since the 1968 IFIP conference on simulation programming languages [Buxton (1968)]. This observation holds in spite of the fact that

Table 4.1 General information

	Language characteristics	*Textbooks*	*Availability*	*Genealogy*
ACSL	Continuous scenario language for differential-equation models	Mitchel & Gauthier (1981)	FORTRAN-based; portable and widely used	Implementation of the SCSC CSSL standard; derived from CSSL III
CSMP	Continuous scenario language for differential-equation models	Speckhart & Green (1976)	FORTRAN-based; widely used on IBM machines	First version CSMP/360; latest version CSMP III; not based on CSSL standard
DYNAMO	Continuous scenario language for system-dynamics models	Pugh (1973)	FORTRAN-based; widely used and implemented (micro-versions)	Developed to support J. W. Forrester's system-dynamics scenarios
GASP	Procedure library for combined simulations	Pritsker (1974, 1979) Pritsker & Young (1975)	FORTRAN-based; portable and widely used	Developed, supported and distributed by A. A. B. Pritsker: GASP, GASP II–IV
SIMPAS	Procedural discrete-event simulation language	Bryant (1981a, 1981b)	PASCAL-based; Portable and easy to install	Developed by Bryant (Univ. of Wisconsin): SIMPAS 2.0, SIMPAS 5.0

Table 4.1 (cont.)

	Language characteristics	*Textbooks*	*Availability*	*Genealogy*
SIMSCRIPT	Procedural programming language with special features for discrete-event simulation	Kiviat *et al.* (1968) Wyman (1970) Russel (1976) Markowitz (1979a, 1979b)	Available for most equipment from major manufacturers; widely used	Designed and developed by RAND Co.; further developed and maintained by CACI Inc.: SIMSCRIPT, SIMSCRIPT I.5, II, II.5
SIMULA	Procedural programming language with special features for discrete-event simulation	Birtwistle *et al.* (1973) Franta (1977) Lamprecht (1982) Nygaard & Dahl (1978) Birtwistle *et al.* (1985) Papazoglou *et al.* (1984b)	Available for a variety of machines; a portable version (S-PORT) is widely implemented	Designed and developed by O. J. Dahl, Nygaard *et al.* at the Norwegian Computing Center; SIMULA I was purely an ALGOL-based simulation language; SIMULA 68 is a completely newly designed system
DEMOS	SIMULA extension for building queueing network scenarios	Birtwistle (1979) Birtwistle *et al.* (1984)	Available as a (*external* or *included*) SIMULA context (class)	Designed and written by G. Birtwistle (Univ. of Bradford, Univ. of Calgary); based on a generalization of GPSS, SIMON 75

Table 4.1 (cont.)

	Language characteristics	*Textbooks*	*Availability*	*Genealogy*
DISCO	SIMULA extension for building combined discrete/continuous scenarios	Helsgaun (1980)	Available as a (*external* or *included*) SIMULA context (class)	Developed by K. Helsgaun at Roskilde Univ. (Denmark); based on CADSIM, COMBINED-SIMULATION
GPSS	Language for building queueing network scenarios	Greenberg (1972) Schriber (1974) Dunning (1981) Gordon (1975, 1978a, 1979) Bobillier *et al*. (1976) O'Donovan (1979) Henriksen (1983) Schmidt (1980) Uyeno (1980)	Available (various dialects) for equipment of practically any major computer manufacturer; very widely used. FORTRAN- and PASCAL-based versions have been implemented	Developed by G. Gordon at IBM for analyzing telecommunication systems: GPSS, GPSS II, GPSS III, GPSS/360, GPSS V; a more efficient, compiled version (GPSS/H) is available from a consultancy
SLAM	Procedure library for building combined discrete/continuous scenarios	Pritsker & Pegden (1979) Pritsker (1983a)	FORTRAN-based; portable and easy to install; widely used	Developed and maintained by A. A. B. Pritsker; based on generalization of GPSS and GERTS

Table 4.1 (cont.)

	Contacts
ACSL	Mitchel & Gauthier Assoc., 1337 Old Marlboro Rd., Concord, MA, USA
CSMP	Nearest IBM rep.
DYNAMO	Sloan Sch. Management, MIT, Cambridge, MA 02139, USA
GASP	Pritsker & Assoc., 1201 Wiley Drive, West Lafayette, IN 47907, USA
SIMPAS	Madison Acad. Computing Center, Univ. Wisconsin, Madison, WI 53706, USA
SIMSCRIPT	CACI Inc., 12011 San Vincente Blvd., Los Angeles, CA 90049, USA
SIMULA	SIMULA Secretary, Norwegian Computing Cent., Forsknigsveien 1B, Blindern, Oslo 3, Norway
DEMOS	Prof. G. Birtwistle, Univ. Calgary, Computer Sci. Dep., Calgary, Canada
DISCO	Dr. K. Helsgaun, Dep. Computer Sci., Roskilde Univ. Cent., 4000 Roskilde, Denmark
GPSS	Manufacturer's nearest rep. For GPSS/H: Wolverine Software Co., 7630 Little River Turnpike, Suite 208, Annandale, VA, USA
SLAM	Pritsker & Assoc., 1201 Wiley Drive, West Lafayette, IN 47907, USA

Table 4.2(a) World models

	World Views	*Model Structuring*	*Random Sampling*
ACSL	Differential equations; lumped-parameter systems; self-driven models	State variables and equations	Some procedures for sampling from theoretical distributions
CSMP	Differential equations; lumped-parameter systems; self-driven models	State variables and equations	Some procedures for sampling from theoretical distributions
DYNAMO	Difference equations; feedback systems; self-driven models	Levels, rates, constants, auxiliaries, . . . , equations	Some procedures for sampling from theoretical distributions
GASP	Open and closed queueing networks; machine- and material-oriented models with event orientation; trace or distribution driven	Entities, event routines, sets; imperative scheduling	Comprehensive library of sampling procedures from theoretical and empirical distributions
SIMPAS	Open and closed queueing networks; machine- and material-oriented models; event orientation; trace or distribution driven	Records, queue members, queues, event procedures; imperative scheduling	Comprehensive library of sampling procedures from theoretical and empirical distributions [adapted from Fishman (1976)]
SIMSCRIPT	Open and closed queueing networks; machine- and material-oriented models; event or process orientation; trace or distribution driven	Entities (permanent and transient), event routines, sets; imperative scheduling	Comprehensive library of sampling procedures from theoretical and empirical distributions

Table 4.2(a) (cont.)

	World Views	*Model Structuring*	*Random Sampling*
SIMULA	Class SIMULATION; open and closed queueing networks; machine- and material-oriented models; activity, event or process orientation; trace or distribution driven	Classes, objects, lists; imperative scheduling	Class SIMULATION provides procedures for sampling from the most common theoretical and empirical distributions
DEMOS	Open or closed queueing networks; machine- or material-oriented scenarios; process or object orientation; trace or distribution driven	Entities; **res**, **bin**, **condq**, **waitq**; imperative (**hold**) and wait/signal based scheduling	Extensive set of real, integer and Boolean objects for sampling
DISCO	Combined difference equations, differential equations, process object scenarios; open or closed queueing networks; machine- or material-oriented models; trace or distribution driven	Entities, continuous classes and variables; imperative (time event), conditional (state event) and 'next dt' based scheduling	Same as for SIMULA class SIMULATION
GPSS	Open or closed queueing networks; material-oriented scenarios; process orientation; distribution driven	Transactions, facilities, storages, blocks, assembly sets, groups, user chains; imperative (**advance**) and conditional (**seize**) scheduling	Procedures (blocks) to sample from (a few) predefined theoretical and empirical distributions; some versions approximate all distributions by x/y co-ordinate values
SLAM	Open or closed queueing networks; machine- or material-oriented scenarios; process orientation; trace or distribution driven; combined discrete/continuous scenarios	Nodes, branches, entities, queues, resources, processes, state variables, equations; event subroutines; time events, state events and quasi-continuous sequencing	A number of predefined procedures for sampling from theoretical and empirical distributions

Table 4.2(b) Utilities

	Statistical Instrumentation	*Utilities for Report Generation*	*Utilities for Experimental Design*	*Debugging Aids*
ACSL	Automatic statistics	Automatic reports and plots	—	—
CSMP	Automatic statistics	Automatic reports and plots	—	—
DYNAMO	Automatic statistics	Automatic reports and plots	—	—
GASP	Some automatic statistics	Automatic and 'manual' reports under user control	'Do it yourself'	Snapshots and event traces
SIMPAS	Some automatic statistics and 'observed variables' (tallied and accumulated)	**display** statement for printing standard statistics and observed variables	Some support for implementing the regenerative method; **reset** statistics	Selective and global snapshots and event traces
SIMSCRIPT	As for SIMPAS (**tally** and **accumulate**)	Flexible report-generation statements	'Do it yourself'	Snapshots and event traces

Table 4.2(b) (cont.)

	Statistical Instrumentation	*Utilities for Report Generation*	*Utilities for Experimental Design*	*Debugging Aids*
SIMULA	Some procedures for histograms	'Do it yourself'	'Do it yourself'	'Do it yourself'
DEMOS	Automatic statistics for some standard QN measures; statistics collection objects (**count**, **tally**, **accumulate**, **histogram**)	Automatic, user modifiable and extensible reports	**reset** statistics	Snapshots and event traces
DISCO	DEMOS's statistics collection devices; automatic instrumentation of continuous processes	Automatic, user modifiable and extensible reports and plots (Class **plotenvironment**)	'Do it yourself'	Snapshots and process traces
GPSS	Automatic statistics and 'Savevalues'	Automatic reports only	**reset** statistics	Snapshots and event traces
SLAM	Some automatic statistics	Some automatic reports	'Do it yourself'	Snapshots and event traces

Table 4.3 Implementation

	Processor Type	*Storage Management*	*Time Management*	*Man–Machine Interface*
ACSL	FORTRAN precompiler into FORTRAN	Static allocation	Next *dt* evaluation loop	Batch and interactive modes; graphics interface
CSMP	FORTRAN precompiler into FORTRAN	Static allocation	Next *dt* evaluation loop	Batch and interactive modes; graphics interface
DYNAMO	FORTRAN precompiler into FORTRAN	Static allocation	Next *dt* evaluation loop	Batch and interactive modes
GASP	FORTRAN- or PL/1-based subroutine package	Static allocation	1 event list (array) and clock	Batch and interactive modes; graphics interface
SIMPAS	PASCAL precompiler into PASCAL	Dynamic, pointer-based allocation	1 event list (linked list) and clock	Batch and interactive modes
SIMSCRIPT	Self-contained compiler	Dynamic allocation with garbage collection	*n* event lists (linked lists); one for each class of events; clock	Batch and interactive modes

Table 4.3 (cont.)

	Processor Type	*Storage Management*	*Time Management*	*Man–Machine Interface*
SIMULA	Self-contained compiler	Dynamic allocation with garbage collection	1 event list (linked list) and clock	Batch and interactive modes
DEMOS	External SIMULA context	Dynamic allocation; as SIMULA	1 event list (binary tree) and clock	Batch and interactive modes
DISCO	External SIMULA context	Dynamic allocation; as SIMULA	Next *dt* evaluation loop and event list; (linked list); clock	Batch and interactive modes
GPSS	Interpreter with precompilation phase; self-contained compiler for GPSS/H	Static allocation; fixed-sized (large) tables	CEC, FEC [linked lists (simulated)]; integer clock	Batch and interactive modes; some versions (GPSS-Norden) with graphics interface
SLAM	FORTRAN precompiler into FORTRAN	Static allocation	Next *dt* evaluation loop and event list (array); clock	Batch and interactive modes; graphics interface

many important ideas about model design, conceptual closure and program reliability have still not gained sufficient foothold within the simulation community. The advantages of process-oriented program structuring and combined models are only now becoming universally accepted.

We may well see a re-emergence of simulation programming among topics at the forefront of computer-science research, spawned by the availability of powerful and inexpensive personal work stations. Research into user-oriented design of programming environments with easy-to-use graphics will provide considerable spin-offs for the construction of flexible and reactive simulation toolboxes.

Simulation analysts are tied into a repetitive cycle encompassing system identification, model conceptualization and design, program construction, program verification, model calibration and validation, simulation, evaluating statistical measures, parameter and model modification.

Methodologies for establishing model credibility, statistical analysis and experimental designs are currently largely unsupported by simulation software. More guidance could be given, particularly in interactive environments. A small body of well-established techniques is available, although mostly scattered through journals and conference proceedings. It is to be hoped that some of the more robust of these will be integrated with new simulation systems, possibly augmented by some 'intelligent' assistance.

The construction of intelligent advisers (expert systems) is a subfield of artificial intelligence and is based on some reasonably well-established techniques for knowledge bases. Special tools, mostly hosted by either LISP or PROLOG [Bundy (1984)] support the application of this programming technology. While it is true that this technology has successfully helped to solve some problems of considerable complexity (in medicine, chemistry, mineral exploration, computer configuration), one must be careful not to overestimate its potential.

There is a growing interest throughout the simulation community in exploring this potential. A conference on simulation and artificial intelligence was hosted by the SCS in early 1985, and some of the above-mentioned expert system tools, such as Knowledge Engineering Environment [Intellicorp (1984)] and LOOPS [Bobrow and Stefik (1983)], offer quite impressive simulation capabilities. Although they were originally developed for rather expensive LISP machines, similar, slightly more modest environments can well be supported on much cheaper hardware (such as Apple's MacIntosh).

Interactive programming environments for discrete-event simulation have received relatively little attention in the past. Nelson (1979) showed how such a capability can be built into an existing language (SIMULA) and greatly enhance its quality. The above-mentioned shift in price–performance ratios for personal work stations makes such facilities very attractive.

The problems of developing simulation software include presentation and debugging of complex, time-dependent behaviour. SMALLTALK-like systems running on personal work stations with window-based graphics may therefore well prove to be a major breakthrough in terms of effectiveness and user acceptance. To further increase the utility of such systems the bandwidth of the user–tool interface should be as wide as possible. A good modelling and programming environment should be capable of high-resolution graphics and provide meaningful and extensible symbol sets. There should also be some easy-to-use pointing device, such as a mouse with buttons for feature selection. It can be most profitably used if processes are menu driven. This is faster and more accurate than any known alternative.

It should also be possible to view different processes in different stages of execution (a graphical model documentation, a view into the source program's execution, various statistical performance measures). Each of these can reside in its own separate window, overlapping other windows or be partly hidden by them. The mouse may again be used to select and activate processes, with control options provided by context-dependent pop-up menus triggered through keys on its back. These may start a simulation, reset it to a previous state,...; edit models and graphic scenarios,...; trace an object's state or a variable's values; obtain a cross-reference listing; perform statistical analyses on an output file,....

We should be free to specify, design, implement and experiment with a model, alternating between different activities at will and, if necessary, leaving processes in some intermediate stage of execution. This is typical for the style most people use when working at their desks. Several tasks may be relevant to a particular project and they need to be interleaved at will; otherwise the creative process may be inhibited. A modelling environment supporting this idea may increase both the productivity and quality of the model-building process by an order of magnitude.

First steps in this direction have already been made. Stella [Richmond (1985)] offers interactive exploration of graphical Systems Dynamics scenarios on the MacIntosh. SimLab is a SIMSCRIPT implementation embedded in a relatively simple interactive environment for the IBM PC. Goldberg and Robson (1979) report on an experiment linking queueing-network simulation and graphical animation, implemented through a SMALLTALK toolkit. This was used successfully by Xerox executives without much prior computer experience. Brown and Sedgewick (1985) show how different graphical symbol structures can be very effectively employed in a programming laboratory to demonstrate the dynamics of algorithms.

Birtwistle *et al.* (1982, 1984), Unger *et al.* (1984), Birtwistle and Luker (1984), Sol (1984) and Stanridge *et al.* (1984) argue convincingly for the merits of computer-based simulation laboratories with graphical animation. These should integrate collections of general-purpose and simulation-specific tools, accessed by a command language presenting a uniform and friendly interface to all types of user [Birtwistle (1984)]. Such systems will eventually also allow end users to directly build and experiment with a model. This is particularly important when using simulations to test implications of theories (see Chapter 1), where 'the purpose of modelling is insight, not numbers' [Hamming (1962)].

Sheppard (1984), Birtwistle (1984) and Vaucher (1984) list some of the likely attributes of simulation work stations of the future. In implementing such systems, methodologies and tools will be needed from the field of artificial intelligence programming. Some of the research in that area promises interesting results [see Klahr and Faught (1980), McArthur and Sowizral (1981), Reddy and Fox (1982), deSwaan (1983), Lemon and Lindstrom (1983), SCS (1985a), Birtwistle (1985), Lavery (1985), O'Keefe (1985, 1986), and Kerckhoffs and Vansteenkiste (1986)].

Chapter 5
Simulation Problems

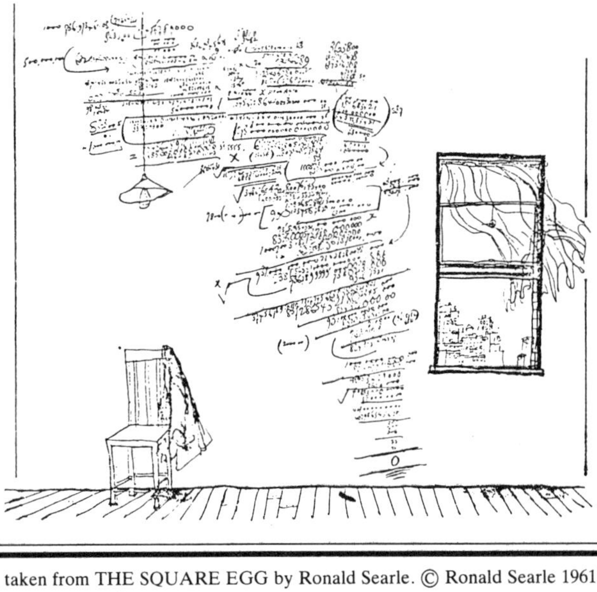

(Solutions to starred exercises are provided in Appendix 4.)

5.1 Search in state spaces

Although the following problems are usually not considered as simulation problems in a strict sense (algorithmic solutions do exist), they make good exercises for a first attempt at writing simulation-style programs. The same basic programming style applies: a symbolic representation of some state space is systematically searched for solutions. State spaces are simpler than in typical simulation problems. There are no time dimensions or stochastic elements. The problems are well defined with clearly described states, actions and goals.

5.1.1 The eight queens problem

The eight queens problem has received much attention by computer scientists, with Dijkstra (1976) and Wirth (1971) as two of the most prominent. Our description follows Kieburtz (1978).

Is it possible without any other pieces on an 8-by-8 chessboard, to position eight queens so that none can reach any other queen in a single move?

Humans solve problems such as this one by visualization and trial and error, taking no more than a few minutes. A trial-and-error method for solving this problem must systematically generate candidates for a solution, testing each candidate to see whether or not it satisfies the specification. This continues until either a satisfactory one is found or all moves have been tried.

The number of possible configurations is very large, although some of these are equivalent to one another by rotating the chessboard. Since no configuration having two or more queens in a single column can succeed, we could restrict the candidate configurations to include only those in which each queen occupies a separate column.

It is not even necessary to generate all of the configurations having one queen per column. A board configuration can be generated by placing the queens one at a time. Obviously, the generation of a particular candidate configuration can stop as soon as it is discovered that one of its partial configurations is unsatisfactory.

Try writing such a 'generate-and-test' model to solve the eight queens problem.

5.1.2 The missionary/cannibals problem

'Crossing problems' like this one are quite popular puzzles for demonstrating different problem-solving and programming strategies in cognitive psychology and artificial intelligence. Amarel (1968) gives a thorough analysis of the topology of this miniworld's state space. The problem's main attraction is again its simple structure and the well-defined set of possible states and moves. The following description is taken from Bundy (1978).

Three missionaries and three cannibals seek to cross a river from its left to its right bank. A boat is available, which will hold only two people and which can be navigated by any combination of missionaries and cannibals. If the missionaries on either bank of the river are ever outnumbered by cannibals, the cannibals will eat the missionaries. When the boat is at a bank, it is counted as part of the bank. The goal is to find the simplest schedule of crossings that will permit all the missionaries and cannibals to cross the river safely.

In writing a program to solve this problem you will need to find an appropriate representation for the entities (missionaries, cannibals, boat) and their actions (river crossings). You will also need to apply moves to produce new states and detect fatal or goal constellations. Be careful not to get caught in a cycle.

5.1.3 The Lineland problem

In this exercise we visit the one-dimensional world of Lineland, on the island of Flattiti. Its description is adapted from Wickelgren (1974).

In Lineland there are two races of people, whites and blacks. As in our three-dimensional world, the whites have for a long time discriminated against the blacks. Lately, however, the blacks are making some progress. For instance, a court decision ordering discrimination reversal in coach travel has just been announced. In the past whites have always ridden in the front of coaches and blacks in the back. Now the court has ordered that for a time blacks will ride in the front and whites in the back.

When the order takes effect, there is one seven-passenger coach already loaded with three blacks in the last three seats, three whites in the next three seats and the front seat empty. Coaches in Lineland are automatic, requiring no driver (steering is obviously unnecessary).

Since the order has already gone into effect, the police stop the bus and insist that blacks and whites must reverse positions immediately. Of course people in Lineland are able to move to adjacent positions in their linear world, so a person can move to an empty adjacent seat in the bus. However, in addition, a clever Linelander called Monorock has invented a special device (a black-gap synthesizer) that allows them to pass through two-dimensional space for a very limited distance, hopping over any intervening persons and objects in either direction. In the coach this hopping ability has a limit of two seats. Thus either a white or a black can hop over one or two adjacent seats, provided the target seat is empty. For example, the first white can move into the front seat, or the second white can hop over one white into the front seat, or the third white can hop over two whites into the front seat. But the first black cannot hop over all three whites.

Using these movement properties of whites and blacks in Lineland, reverse the relative positions of the three whites and three blacks, so that all blacks are in front of all whites in the bus. Note that it is irrelevant where the empty seat occurs in the bus, so long as all three blacks are sitting in front of all three whites.

Your model may again use a simple 'generate-and-test' strategy, but you may want to provide some measure of achievement for a given constellation. This can give more direction to your program, which should try to increase that value for each move.

5.2 Problems for Monte Carlo simulation

These exercises add a stochastic aspect to a model. They can be solved with any appropriate simulation system or using the MC and STATS toolboxes. Dewdney (1985) describes some additional Monte Carlo problems of particular interest.

5.2.1 A simple crew-scheduling problem

This problem is adapted from Schriber (1974).

Albatross Airways, which serves DreamWorld's polar regions, is based at Ragnaroek. There are 15 flights out of Ragnaroek per day, each with one pixie as stewardess. The airline has a policy of keeping three pixies on call to replace sick stewardesses. A probability distribution for the daily number of sick stewardesses is given in the table below. It can be applied each day, independent of yesterday's number of sick pixies.

Number sick		**Probability**	
	0		20%
	1		25%
	2		20%
	3		15%
	4		10%
	5		10%

Model the operations of Albatross Airways to estimate the utilization of reserve stewardesses and the distribution of the random variable 'number of flights cancelled each day because no stewardess is available'. Simulate this scenario for 1000 days of operation.

5.2.2 Louis' destiny*

Louis is a leprechaun who earns his living in an icicle quarry. He owns two tame centipedes, XXY and YYY, whom he uses for transport to and from work. XXY is about 3000 and YYY close to 4000 years old, so they tire easily and sleep a lot. The probability that Louis will succeed in trying to wake up one of them is about 50% if Louis has eaten fewer than 10 bags of fish and chips the previous night. It is 90% otherwise (old centipedes are very experienced). We can also assume that both of them react independently. Luckily once a centipede has started it will, without fail, continue the ride all the way from home to work or back home without falling asleep once!

Louis has long been engaged to Wendelin, a water nymph, who sometimes (30%) drops by to give him a lift on her butterfly, either from home to work or from work to home. He then does not even try to wake one of his centipedes, who stay where they are.

It is, therefore, quite possible that Louis will find himself at home in the morning, without any transportation (because Wendelin did not come by, he cannot get the centipedes started, or both are parked at work). In this case he will call his boss, a grumpy old wizard who will immediately cast a spell, turning poor Louis into a toad. Louis will then call Wendelin for consolation, whereupon they will marry and live in eternal bliss. Alternatively Louis may be stranded at work in the evening. He will then elope to far away Morven with Harold, the bookkeeper (note that leprechauns can switch sex if they get *very* excited).

If Louis arrives home safely he will invariably eat fish and chips (normally distributed with a mean of 9 and a standard deviation of 3 bags) before he goes to bed, ready for a new day of back-breaking work.

Given that Louis starts with both centipedes at home, write a Monte Carlo model and take 150 samples of three consecutive days to find the expected frequencies with which Louis will continue to mine icicles, live in eternal bliss or get to Morven. You may also want to run the same experiment with an exponential 'fish and chips' distribution (mean of 9).

5.2.3 The clockmasters of Tick and Tock*

This exercise was inspired by a problem reported by F. Rubin as described in *Journal of Recreational Mathematics*, **12**(1), 53–54.

The twin towns of Tick and Tock in the remote and inaccessible Oriab mountains are jealously proud of their ancient clockmaking traditions. Both towns are equally skilled, their clocks running from one minute slow to one minute fast per day, with uniform and independent probabilities.

The master clockmakers, of whom there are exactly 100 in each town, are so proud that each wears one of her own clocks on a gold chain around her neck. Whenever two masters meet in the town square, which happens frequently, they perform the 'masters' matching ceremony'. In Tick, each master sets her clock to the mean of their two clocks; in Tock, each sets her clock to the time shown on the other's clock.

On New Year's Eve, all of the clockmasters meet in the town square to compare their clocks with the one on the town hall tower, which is always exactly correct. Any master whose clock is more than one hour in error is dropped from the guild and exiled forever. Her place is then taken by an aspiring young apprentice and everyone sets their clocks to the correct time.

Design and implement a model for 100 clocks (masters), whose rates of deviation are evenly distributed from -0.99 to $+0.99$ minutes per day. At

*Adapted by permission from Rubin Frank (1979–80), 'The Clocks of Klotz and Klutz', *J. Recreational Mathematics* **12(1)**, pp. 53–54.

the time of each meeting all clocks are updated and then two are selected at random. Then either the Tick or the Tock transformation is performed. Assume that meetings occur equally spaced throughout a year. At the end of the year the number of clocks with errors greater than one hour should be printed.

Perform sampling experiments with 150, 300 and 600 meetings, respectively, for both Tick and Tock. In which town can we expect the most tragedies?

Note: You will need to sample from two streams of random numbers:

1. continuous uniform distribution −0.99 to +0.99 (daily rate of deviation from the correct time);
2. integer uniform distribution 1 to 100 (selection of clocks; don't compare a clock with itself!).

5.2.4 The snark and snork problem

This problem is adapted from Dyck *et al.* (1982), p. 513.

On the island of Atuan, in the heart of the Kargadian empire, annual games are held in the Colosseum. One of these contests is conducted according to the following ancient ritual.

In the arena there is a line-up of gladiators. The arena has seven doors. Four of them hide fierce and hungry snarks, while beautiful, blue-furred snork maidens are imprisoned behind the other three. The first gladiator chooses a door. If there is a snark behind it, he is eaten and dies. The snark is put back in her cage. If he chooses a door with a snork, the gladiator is set free, rewarded with riches, the snork is put back and one of the snarks is sacrificed to the gods (randomly). There is then one fewer door to choose from. The gladiators will keep on choosing doors until all the snarks or all the gladiators are killed. After that, any remaining gladiators are kept as slaves.

Simulate this scenario 100 times. How many gladiators can expect to be eaten and how many can expect to be free after each contest?

5.2.5 The 'intoxicated beetle': A random walk

This problem is adapted from Goodman and Hedetniemi (1977), p. 168.

After a few jugs of beer in her local pub, a silly beetle has foolishly made the following bet. She claims that she can walk randomly around a ten-by-ten grid square, suspended over a cauldron filled with boiling hot love potion. Each step she takes is only one unit in the direction of one of the principal points of the compass (north, south, east, west). She will start from the centre of the square and take ten steps.

Simulate her predicament 100 times to determine the chances of staying alive or falling off the edge.

5.3 Problems for quasi-continuous simulation

These problems impose a time dimension on a microworld. They can be defined by both difference (step variables) and differential (continuous processes) equations. DYNAMO, any continuous-system simulation language or our CS toolbox may be used.

5.3.1 The bouncing ball problem

The following problem is taken from Dyck *et al.* (1982), pp. 451–6.

Calculate and plot the path of a rubber ball for several bounces along a level floor, given its initial location and velocity. Assume that the only external force is gravity and that the acceleration due to gravity is 9.8 m/s^2. The bouncing characteristics of the ball follow the rule that 10% of the vertical velocity is lost when the ball bounces, but no loss in horizontal velocity occurs. A state-space description for such a model will need the x and y co-ordinates of the ball and its velocity components in the x and y directions. A natural choice of co-ordinate system is the direction along the floor as x and the vertical height as the y co-ordinate. The initial state of the simulation will be defined by the ball's initial co-ordinates and velocity components.

State changes are described by the physical laws governing motion under acceleration. Since no forces act to alter the motion in the x direction, the velocity will remain constant at its initial value. If the ball is projected at v m/s at an angle θ to the horizontal, this initial velocity in the x direction is $v_x = v \cos \theta$. Similarly, the motion in the y direction is initially $v_y = v \sin \theta$, but this velocity changes as the ball is accelerated under the force of gravity. Since acceleration is the rate of change of velocity, the new velocity component will be:

$$v_y(t + \delta t) = v_y(t) + a_y * \delta t$$

In this case, the acceleration a_y is due to gravity in the negative y direction, so $a_y = -9.8$ m/s^2. This velocity may be used to calculate the new y co-ordinate. For motion under constant acceleration we have the formula:

$$y(t + \delta t) = y(t) + \tfrac{1}{2}[v_y(t + \delta t) + v_y(t)] * \delta t$$

The new x co-ordinate follows from:

$$x(t + \delta t) = x(t) + v_x * \delta t$$

There remains the problem of what to do when the ball hits the floor. The assumption that 10% of the velocity is lost when the bounce occurs is implemented as:

$$\text{if } y(t + dt) < 0 \text{ then}$$
$$v_y(t + dt) := |\, 0.9 * v_y * (t + dt) \,|$$

This assumption is not quite correct, since it allows the ball to move slightly below floor level before its direction is changed. The error caused by this assumption is negligible if δt is small enough.

5.3.2 Salmon farming on Lorbanery*

This problem is adapted from Payne (1982).

A certain, very isolated lake on Lorbanery can support a maximum population of 500 000 glowfish, which feed on various water plants and other marine life. With no competition glowfish would triple each year. Lorbanery is a poor island, depending very much on tourism for most of its income. It therefore wants to promote a sport fishing industry by introducing salmon to the lake. The salmon prey on the glowfish and would increase at the rate of 50% per year in the presence of abundant supply. Their growth rate drops to zero as their numbers exceed 10% of the glowfish population. Assume that a typical salmon consumes 10 glowfish per month.

1. Develop a model to simulate the growth of the glowfish from an initial level of 5000, with no salmon.
2. Three years after the start of the simulation, introduce salmon to the lake and observe the system for 20 years. Experiment with initial salmon populations of 200, 500, 750, 1000 and 2000. How long does the glowfish resource last?
3. In the previous glowfish–salmon system, assume that each of the populations is reduced at a relative rate of 0.2% per day for a six-month fishing season each year; that is, 2 fish per 1000 of each species are caught each day. Start the fishing season two years after the introduction of the salmon and follow the model's behaviour for 18 years. How do the numbers compare to the previous situation?

5.3.3 A simple 'government game'

Write a program to simulate the decision-making behaviour of a benevolent tyrant ruling some ancient city state.

Each year current information about population, area in use for agriculture and food supplies is obtained, and decisions have to be made for the next period on how much land to plough, how much grain to eat or sell, and how much land to buy or sell. Assume that grain is the only means of payment. All state variables and policy parameters are interdependent; the food supply for the population will influence its death rate, the population density its birth rate, and so on. Random factors (such as weather) can be introduced to vary the land's yield and price and other characteristics.

Draw a system-dynamics diagram of such a scenario and simulate the model with suitably chosen initial values. You may want to evaluate a particular strategy and try to keep the population relatively constant while increasing your wealth (grain per head, land per head). You can design the same style of decision-making game for a variety of other contexts (markets, ecologies, etc.).

5.4 Problems for discrete-event simulation

The following problems lead to typical discrete-event models. The first two are taken from the computing folklore. The others are based on queueing network scenarios. Any discrete simulation system or the MC, STATS, DES and QN toolboxes may be used.

5.4.1 The game of life*

The game of life, invented by John Conway, is supposed to model the genetic laws of birth, survival and death [see Gardner (1984)] of some colony of creatures.

It is played in a square universe (board) of arbitrary dimensions. Each square (apart from the border squares) has eight neighbours. It may be empty or may contain an organism of unspecified shape. A new generation is determined by the following rules:

1. BIRTH — An organism will be born in an empty location that has exactly 3 neighbours;
2. DEATH — An organism with 4 or more neighbours will die from overcrowding;
3. SURVIVORS — An organism with exactly 2 or 3 neighbours will survive to the next generation.

This scenario requires only a simple time-slicing procedure.

Design an appropriate representation for the universe and the organisms who populate it. Starting with some well-chosen [see Gardner (1984) for guidance] initial state you can then simulate the life of a colony for a number of generations.

Although the rules are simple, this game is quite an absorbing one with many interesting patterns. It is particularly intriguing if you watch it in 'slow motion' on a terminal screen.

5.4.2 The dining philosophers*

The following scenario is a good example of mutual exclusion, where a number of entities (processes) are waiting on certain conditions. It has served as a classical vehicle for studying process synchronizations in operating systems. A number of variants [such as 'drinking philosophers'; see Chandy and Misra (1984)] have also been proposed.

Five philosophers are seated around a circular table containing an inexhaustible supply of spaghetti, within easy reach of all. Between each pair of adjacent philosophers is a single fork. Philosophers have a rather simple life style:

loop: (*1*) *think*; (*2*) *sleep*; (*3*) *eat*;

Eating requires both the fork on the left- and the fork on the right-hand side. Each philosopher thus competes for resources with his or her immediate neighbours.

Simulate a 30-hour orgy, using the following timings:

Thinking	Uniform 20–30 minutes. (exhausting)
Sleeping	Negative exponentially 180 minutes.
Eating	Normally (20, 5) minutes.

At the end you should obtain tallies (*min*, *max*, *mean*) of the times each of the five philosophers' had to wait between sleeping and eating. Try to avoid deadlocks.

5.4.3 Manuel revisited

Manuel, as you will recall, engages in the production and marketing of love potions on the island of Gont. His current premises are rather small, and, since his fame has spread all over DreamWorld, long queues of customers have lately been observed in front of his stall. Manuel has therefore commissioned a well-known consultancy (Auercraft) to analyze his business procedures and suggest ways to boost his productivity.

According to their report Manuel must rationalize his product line. Hitherto he has used a wide range of individualized concoctions (toad slimes, leg of tarantula, snake poisons, Eye of Newt) for his potions, which he prepared fresh on the premises. In future he is to sell only two varieties of precast love spells, stored in two big cauldrons. He must also stop chatting to his customers and streamline his diagnostic procedures (determine whether a weaker or stronger potion is required), so that customers are funnelled through the shop as shown in Figure 5.1.

Manuel is somewhat suspicious of Auercraft's recommendations. He therefore first wants to model and simulate the new set-up in order to determine whether the queue can ever be expected to reach 13 customers (Manuel is very superstitious) and what might be his average utilization (he is also quite lazy and hates to be busy for more than 70% of the time).

Design, code and run a simulation to help Manuel. A sample of 500 customers should be sufficiently informative.

5.4.4 The ring of Avalon*

Avalon, a tranquil community on DreamWorld's southern fringe, is populated by a coven of witches and wizards engaged in the ancient but as yet unsuccessful enterprise of gold making. Three of its most prominent research dungeons are located in Kadatheron, Sona-Nyl and near the Towers of Koth

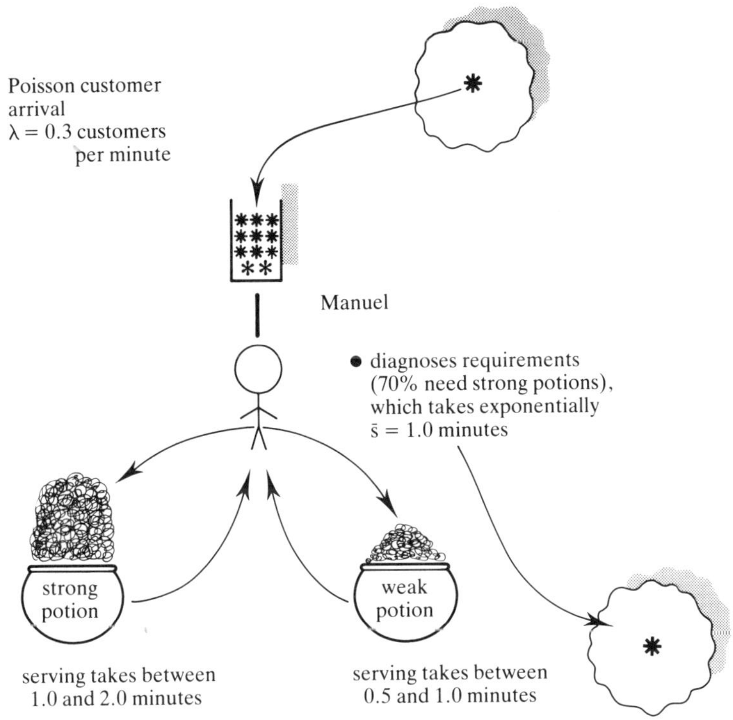

Figure 5.1 Manuel revisited

in Gug Kingdom. In order to exchange new spell mixes (mostly failures) the coven plans to install a local area communications network (Spellnet).

It has been decided to use specially designed sailing boats, piloted by highly skilled and experienced leprechauns. The boats are not terribly fast, but quite sturdy, as they need to be in view of the nature of the documents they are designed to carry. DreamWorld still exists in a paperless society and all records are chipped in stone. Such petrified spells are called 'packets' and each boat can carry up to three of them in its pouch. The boats will be in continuous use (in Avalon, leprechauns are not entitled to holidays), circling through Kadatheron, Sona-Nyl and Morven (the harbour town nearest to the Tower of Koth).

The southern sea is dangerous, teeming with sea serpents, dragons and other fearsome creatures. For each sea voyage there is a 30% probability that a boat will be lost. It can then be replaced only in Sona-Nyl, where it takes between 2 and 5 days to equip a new boat and round up a crew of leprechauns. It is assumed that it is immediately known (through witchcraft!) if a boat has been lost.

Each of the three stations has a spell-generation pattern of:

Kadatheron   1 spell every 7 days;

Sona-Nyl	1 spell every 3 days;
Morven	1 spell every 5 days.

These timings are means of exponential distributions. Spells must wait at each station until a boat comes by. When a boat enters a station it is unloaded (but only those spells which have reached their destinations) and all free slots are filled with waiting spells (if any). Then the boat is off to the next port.

There is a uniform probability that a spell's destination is any other particular station. For simplicity's sake we ignore the time it takes to unload and load the boats. The travelling times between stations are:

Kadatheron	to	Morven	3 days;
Morven	to	Sona-Nyl	4 days (high seas);
Sona-Nyl	to	Kadatheron	2 days.

All timings are subject to a normal distribution ($\pm 10\%$ standard deviation).

The coven would like to know the amount of traffic with which Spellnet can cope without queues at any station exceeding three spells.

Build and explore a model of this network, running it for an initial period of 15 and a statistics-gathering period of 60 days. Assume that the model starts in an empty state, with one of the three boats moored at each of the three harbours. Gather statistics about the boats' utilizations, the lengths of the spell queues at each station and the average time a spell is delayed in transit between its origin and its destination (across all spells).

5.4.5 The busy barmaid model

The Blue Unicorn, in the port city of Atuan, is one of DreamWorld's most infamous establishments. Travellers with dubious credentials gather in its smoke-filled tavern each night, to exchange secret information and plot nefarious deeds. Recent complaints about the standard of service have prompted the inn's management to explore alternative operating procedures.

It has been observed that guests arrive randomly at a rate (negative exponential) of 0.03 per minute, their average thirst is uniformly distributed between 1 and 10 jugs, and that each customer takes about 10 minutes (normally distributed with a deviation of 2.5) to empty his jug.

The Blue Unicorn's staff consists of a number of highly skilled fairies who invariably take exactly 5 minutes to fill a jug. The inn keeps ample stocks of ale and spirits, which means that for all practical purposes we can assume an unlimited supply of these.

All of the four scenarios described below should be simulated for 180 minutes of the tavern's operation. This may provide the inn's management with some data on whose basis they can evaluate their reactions to future customer complaints.

1. The first scenario reflects the Blue Unicorn's current operating procedure. It suggests a material-oriented framework. After customers

enter the tavern they queue for the single barmaid's services. Jugs are not considered as bottlenecks; therefore they are not explicitly modelled. The barmaid should be represented by a passive resource, who can be seized and released by waiting customers. Filling glasses is the only duty performed by the bar staff. After obtaining and drinking a jug, customers either leave the inn or rejoin the queue for a refill (depending on their thirst). The model should gather queue and maid-utilization statistics and count the number of jugs filled during its operation.

2. Jugs may not be as irrelevant to service delays as the first model suggests. Our next scenario will therefore explore their effects on customer's queueing delays.

   The tavern owns a bin of 10 wooden jugs, all initially empty, which the barmaid fills in advance and lines up on the counter. If there are no more jugs to fill, the maid has to idle until an empty one arrives (note that jugs are never washed in the Blue Unicorn). Incoming customers now queue in front of the bin of full jugs, drink their ale and always return the empty jug to the barmaid before leaving or rejoining the queue. This set-up suggests a mixed machine/material-oriented model with both customers and barmaid as active processes, synchronized by the jugs they pass between them.

   Apart from the usual statistics we now also want to measure the average level of ale in our guests at the time they are leaving.

3. At another level of refinement we may want to treat the 10 jugs as active entities with life cycles of their own.

   Customers will co-opt full ones on arrival and relinquish them after drinking, whereupon the barmaid co-opts them for a filling activity. Starting with an array of full jugs at opening time, we now want to obtain all the previous standard statistics as well as the average number of drinkers in the tavern (its utilization).

4. A final variation features a leprechaun, employed as a waiter collecting empty jugs from tables at random intervals. We can assume that he performs this task each time he awakes, instantaneously moving all empty jugs to a pile of dirty ones; but he sleeps for 20–40 minutes in between.

   This time we again represent the barmaid by a passive resource and dirty and empty jugs as bins or counters. Customers and the leprechaun are the active processes which drive the simulation. Upon arrival a customer now has to queue until two conditions are met simultaneously: he can grab an empty jug and the barmaid is available to fill it. After drinking he may leave the empty jug on the table, from where it will later be collected.

   Statistics should now include a frequency diagram of the times spent in the pub by each customer and a counter of lost business. This counter is incremented each time customers leave without drinking, which invariably happens whenever there are more than 13 guests in the tavern.

It is a good exercise to draw activity cycle diagrams for each of these four versions.

The scenarios can be simulated with any discrete-event simulation system or using our toolboxes, although the specifications and the way they are worded suggest a DEMOS implementation.

5.4.6 An airport model for Albatross Airways

This problem is adapted from Reitman (1971).

Due to the unfaulted safety record of its birds Albatross Airways has recently experienced a marked increase in demand for its services. Although this is of course pleasing to the company's accountants, it also creates some problems for the airline's management. While it is quite feasible to contract and train additional albatrosses and pixies, the company's old airport at Ragnaroek is already operating close to capacity limits. A new airport has to be built to cope with the projected increase in services. After careful deliberation Albatross Airways has decided on a modern design with two crossing runways, for which it expects to use the following takeoff and touchdown procedures.

1. Bird departure schedule: flocks of birds to depart shortly after the hour and half-hour.
2. Ground servicing: during the busy hours of the day maximum time for a bird on the ground is 1.5 hours, with 0.75 hours as average. There is a minimum 0.33 hours rest interval between flights.
3. Runway scheduling: both runways are used when wind conditions permit; otherwise one runway is used for both takeoff and landing.
4. Albatross landing: under visual flight rules (VFR) the interval between albatrosses ranges from 60 to 75 seconds. All of Albatross Airlines' birds will be equipped with infrared goggles. Landings under instrument flight rules (IFR) require from 75 to 110 seconds. Weather conditions determine VFR and IFR.
5. Crossing runway scheduling: when both runways are in use one is used only for landings and the other for takeoffs. At least 20 seconds must pass from the time a landing bird has run past the runway intersection point and the time an albatross will reach the intersection on a takeoff sprint. Forty seconds are required until a landing may occur, once the departing bird has passed the intersection.

Select some appropriate meteorological data (remember that Ragnaroek lives up to its name; its weather is usually ferocious) for both wind and visibility conditions and simulate the proposed airport to determine:

1. Hourly runway acceptance rates; hourly and daily utilization statistics.
2. Runway utilizations under both VFR and IFR conditions.

3. The effects of a storm crossing the airport and the exploration of alternative strategies to return the airport to normal operation as quickly as possible.
4. Evaluation of grouping flocks of birds into landing and takeoff queues during single-runway operation.

5.5 Problems for combined simulation

The following scenarios combine elements of both continuously fluctuating states and sequences of discrete-state transitions. They can be simulated using either a special-purpose language for combined simulation or the MC, STATS, CS, DES, QN and COMBINED toolboxes.

5.5.1 The fire dragons of Ismay*

This example simulates a fire station. It was first proposed by R. W. Sierenberg of Delft university. Helsgaun (1980) wrote a DISCO implementation.

The small town of Ismay, on Iffish, owns a well-equipped fire station and employs three dragons on a contract basis (DreamWorld's big southern dragons can spit water as well as fire). Fire alarms occur randomly in exponentially distributed intervals, with a mean of six hours. Most houses on Iffish are built out of driftwood, containing a large amount of flammable material. When a fire is discovered, it is normally already of a certain size, increasing at a rate proportional to this size as long as no extinguishing activity takes place. At the moment an alarm is given, one fire dragon, if available, is sent to the fire. When he reaches the fire and finds out that his extinguishing capacity is smaller than its rate of increase, he requests further assistance by sending a second alarm referring to the same fire. Fortunately Iffish's dragons are telepathic and alarms can therefore be transmitted instantaneously. Dragons do, however, take some time to reach the origin of an alarm. After the fire is put out or when all inflammable material is consumed all fire dragons return to the station.

Simulate this scenario for 1 month, gathering statistics about all successful and unsuccessful fire-fighting events. You may use the following data:

Flammable material per house	uniform 100–600 cm^3;
Travelling time per dragon	uniform 5–15 minutes;
A fire's size at the time it is detected	uniform 0–5 cm^3;
Intensity of fire (rate)	uniform 0.01–0.06 cm^3/min.;
Extinguishing capacity of dragons	deterministic 10 cm^3;
Number of fires (rate!)	negative exponential 1 per hour.

You may also want to explore the effect of different distributions and parameters on the effectiveness of Ismay's fire-fighting strategy.

5.5.2 Kadatheron's rocket

This problem is described in Helsgaun (1980).

A research dungeon at Kadatheron has recently deciphered the long lost secret of how to build multi-stage solid fuel rockets. Avalon's wizards now hope to use this discovery to visit one of DreamWorld's four moons, where they suspect large deposits of lead, an ore which is essential for their research into gold making.

The following scenario has been proposed to analyze the trajectory of a three-stage rocket launched from DreamWorld's surface. During the rocket's flight its mass, velocity and altitude change according to well-known physical laws.

The change in mass is due to the expulsion of burnt fuel and takes place at a constant rate: $-MassFlow$.

$$Mass.rate = - MassFlow;$$

From Newton's second law of motion we can determine the rocket's acceleration, that is its *velocity rate*, by summing the forces acting upon the rocket and dividing the result by the rocket's mass. In this example three forces act upon the rocket:

Thrust denotes the force generated by the expulsion of burnt fuel.

Drag is the air resistance. It is dependent on the rocket's cross-sectional area.

Gravity is the earth's gravitational attraction acting upon the rocket. Its effect depends on its mass and altitude.

$$\begin{aligned} Thrust &= MassFlow * FlowVelocity; \\ Drag &= c_1 * Area * EXP\,(-c_2 * Altitude.state) \\ &\quad * Velocity.state^2; \\ Gravity &= Mass.state \,/\, (c_3 + c_4 * Altitude.state)^2; \\ Velocity.rate &= (Thrust - Drag - Gravity) \,/\, Mass.state; \end{aligned}$$

The rocket's change in altitude per unit of time, *altitude rate*, is equal to its velocity:

$$Altitude.rate = Velocity.state;$$

This description defines a single continuous process through three differential equations. In this example and in most programming systems it is the user's responsibility to make sure that the equations involved are evaluated in the correct sequence. This means that the variables occurring on the right-hand side of an assignment statement must have values which reflect the current state of the system. Here *Thrust*, *Drag* and *Gravity* must be computed prior to *Velocity. rate*.

Use these specifications together with suitably chosen constants (c_1 to c_4) and initial values (for *Mass*, *Thrust*, *Gravity*) to simulate the flight of a three-stage rocket. Assume that each stage, k, carries enough fuel to last for $t(k)$ seconds of flight. Stage separation can be modelled by time events. You may want to plot the rocket's flight path.

5.5.3 A tanker fleet simulation

(The following problem was adapted from Pritsker and Young (1975).)

The rainbow fountain on the island of Osskil, so the saying goes, will give eternal youth to whoever drinks from it. This well-known myth has recently prompted a group of Havnor businessmen to purchase a leasehold and exploit its commercial potential. In order to process large amounts of the fountain's water, they must use a fleet of tankers to transport it from Neshum, Osskil's port, to Great Port on Havnor. In Havnor the water will then be expensively bottled, lavishly labelled and exported to all who can afford to pay its price.

Market research has indicated a substantial demand for the product. The initial fleet will therefore consist of 15 tankers, each with a nominal carrying capacity of 150 tb (thousand barrels of water). A pipeline and sufficient dock space has already been built in Neshum to accommodate all tankers simultaneously, if necessary. The average loading rate for a tanker is 50 tb/day. A tanker is loaded to its nominal carrying capacity ± 10 tb. The time to load it is exponentially distributed with a mean equal to the amount to be loaded, divided by the average loading rate. All the tankers unload at the same dock in Great Port with an average unloading rate of 200 tb/day. The amount left after unloading is uniformly distributed between zero and 7.5 tb.

Rowing time from Neshum to Great Port is normally distributed with a mean of 5 and a standard deviation of 1.5 days, when a tanker is loaded. However, a tanker never takes less than 3.5 days or more than 8 days in travelling to Great Port. The time to travel from Great Port to Neshum, unloaded, is also normally distributed with a mean of 4 days, a standard deviation of 1 day, a minimum of 3 days and a maximum of 7 days. The random variations in tankers' travel time are due to weather conditions, the state of fatigue of their leprechauns, and other uncertainties associated with ocean travel.

Simulate the tanker fleet for 365 days to obtain estimates of the following quantities: *dock utilization*, *tanker round-trip time*, *tanker waiting time* and *number of tankers waiting to be unloaded*.

1. Initial conditions for the first simulation specify that the tankers are to arrive at their loading points at half-day intervals starting with the first at time zero, at which time the unloading dock is idle and no tankers are waiting to be unloaded.
2. So far the model is purely a discrete-event one. It is still necessary to

specify the continuous component. It has been assumed that all tankers can be loaded simultaneously if necessary.

In Great Port there is only one unloading dock; this supplies a storage tank, feeding a bottling plant through a pipeline. The storage tank receives well water from a tanker at a constant rate of 300 tb/day and continuously supplies water to the bottling plant at a constant rate of 150 tb/day. A two-shift day is used for operating the unloading dock. It is closed at 8 P.M. and opens at 4 A.M. Safety considerations require that unloading be stopped when the dock is shut down for the night. The storage tank has a capacity of 2000 tb. When it is full, unloading is halted until the amount in the tank decreases to 80% of capacity. When the storage tank is empty, supply to the bottling plant is halted until a level of 50 tb is reached. This should avoid frequent start-ups and shut-downs.

The main characteristics associated with the tankers are summarized below:

1. Nominal carrying capacity is 150 tb.
2. Travel time loaded is normally distributed with a mean of 5.0 days and a standard deviation of 1.5 days.
3. Travel time unloaded is normally distributed with a mean of 4.0 days and a standard deviation of 1 day.
4. Time to load is $(150 + X)/50$ days, where X is uniformly distributed in the interval $(-10, 10)$.
5. Unloading rate is 300 tb/day.

Initial conditions for this second simulation are that the storage tank is half-full and the tankers are to arrive at their loading points at half-day intervals, starting with the first at time 0.0. The objective of this study is to simulate the above system for 365 days to obtain estimates of the following quantities:

1. Dock utilization for unloading.
2. Dock utilization for loading.
3. Time when bottling plant has well water to process.
4. Amount of well water in the storage tank.
5. Tanker round-trip time.
6. Tanker waiting time.
7. Number of tankers waiting for unloading.

Appendix 1
Some Utility Programs

1.1 Shells for model executives

1.1.1 Monte Carlo simulation

- declare CONST: MaxHistoClasses, MaxPlotPts;
- Include declarations and definitions for MC and STATS toolboxes;
- {declare, instantiate and initialize aStaticClock, aStateSamplingMonitor, distributions and data-collection devices};
- ReadAndInitialize;
- WHILE AnotherSampleP DO
  (1) sample from distributions;
  (2) record results;
- WriteReport; {report results (frequencies)}

1.1.2 Quasi-continuous simulation

- declare CONST: MaxPlotPts, MaxHistoClasses;
- Include declarations and definitions for CS and STATS toolboxes;
- {declare, instantiate and initialize aSynchClock, aTimeSlicingMonitor, step or continuous variables and data-collection devices};
- ReadAndInitialize;
- WHILE AnotherTickP DO
  step or integrate variables; record results;
- WriteReport; {report results (chronological graphs)}

1.1.3 Discrete-event simulation (of queueing networks)

The event-oriented modelling frame

- declare CONST: MaxHistoClasses, MaxPlotPts;
  and TYPE : anEventType = {all event names used in model};
- Include declarations and definitions for MC, STATS, DES and QN toolboxes;
- {declare, instantiate and initialize anAsynchClock, anEventSchedulingMonitor, sources, sinks, distributions, servers, queues and data-collection devices};
- ReadAndInitialize;
- {define an event procedure for each class of event}.
- WHILE MoreSamplesToTakeP (TheMonitor) AND
  MoreEventsP (TheMonitor) DO
  {or MoreTimeToSimulate . . .}
  (1) GetNextEventNotice (TheMonitor);
  (2) CASE KindOfEventQ (TheMonitor.currentEvent) OF
  {a case list of event names associated with calls on event procedures}
  END;
  (3) FlushOutEvent (TheMonitor.currentEvent);
- WriteReport; {report results (server utilizations,
  queue statistics, transit times)}

The process-oriented modelling frame (single process class !)

- declare CONST: MaxHistoClasses, MaxPlotPts;
  and TYPE : anEventType = {list of all names of process phases};
  and VAR : FirstPhase, LastPhase; {eventType names};
- Include declarations and definitions for MC, STATS, DES, PROCESS and QN toolboxes;
- (Note: the TAPtrType declaration in DES_DCLS.LIB must be changed from ↑ aTarecordType to ↑ aTAprocessType);
- {define, instantiate and initialize FirstPhase, LastPhase, anAsynchClock, anEventSchedulingMonitor, sources, sinks, distributions, servers, queues and data-collection devices};
- ReadAndInitialize;
- {define a TAprocessClass {may assume a reference variable *"self"*: aTAPtrType} procedure as a shell for a CASE statement indexed by the names of all of a process' phases; each associated with a block of code};
- WHILE MoreSamplesToTakeP (TheMonitor) AND
  MoreEventsP (TheMonitor)
  {or MoreTimeToSimulateP (TheMonitor)} DO
  ResumeNextTAprocess;
- WriteReport; {report results (server utilizations, queue statistics, transit times)}

The activity-oriented modelling frame

- declare CONST: MaxHistoClasses, MaxPlotPts;
  and TYPE : anEventType = {dummy to compile DES_PROCS.LIB};
  and TYPE : anActivityType = {list of activity names};
- Include declarations and definitions for MC, STATS, DES, ACTIVITY and QN toolboxes;
- (Note: *"anEventPtrType"* and *"anEventNoticeType"* in DES_DCLS.LIB are replaced by *"anActivityPtrType"* and *"anActivityNoticeType"*);
- {declare, instantiate and initialize anAsynchClock, anActivityScanMonitor, distributions, servers, queues and data-collection devices};
- ReadAndInitialize;
- {define one activity class procedure for each activity. These procedures need references to their state records (*self*) and indicators (*state*). Their body is a CASE statement with two entry points; one for *"start"* and one for *"finish"* phases};
- {define a *"StartActivities"* procedure which tries to trigger all activities in *"start"* states willing to go, by cycling through the agenda. Conditions associated with the start of each activity are checked. Each successful start is signalled to the monitor by invoking procedure *"ActivityStarted"*. Define *"FinishActivities"* procedure which tries to trigger all activities in *"finish"* states ready to go, checking their *"finishTime"* slots.};
- WHILE StillMoreSamplesP (TheMonitor)
  {or StillMoreTimeP (TheMonitor)} DO

```
Phase   I:  REPEAT
               StartNewScan; StartActivities;
            UNTIL ScanWithoutAnyStarts;
Phase  II:  SetClock (TheClock); {to smallest finish
                                   time on agenda}
Phase III:  FinishActivities; {which finish at this time}
```

- WriteReport; {report results (server utilizations, queue statistics, transit times)}

1.1.4 Combined discrete/continuous simulation

- (Note: this can be predominantly continuous, driven by a time-slicing loop with some events superimposed on it; or it can be predominantly event based, driven by a next event monitor with continuously modelled subprocesses};
- declare CONST: MaxHistoClasses, MaxPlotPts;
  and TYPE : anEventType = {list of event names};
- Include declarations and definitions of MC, STATS, CS, DES, QN and COMBINED toolboxes;
- {declare, instantiate and initialize aSynchClock or anAsynchClock, aTimeSlicingMonitor or anEventSchedulingMonitor, state variables, distributions, sources, sinks, servers, queues and data-collection devices};
- ReadAndInitialize;
- {provide integration procedures and event procedure definitions. Event procedures may by implement time events (triggered by a clock) or state events (triggered by model states)};
- WHILE AnotherTickP DO
  (1) integrate all continuous variables (see CS shell);
  (2) check whether discrete events can occur now (see DES shell);
  (3) record the state of the model;
- WriteReport; {report results (frequencies, time plots (server utilizations, queue statistics, transit times))}

1.2 Monte Carlo methods

The Monte Carlo library

Object : *aStaticClockType*;

Methods : none {dummy declaration};

Object : *aStateSamplingMonitorType*;

Methods : *InitStateSampling Monitor* (*aStateSamplingMonitor*, *samples*);
AnotherSampleP (*aStateSamplingMonitor*): *BOOLEAN*;
ReportOnSamplesTaken (*aStateSamplingMonitor*);

{all distribution objects share the following internal method:
GenerateZeroToOneNumbers (*seed*): *REAL*;}

Object : *aDrawGenType*;

Methods : **private**
MeanOfDrawGenQ (*aDrawGen*): *REAL*;

public
InitDrawGen (*aDrawGen*, *seedling*, *percentTrue*);
Draw (*aDrawGen*): *BOOLEAN*;
ResetDrawStats (*aDrawGen*);
ReportOnDrawGen (*aDrawGen*);

Object : *aRandIntGenType*;

Methods : **private**
MeanOfRandIntGenQ (*aRandIntGen*): *REAL*;

public
InitRandIntGen (*aRandIntGen*, *seed*, *low*, *high*);
RandInt (*aRandIntGen*): *INTEGER*;
ResetRandIntStats (*aRandIntGen*);
ReportOnRandIntGen (*aRandIntGen*);

(cont.)

Object	: *aUniformGenType*;
Methods	: **private**
	MeanOfUniformGenQ (*aUniformGen*): *REAL*;
	public
	InitUniformGen (*aUniformGen*, *seed*, *low*, *high*);
	Uniform (*aUniformGen*): *REAL*;
	ResetUniformStats (*aUniformGen*);
	ReportOnUniformGen (*aUniformGen*);

Object	: *aNegExpGenType*;
Methods	: **private**
	MeanOfNegExpGenQ (*aNegExpGen*): *REAL*;
	public
	InitNegExpGen (*aNegExpGen*, *seed*, *meanRate*);
	NegExp (*aNegExpGen*): *REAL*;
	ResetNegExpStats (*aNegExpGen*);
	ReportOnNegExpGen (*aNegExpGen*);

Object	: *aNormalGenType*;
Methods	: **private**
	MeanOfNormalGenQ (*aNormalGen*): *REAL*;
	public
	InitNormalGen (*aNormalGen*, *seed*, *mean*, *sigma*);
	Normal (*aNormalGen*): *REAL*;
	ResetNormalStats (*aNormalGen*);
	ReportOnNormalGen (*aNormalGen*);

{UTILITIES FOR MONTE CARLO METHODS}

```
{****************************************}

{******************************************}
{* TheMonitor is assumed to be a GLOBAL of  *}
{* aStateSamplingMonitorType                *}
{******************************************}

{Data Type Declarations}
{  ===========  }

{***** EXECUTION MONITOR for conducting sampling experiments        }
{this data type serves no purpose here. It is only included to allow }
{declarations of statistics collection classes to compile as one unit }

aStaticClockType = record
                     NOW : REAL
                   end;

aStateSamplingMonitorType = record
                              samplesTaken, toBeTaken : 0..MAXINT
                            end;

{***** RANDOM GENERATORS to sample from probability distributions}

aDrawGenType =     record
                     seed          : INTEGER;
                     samplesTaken  : 0..MAXINT;
                     integral      : REAL;
                     percentTRUE   : REAL
                   end;

aRandIntGenType =  record
                     seed          : INTEGER;
                     samplesTaken  : 0..MAXINT;
                     integral      : REAL;
                     low, high     : INTEGER
                   end;

aUniformGenType =  record
                     seed          : INTEGER;
                     samplesTaken  : 0..MAXINT;
                     integral      : REAL;
                     low, high     : REAL
                   end;

aNegExpGenType =   record
                     seed          : INTEGER;
                     samplesTaken  : 0..MAXINT;
                     integral      : REAL;
                     lambda        : REAL
                   end;
```

```
aNormalGenType =  record
                    seed          : INTEGER;
                    samplesTaken : 0..MAXINT;
                    integral      : REAL;
                    mean, sigma   : REAL
                  end;
```

```
{Utility Functions and Procedures for Monte Carlo Methods }
{ ================================= }

{a "static" clock must be instantiated to allow the declarations}
{of some statistics gathering devices to compile !              }

procedure InitStaticClock (var clock : aStaticClockType);
begin
  clock.NOW := 1.0;
end;

{***** Messages understood by a state SAMPLING MONITOR *********}

procedure InitStateSamplingMonitor (var sampler : aStateSamplingMonitorType);
                                        samplesToTake : INTEGER);
begin
  if (samplesToTake > 0) then
    sampler.toBeTaken := samplesToTake
  else
    WRITELN ('only positive number of samples please !');
  sampler.samplesTaken := 0;
end;

function AnotherSampleP (var sampler : aStateSamplingMonitorType) : BOOLEAN;

{NOTE:  this function has a side effect on the monitor's <samplesTaken>}
{       field !!!    Each time it is used it increments it by one.     }
begin
  sampler.samplesTaken := sampler.samplesTaken + 1;
  AnotherSampleP := sampler.samplesTaken <= sampler.toBeTaken;
end;

function ReportOnSamplesTakenQ (sampler : aStateSamplingMonitorType) :
INTEGER;
begin
  ReportOnSamplesTakenQ := sampler.samplesTaken - 1;
end;

{***** Methods for RANDOM SAMPLING ****************************}
{these methods are based on descriptions in Fishman (1978) p.345 ... where  }
{generators for other distributions can also be found                       }

{"private" method used by other functions; adopted from Sedgewick (1982).    }

function GenerateZeroToOneNumbers (var seed : INTEGER) : REAL;
const
  m  = 100000000;
```

```
  m1 =       10000;
  b  =  31415821;

  function Mult (p, q : INTEGER) : INTEGER;
  var
    p1, p0, q1, q0 : INTEGER;
  begin
    p1 := p div m1; p0 := p mod m1;
    q1 := q div m1; q0 := q mod m1;
    mult := ((( p0 * q1 + p1*q0) mod m1) * m1 + p0*q0) mod m;
  end {Mult function};

  function Random : INTEGER;
  begin
    seed := (mult (seed, b) + 1) mod m;
    random := seed;
  end {Random function};

begin
  GenerateZeroToOneNumbers := ABS (Random) / m;
end; {GenerateZeroToOneNumber function}

procedure InitDrawGen (var gen       : aDrawGenType;
                           seedling  : INTEGER;
                           percentP  : REAL);
begin
  if(percentP > 1.0) or (percentP < 0) then
    WRITELN ('Percentage parameter out of range ! (0 to 1 only)')
  else
    with gen do
      begin
        seed         := seedling;
        samplesTaken := 0;
        integral     :- 0.0;
        percentTrue  := percentP;
      end;
  end; {of procedure InitDrawGen}

procedure InitRandIntGen (var gen          : aRandIntGenType;
                              seedling     : INTEGER;
                              lowP, highP  : INTEGER);
begin
  if lowP > highP then
    WRITELN ('bounds parameters out of range ! (low must be <= high)')
  else
    with gen do
      begin
        seed         := seedling;
        samplesTaken := 0;
        integral     := 0.0;
        low          := lowP;
        high         := highP;
      end;
end; {of procedure InitRandIntGen}
```

```
procedure InitUniformGen (var gen          : aUniformGenType;
                              seedling     : INTEGER;
                              lowP, highP: REAL);
begin
  if lowP > highP then
    WRITELN ('bounds parameters out of range ! (low must be <= high)')
  else
    with gen do
      begin
        seed          := seedling;
        samplesTaken := 0;
        integral      := 0.0;
        low           := lowP;
        high          := highP;
      end;
end; {of procedure InitUniformGen}

procedure InitNegExpGen (var gen          : aNegExpGenType;
                             seedling     : INTEGER;
                             meanRateP : REAL);
begin
  with gen do
    begin
      seed          := seedling;
      samplesTaken := 0;
      integral      := 0.0;
      lambda        := meanRateP;
    end;
end; {of procedure InitNegExpGen}

procedure InitNormalGen (var gen              : aNormalGenType;
                             seedling         : INTEGER;
                             meanP, sigmaP: REAL);
begin
  if sigmaP < 0 then
    WRITELN ('the sigma value of a normal distribution must be positive !')
  else
    with gen do
      begin
        seed          := seedling;
        samplesTaken := 0;
        integral      := 0.0;
        mean          := meanP;
        sigma         := sigmaP;
      end;
end; {of procedure InitNormalGen}

function Draw (var gen : aDrawgenType) : BOOLEAN;

{NOTE:  this function produces a side effect ! It changes the <seed> and   }
{       statistics fields of the generator                                  }
var
  temp : BOOLEAN;
```

```
begin
  gen.samplesTaken := gen.samplesTaken + 1;
  if GenerateZeroToOneNumbers(gen.seed) <= gen.percentTRUE then
    temp := TRUE
  else
    temp := FALSE;
  if temp then
    gen.integral := gen.integral + 1;
  Draw := temp;
end; {of DRAW function}

function MeanOfDrawGenQ (gen : aDrawGenType) : REAL;
begin
  MeanOfDrawGenQ := gen.integral / gen.samplesTaken;
end;

procedure ResetDrawStats (var dist : aDrawGenType);
begin
  WRITELN ('Draw generator statistics reset at ', TheClock.NOW:7:2);
  dist.samplesTaken := 0;
  dist.integral      := 0.0;
end;

procedure ReportOnDrawGen (dist : aDrawGenType);
begin
  WRITE    (dist.samplesTaken:5, ' samples were taken by this binary ');
  WRITELN ('distribution. The sample mean was ', MeanOfDrawGenQ (dist):7:2);
end;

function RandInt (var gen : aRandIntGenType) : INTEGER;
{NOTE:  this function produces a side effect. It changes the <seed> and    }
{       statistics fields of the generator                                 }
var
  temp : INTEGER;
begin
  gen.samplesTaken := gen.samplesTaken + 1;
  temp := TRUNC( (gen.high - gen.low + 1) *
                  GenerateZeroToOneNumbers (gen.seed) + gen.low );
  gen.integral := gen.integral + temp;
  RandInt := temp;
end; {of RandInt function}

function MeanOfRandIntGenQ (gen : aRandIntGenType) : REAL;
begin
  MeanOfRandIntGenQ := gen.integral / gen.samplesTaken;
end;

procedure ResetRandIntStats (var dist : aRandIntGenType);
begin
  WRITELN ('RandInt generator statistics were reset at ', TheClock.NOW:7:2);
  dist.samplesTaken := 0;
  dist.integral      := 0.0;
end;
```

```
procedure ReportOnRandIntGen (dist : aRandIntGenType);
begin
  WRITE    (dist.samplesTaken, ' samples were taken by this uniform integer ');
  WRITELN ('distribution. The sample mean was ',
               MeanOfRandIntGenQ (dist):7:2);
end;

function Uniform (var gen : aUniformGenType) : REAL;
{NOTE:  this function produces a side effect. It changes the <seed> and   }
{          statistics fields of the generator                               }
var
  temp : REAL;
begin
  gen.samplesTaken := gen.samplesTaken + 1;
  temp:= ((gen.high - gen.low) * GenerateZeroToOneNumbers (gen.seed) +
           gen.low);
  gen.integral := gen.integral + temp;
  Uniform := temp;
end; {of Uniform function}

function MeanOfUniformGenQ (gen : aUniformGenType) : REAL;
begin
  MeanOfUniformGenQ := gen.integral / gen.samplesTaken;
end;

procedure ResetUniformStats (var dist : aUniformGenType);
begin
  WRITELN ('Uniform generator statistics reset at ', TheClock.NOW:7:2);
  dist.samplesTaken := 0;
  dist.integral        := 0.0;
end;

procedure ReportOnUniformGen (dist : aUniformGenType);
begin
  WRITE    (dist.samplesTaken:5, 'samples were taken by this uniform ');
  WRITELN ('distribution. The sample mean was ',
               MeanOfUniformGenQ (dist):7:2);
end;

function NegExp (var gen : aNegExpGenType) : REAL;
{NOTE:  this function produces a side effect. It changes the <seed> and   }
{          statistics fields of the generator                               }
var
  temp : REAL;
begin
  gen.samplesTaken := gen.samplesTaken + 1;
  temp                 := - LN (generateZeroToOneNumbers (gen.seed)) /
                             gen.lambda;
  gen.integral        := gen.integral + temp;
  NegExp               := temp;
end; {ofNegExp function}

function MeanOfNegExpGenQ (gen : aNegExpGenType) : REAL;
begin
  MeanOfNegExpGenQ := gen.integral / gen.samplesTaken;
end;
```

```
procedure ResetNegExpStats (var dist : aNegExpGenType);
begin
  WRITELN ('Neg. Exp. generator statistics reset at ', TheClock.NOW:7:2);
  dist.samplesTaken := 0;
  dist.integral     := 0.0;
end;

procedure ReportOnNegExpGen (dist : aNegExpGenType);
begin
  WRITE   (dist.samplesTaken:5, ' samples were taken by this neg. expon. ');
  WRITELN ('distribution. The sample mean was ',
           MeanOfNegExpGenQ (dist):7:2);
end;

function Normal (var gen : aNormalGenType) : REAL;
{NOTE:  this function produces a side effect. It changes the <seed> and   }
{       statistics fields of the generator                                }
var
  temp               : REAL;
  firstRN, secondRN : REAL; {the function uses a number of RNs to generate   }
                            {one normally distributed RV (see Fishman (1978))}
begin
  gen.samplesTaken := gen.samplesTaken + 1;
  repeat
    firstRN   := - LN (GenerateZeroToOneNumbers (gen.seed));
    secondRN := - LN (GenerateZeroToOneNumbers (gen.seed));
  until (2 * firstRN) >= (SQR (secondRN - 1));
  if (GenerateZeroToOneNumbers (gen.seed) < 0.5) then
    secondRN := - secondRN;
  temp := gen.sigma * secondRN + gen.mean;
  gen.integral := gen.integral + temp;
  Normal := temp;
end; {of Normal function}

function MeanOfNormalGenQ (gen : aNormalGenType) : REAL;
begin
  MeanOfNormalGenQ := gen.integral / gen.samplesTaken;
end;

procedure ResetNormalStats (var dist : aNormalGenType);
begin
  WRITELN ('Normal generator statistics reset at ', TheClock.NOW:7:2);
  dist.samplesTaken := 0;
  dist.integral     := 0.0;
end;

procedure ReportOnNormalGen (dist : aNormalGenType);
begin
  WRITE   (dist.samplesTaken:5, ' samples were taken from this normal ');
  WRITELN ('distribution. The sample mean was ',
           MeanOfNormalGenQ (dist):7:2);
end;
```

1.3 Statistical instrumentation

The statistics library

Object	: *aCounterType*;
Methods	: **private** *CounterValueQ* (*aCounter*): *INTEGER*; **public** *InitCounter* (*aCounter*); *UpdateCounter* (*aCounter*); *ResetCounterStats* (*aCounter*); *ReportOnCounter* (*aCounter*);

Object	: *aTallyType*;
Methods	: **private** *TalliedMeanQ* (*aTally*): *REAL*; *TalliedSigmaQ* (*aTally*): *REAL*; *TalliedMinQ* (*aTally*): *REAL*; *TalliedMaxQ* (*aTally*): *REAL*; **public** *InitTally* (*aTally*); *UpdateTally* (*aTally*, *obs*); *ResetTallyStats* (*aTally*); *ReportOnTally* (*aTally*);

Object	: *aHistogramType*;
Methods	: **public** *InitHistogram* (*aHistogram*, *low*, *high*, *noOfClasses*); *UpdateHistogram* (*aHistogram*, *obs*); *ResetHistogramStats* (*aHistogram*); *ShowHistogram* (*aHistogram*);

Object : *aMeasureType*;

Methods : **private**

MeasuredMeanQ (*aMeasure*): *REAL*;
MeasuredSigmaQ (*aMeasure*): *REAL*;
MeasuredMaxQ (*aMeasure*): *REAL*;

public

InitMeasure (*aMeasure*);
UpdateMeasure (*aMeasure*, *aStateChange*);
ResetMeasureStats (*aMeasure*);
ReportOnMeasure (*aMeasure*);

Object : *aChronographType*;

Methods : **public**

InitGraph (*aChronograph*, *maxXobs*, *maxYobs*, *maxXcoord*,
maxYcoord);
StorePoint (*aChronograph*, *xObs*, *yObs*, *printSymbol*);
ResetGraphStats (*aChronograph*);
ShowGraph (*aChronograph*);

```
{UTILITIES FOR STATISTICS COLLECTION}
{ *************************************** }

{ ************************************************* }
{* TheClock & TheMonitor are assumed to be GLOBALS     *}
{ ************************************************* }

{the following CONSTants must always be included in programs using these}
{utilities:   MaxHistoClasses  = <whatever seems appropriate>;             }
{             MaxPlotPts       = <whatever seems appropriate>;             }

{***** DATA COLLECTORS}

{gather statistics about series of events}

aCounterType = record
                  startTime : REAL;
                  aValue    : INTEGER
               end;

aTallyType = record
                startTime           : REAL;
                noOfObs             : 0..MAXINT;
                integral, sumOfSq   : REAL;
                minObs, maxObs      : REAL
             end;

aHistogramType = record
                    startTime           : REAL;
                    NoOfClasses         : 0..MaxHistoClasses;
                    classWidth          : REAL;
                    low, high           : REAL;
                    tally               : aTallyType;
                    counters            : array [0..MaxHistoClasses]
                                          of 0..MAXINT;
                    overflow, underflow : 0..MAXINT
                 end;

{statistical instrumentation of state changes}

aMeasureType = record
                  startTime           : REAL;
                  noOfObs             : 0..MAXINT;
                  state               : REAL;
                  timeOfLastChange    : REAL;
                  integral, sumOfSq   : REAL;
                  minState, maxState  : REAL
               end;
```

```
aChronographType = record
                       startTime                : REAL;
                       noOfObs                  : INTEGER;
                       maxXcoord, maxYcoord     : 0..MaxPlotPts;
                       xCoordUnit, YCoordUnit   : REAL;
                       minXvalue, maxXvalue     : REAL;
                       minYvalue, maxYvalue     : REAL;
                       grid                     : array [0..MaxPlotPts,0..
                                                  MaxPlotPts] of CHAR
                   end;
```

```
{Utility Functions and Procedures for Statistics Collection}
{ ================================= }

{***** Methods for MODEL INSTRUMENTATION  ************************}


{Messages to COUNTER objects: can be used to observe frequencies}
{               *******        of events.                       }

procedure InitCounter (var counter : aCounterType);
begin
  counter.startTime := TheClock.NOW;
  counter.aValue    := 0;
end;

procedure UpdateCounter (var counter : aCounterType);
begin
  counter.aValue := counter.aValue + 1;
end;

function CounterValueQ (counter : aCounterType) : INTEGER;
begin
  CounterValueQ := counter.aValue;
end;

procedure ResetCounterStats (var counter : aCounterType);
begin
  WRITELN ('Counter statistics were reset at ', TheClock.NOW:7:2);
  InitCounter (counter);
end;

procedure ReportOnCounter (counter : aCounterType);
begin
  WRITE    ('Value of this Counter at ', TheClock.NOW:7:2, ' : ');
  WRITELN (counter.aValue:7);
end;

{Messages to TALLY objects: can be used to observe the mean and deviation}
{               ******       of some series of data values               }
```

```
procedure InitTally (var tally : aTallyType);
const
  aLargeNumber = MAXINT;
begin
  with tally do
    begin
      startTime := TheClock.NOW;
      noOfObs  :=   0;
      integral  := 0.0;
      sumOfSq  := 0.0;
      minObs   := aLargeNumber;
      maxObs   :=   0;
    end;
end;

procedure UpdateTally (var tally : aTallyType;
                           obs  : REAL);
begin
  with tally do
    begin
      noOfObs := noOfObs + 1;
      integral  := integral + obs;
      sumOfSq := sumOfSq + SQR (obs);
      if obs < minObs then
        minObs := obs;
      if obs > maxObs then
        maxObs := obs;
    end;
end; {of procedure UpdateTally}

function TalliedMeanQ (tally : aTallyType) : REAL;
begin
  if tally.noOfObs > 0 then
    TalliedMeanQ := tally.integral / tally.noOfObs
  else
    TalliedMeanQ := 0;
end;

function TalliedSigmaQ (tally : aTallyType) : REAL;
begin
  with tally do
    TalliedSigmaQ := SQRT ((sumofSq − SQR (integral) / noOfObs) /
    (noOfObs−1));
end;

function TalliedMinQ (tally : aTallyType) : REAL;
begin
  TalliedMinQ := tally.minObs;
end;

function TalliedMaxQ (tally : aTallyType) : REAL;
begin
  TalliedMaxQ := tally.maxObs;
end;
```

```
procedure ResetTallyStats (var tally : aTallyType);
begin
  WRITELN ('Tally statistics reset at ', TheClock.NOW:7:2);
  InitTally (tally);
end;

procedure ReportOnTally (tally : aTallyType);
begin
  with tally do
    begin
      WRITE    ('Until ', TheClock.NOW:7:2, ' , ', noOfObs:5, ' observations ');
      WRITELN ('were made.');
      WRITELN;
      WRITELN ('This tally''s characteristics were as follows: ');
      WRITELN ('  ----------------------    ');
      WRITELN;
      WRITELN ('Mean      : ', TalliedMeanQ (tally):7:2);
      WRITELN ('Sigma     : ', TalliedSigmaQ(tally):7:2);
      WRITELN ('Min. Obs.: ', TalliedMinQ   (tally):7:2);
      WRITELN ('Max. Obs.: ', TalliedMaxQ   (tally):7:2);
    end;
end;

{Messages to HISTOGRAM objects: can be used to observe the frequency          }
{                *********      with which observed values fall into ranges   }
{NOTE: uses TALLY's declarations and methods !!!                              }

{The CONST MaxHistoClasses must be defined by all programs using a histogram !}

procedure InitHistogram (var histo         : aHistogramType;
                             lowP, highP : REAL;
                             classesP      : INTEGER);
var
  index : 1..MaxHistoClasses;
begin
  if (classesP <= 0) or (classesP > MaxHistoClasses) then
    begin
      WRITELN ('There must be a positive number of frequency classes!');
      WRITELN ('This number must not be greater <MaxHistoClasses>!' );
    end
  else
    if lowP >= highP then
        WRITELN('lower bounds must be < upper bounds !')
    else
        with histo do
          begin
            startTime    := TheClock.NOW;
            noOfClasses := classesP;
            low          := lowP;
            high         := highP;
            classWidth   := (high - low) / noOfClasses;
            InitTally (histo.tally);
            for index :=1 to noOfClasses do
              counters[index] :=0;
```

```
                overflow    := 0;
                underflow   := 0;
              end; {with}
end; {of procedure InitHistogram}

procedure UpdateHistogram (var histo : aHistogramType;
                               obs   : REAL);
var
  class : 1..MaxHistoClasses;
begin
  UpdateTally (histo.tally, obs);
  if obs < histo.low then
    histo.underflow := histo.underflow + 1
  else
    if obs > histo.high then
      histo.overflow := histo.overflow + 1
    else
      begin
        class := TRUNC ((obs - histo.low) / histo.classWidth) + 1;
        if class > histo.noOfClasses then
          class := histo.noOfClasses;
        histo.counters [class] := histo.counters [class] + 1;
      end; {if}
end; {of procedure UpdateHistogram}

procedure ResetHistogramStats (var histo : aHistogramType);
begin
  WRITELN ('Histogram statistics were reset at ', TheClock.NOW:7:2);
  InitHistogram (histo, histo.low, histo.high, histo.noOfClasses);
end;

procedure ShowHistogram (var histo : aHistogramType);
const
  lineLength  =  50; {Number of characters in a bar}
  barWidth    =   3; {Thickness of bars            }
  space       =  ' ';
  printSymbol =  '*';
var
  classLabel                   : REAL;     {label for class boundaries              }
  percent                      : REAL;     {obs.percentage per class                }
  stars                        : REAL;     {number of printsymbols per height unit  }
  index, barLines, barStars: INTEGER;  {loop indices                            }
  maxCount                     : INTEGER;  {maximum frequency count                 }
begin
  with histo do
    begin
      {find the max. frequency count}
      maxCount := counters [1];
      for index := 2 to noOfClasses do
        if counters [index] > maxCount then
          maxCount := counters[index];
      ReportOnTally (tally);
      WRITELN;
      WRITELN ('Overflow  was ', overflow :5);
      WRITELN ('Underflow was ', underflow:5);
```

```
        WRITELN;
        {Compute a scale factor for stars per observation}
        stars := lineLength / maxCount;
        {Set the label for the first class boundary}
        classLabel := low;
        WRITELN;
        WRITELN (classLabel:8:3, '|':2);
        {Print a bar for each class}
        for index := 1 to noOfClasses do
          begin
            percent := counters [index] / tally.noOfObs;
            {print a bar for this frequency count}
            for barLines := 1 to barWidth do
              begin
                WRITE (space:9, '|');
                for barStars := 1 to (ROUND (stars * counters [index])) do
                  WRITE(printSymbol);
                if barLines = 2 then
                  WRITE(' ', percent:5:2, '%');
                  WRITELN;
          end; {barlines for}
        {Print the next bar boundary}
        classLabel := classLabel + classWidth;
        WRITELN (classLabel:8:3, '|':2);
          end; {index for}
      end; {with clause}
end; {of procedure ShowHistogram}

{Messages to MEASURE objects:  can be used to observe the time-weighted      }
{              *********        mean and dcviation of series of states        }

procedure InitMeasure (var measure : aMeasureType);
begin
  with measure do
    begin
      startTime        := TheClock.NOW;
      noOfObs          :=   0;
      state            :=   0;
      timeOfLastChange := 0.0;
      integral         := 0.0;
      sumOfSq          := 0.0;
      minState         :=   0;
      maxState         :=   0;
    end;
end;

procedure UpdateMeasure (var measure     : aMeasureType;
                             stateChange : REAL);
begin
  with measure do
    begin
      noOfObs := noOfObs + 1;
      integral := integral + state        * (TheClock.NOW - timeOfLastChange );
      sumOfSq := sumOfSq+ SQR (state * (TheClock.NOW - timeOfLastChange));
      state    := state + stateChange;
```

```
        timeOfLastChange := TheClock.NOW;
        if state < minState then
          minState := state;
        if state > maxState then
          maxState := state;
      end;
end;

function MeasuredMeanQ (measure : aMeasureType) : REAL;
begin
  MeasuredMeanQ := measure.integral / (TheClock.NOW - measure.startTime);
end;

function MeasuredSigmaQ (measure : aMeasureType) : REAL;
begin
  with measure do
    MeasuredSigmaQ := SQRT (sumOfSq - SQR (integral / TheClock.NOW) /
                            (TheClock.NOW - measure.startTime));
end;

function MeasuredMinQ (measure : aMeasureType) : REAL;
begin
  MeasuredMinQ := measure.minState;
end;

function MeasuredMaxQ (measure : aMeasureType) : REAL;
begin
  MeasuredMaxQ := measure.maxState;
end;

procedure ResetMeasureStats (var measure : aMeasureType);
begin
  WRITELN ('Measure statistics were reset at ', TheClock.NOW:7:2);
  with measure do
    begin
      startTime := TheClock.NOW;
      noOfObs  := 0;
      integral  := 0.0;
      sumOfSq  := 0.0;
      minState  := state;
      maxState  := state;
    end;
end;

procedure ReportOnMeasure (measure : aMeasureType);
begin
  with measure do
    begin
      WRITE    (noOfObs:5, ' measurements were taken over a time period of ');
      WRITELN ((TheClock.NOW - startTime):7:2, ' time units.');
      WRITELN;
      WRITELN ('This measure''s characteristics were :');
      WRITELN ('  -------------------  ');
      WRITELN;
      WRITELN ('Measured mean = ', MeasuredMeanQ (measure):7:2);
```

```
      WRITELN ('Measured sigma = ', MeasuredSigmaQ(measure):7:2);
      WRITELN ('Min. state     = ', MeasuredMinQ  (measure):7:2);
      WRITELN ('Max. state     = ', MeasuredMaxQ  (measure):7:2);
    end;
end;

{Messages to CHRONOGRAPH objects: can be used to plot time series of states   }
{              ***************                                                }

procedure InitGraph (var graph                 : aChronographType;
                         maxXobs, maxYobs       : REAL;
                         maxXcoordP, maxYcoordP : INTEGER);
const
  space = ' ';
var
  rows,cols : 0..maxPlotPts;
begin
  with graph do
    begin
      noOfObs    :=   0;
      minXvalue  := 0.0;
      minYvalue  := 0.0;
      maxXvalue  := maxXobs;
      maxYvalue  := maxYobs;
      maxXcoord  := maxXcoordP;
      maxYcoord  := maxYcoordP;
      xCoordUnit := maxXvalue / maxXcoord;
      yCoordUnit := maxYvalue / maxYcoord;
      for rows := 0 to maxYcoord do
        for cols := 0 to maxXcoord do
          grid[rows, cols] := space;
    end;
end;

procedure StorePoint (var graph      : aChronographType;
                          xObs, yObs : REAL;
                          symbol     : CHAR);
const
  overprintSymbol = '#';
  space           = ' ';
var
  x, y : 0..MaxPlotPts;
begin
  with graph do
    begin
      if (xObs < 0) or (xObs > maxXvalue) then
        WRITELN ('X = ', xObs:8:2, ' out of range for this graph!')
      else
        if (yObs < 0) or (yObs > maxYvalue) then
          WRITELN ('Y = ', yObs:8:2, ' out of range for this graph!')
        else
          begin
            noOfObs := noOfObs + 1;
            x := ROUND (xObs / xCoordUnit);
            y := ROUND (yObs / yCoordUnit);
```

```
            if grid [x,y] <> space then
               grid [x,y] := overprintSymbol
            else
               grid [x,y] := symbol;
         end;{if}
   end;{with}
end; {of procedure StorePoint}

procedure ResetGraphStats (var graph : aChronographType);
begin
  with graph do
    InitGraph (graph, maxXvalue, maxYvalue, maxXcoord, maxYcoord);
end;

procedure ShowGraph (graph : aChronographType);
const
  space = ' ';
var
  rows ,cols      : 0..MaxPlotPts;
  xLable ,yLable : REAL;
begin
  with graph do
    begin
      WRITELN;
      WRITELN (noOfObs:5,' plot points were collected.');
      WRITELN;
      {Print the graph hanging down (with (0,0) at the top left corner) !}
      WRITE (space:9, '0');
      for cols := 0 to maxXcoord do
        WRITE (space);
      WRITELN (maxXvalue:8:2);
      WRITE (space:10);
      for cols := 0 to maxXcoord do
        WRITE ('-');   WRITELN;
      yLable := 0;;
      for rows := 0 to maxYcoord do
        begin
          if (rows mod 5) = 0 then
            WRITE (yLable:8:2, ' |')
          else
            WRITE(space:8, ' |');
          for cols := 0 to maxXcoord do
            WRITE (grid [cols, rows] :1); {X-coord. is horizontal;            }
          WRITELN;                      {Y-coord. is vertical. Hanging down !}
          yLable := yLable + yCoordUnit;
        end;{"rows-for"}
    WRITELN;
    end; {with clause}
end; {of procedure ShowGraph}
```

1.4 Quasi-continuous simulation

The continuous simulation library

Object : *aSynchClockType*;

Methods : **public**
InitSynchClock (*aSynchClockType*, *resolution*);
AnotherTickP (*aSynchClockType*): *BOOLEAN*;
HowLongSimulatedQ (*aSynchClockType*): *REAL*;

Object : *aTimeSlicingMonitorType*;

Methods : **public**
InitTimeSlicingMonitor (*aTimeSlicingMonitorType*, *simLength*);

Object : *aStepVarType*;

Methods : **public**
InitStepVar (*aStepVarType*, *anIntegerQtty*);
StepState (*aStepVarType*, *anIntegerQtty*);
ResetStepVar (*aStepVarType*);

Object : *aContVarType*;

Methods : **public**
InitContVar (*aContVarType*, *aRealQtty*);
UpdateState (*aContVarType*, *aRealQtty*);
ResetContVar (*aContVarType*);

```
{UTILITIES FOR QUASI-CONTINUOUS SIMULATION}
{ ********************************************** }

{*********************************************************}
{* TheClock & TheMonitor are assumed to be GLOBALS of    *}
{* aSynchClockType and aTimeSlicingMonitorType respectively! *}
{*********************************************************}

{Data Type Declarations}
{ =========== }

{***** EXECUTION MONITOR for time-slicing experiments}

aSynchClockType = record
                    NOW             : REAL;
                    resolution      : REAL;
                    timeOfNextEvent : REAL
                    {dummy dcl. to permit cs-procs to compile together}
                    {with des-procs (i.e. for "combined models")      }
                  end;

aTimeSlicingMonitorType = record
                            termCond : record
                                         simLength : REAL
                                       end
                          end;

{***** "CONTINUOUS VARIABLES" for state representation}

aStepVarType = record
                 level : INTEGER
               end;

aContVarType = record
                 level : REAL
               end;
```

```
{Utility Functions and Procedures for Quasi-Continuous Simulations}
{ ==================================== }

{***** Methods associated with a SYNCHRONOUS CLOCK *****}

procedure InitSynchClock (var clock      : aSynchClockType;
                              resolutionP : REAL);
begin
  clock.NOW := 0.0;
  if resolutionP <= 0 then
    begin
      WRITELN ('Time grain can''t be negative or zero! Set to 1.');
      clock.resolution := 1;
    end
  else
    clock.resolution := resolutionP;
end; {InitSynchclock procedure}
```

```
function AnotherTickP (var clock : aSynchClockType) : BOOLEAN;

{NOTE:  this function causes a side effect! It changes the <clock> value of      }
{       the monitor by incrementing it by <resolution> each time it is used.      }
begin
  if clock.NOW >= TheMonitor.termCond.simLength then
    AnotherTickP := FALSE
  else
    AnotherTickP := TRUE;
  clock.NOW := clock.NOW + clock.resolution;
end;

function HowLongSimulatedQ : REAL;
begin
  HowLongSimulatedQ := TheClock.NOW;
end;

{***** Methods associated with the TIME-SLICING MONITOR *****}

procedure InitTimeSlicingMonitor (var monitor     : aTimeSlicingMonitorType;
                                      simLengthP : REAL);
begin
  if simLengthP < 0 then
    WRITELN ('Simulation must last a pos. number of time units !')
  else
    monitor.termCond.simLength := simLengthP;
end;

{***** Messages to quasi-continuous VARIABLES *****}

procedure InitContVar (var variable  : aContVarType;
                           initLevel : REAL);
begin
  variable.level := initLevel;
end;

{the UpdateState method can only be used for SIMPLE INTEGRATION}
{of a state by some constant amount per tick of time            }

procedure UpdateState (var state : REAL; qtty : REAL);
{this body should be replaced by an appropriate integration procedure}
begin
  state := state + (qtty * TheClock.resolution);
end;

procedure ResetContVar (var variable : aContVarType);
begin
  WRITELN ('Cont. variable reset at ', TheClock.NOW:7:2);
  variable.level := 0.0;
end;

procedure InitStepVar (var variable : aStepVarType; initLevel : INTEGER);
begin
  variable.level := initLevel;
end;
```

```
procedure StepState (var state : INTEGER; qtty: INTEGER);
begin
  state := state + ROUND (qtty * TheClock.resolution);
end;

procedure ResetStepVar (var variable : aStepVarType);
begin
  WRITELN ('Step Variable reset at ', TheClock.NOW:7:2);
  variable.level := 0;
end;
```

1.5 Discrete-event simulation

The event-based library

Object : *anEventType*; {defined by the model}
Object : *anEventPtrType*; {to an EventNotice}
Object : *aTAPtrType*; {to a (passive) TArecord}
Object : *anEventNoticeType*;

Methods : **public**
KindOfEventQ (*anEventPtrType*): *anEventType*;
TAreferenceQ (*anEventPtrType*): *aTAPtrType*;
FlushOutEvent (*anEventPtrType*);

Object : *anAsynchClockType*;

Methods : **public**
InitAsynchClock (*anAsynchClockType*);
SetClock (*anAsynchClockType*); {*in ACTIVITY_PROCS.LIB*}

Object : *anEventListType*;

Methods : **public**
ScheduleEvent (*anEventSchedulingMonitorType*, *aRealDelay*, *anEventType*, *aTAPtrType*);
GetNextEventNotice (*anEventSchedulingMonitorType*);
MoreEventsP (*anEventSchedulingMonitorType*): *BOOLEAN*;
ShowAgenda (*anEventSchedulingMonitorType*);

Object : *anEventSchedulingMonitorType*;

Methods : **private**
HowManySamplesQ (*anEventSchedulingMonitorType*): *INTEGER*;

public
InitEventSchedulingMonitor (*anEventSchedulingMonitorType*, *simLength*, *sampleSize*);
TraceMonitor (*anEventSchedulingMonitor*);
UntraceMonitor (*anEventSchedulingMonitor*);
MoreTimeToSimulateP (*anEventSchedulingMonitorType*): *BOOLEAN*;
MoreSamplesToTakeP (*anEventSchedulingMonitorType*): *BOOLEAN*;
RecordAnotherSample;
ResetMonitorStats (*anEventSchedulingMonitorType*, *simLength*, *sampleSize*);
ReportOnMonitor (*anEventSchedulingMonitorType*);

The process-based **library**

Object	: *aTAprocessClass* (*self*: *aTAPtrType*); {*FORWARD declaration*}
Object	: *aTAPtrType*; {*to a TAprocessType*}
Object	: *aTAprocessType*;
Methods :	**public** *CreateAndStartTAprocess* (*anInitEventPhase*, *aDelay*); *FlowToNextEvent* (*aTAPtrType*, *aDelay*); *ResumeNextTAprocess*;

The activity-based **library**

Object	: *anActivityType*; {*list of activity classes in the model*}
Object	: *anActivityPhaseType*; {*start and finish*}
Object	: *anActivityPtrType*; {*to an activitynotice*}
Object	: *anActivityNoticeType*;
Methods :	**public** *InitAndStartActivity* (*anActivityType*); *AdvanceActivityPhase* (*anActivityPtrType*); *SetFinishTime* (*anActivityPtrType*, *aDelay*, *aTAptrType*); *PhaseOfActivityQ* (*anActivityPtrType*): *anActivityPhaseType*; *ClassOfActivityQ* (*anActivityPtrType*): *anActivityType*; *FinishTimeOfActivityQ* (*anActivityPtrType*): *REAL*;

Object	: *anActivityListType*;
Methods :	**public** *EndOfAgendaP*: *BOOLEAN*; *GetNextActivityQ* (*anActivityScanMonitorType*): *anActivityPtrType*;

Object	: *anActivityScanMonitorType*;
Methods :	**public** *InitActivityScanMonitor* (*anActivityScanMonitorType*, *simLength*, *sampleSize*); *AdvanceActivityScan*; *ResetActivityScan*; *ActivityStarted*; *StartNewScan*; *ScanWithoutAnyStartsP*: *BOOLEAN*; *StillMoreSamplesP* (*anActivityScanMonitorType*): *BOOLEAN*; *StillMoreTimeP* (*anActivityScanMonitorType*): *BOOLEAN*;

```
{UTILITIES FOR DISCRETE-EVENT SIMULATION}
{ ******************************************* }

{*****************************************************}
{* TheClock & TheMonitor are assumed to be GLOBALS of        *}
{* anAsynchClockType and anEventSchedulingMonitorType,       *}
{* aProcessSynchMonitorType or anActivityScanMonitorType.    *}
{*****************************************************}

{Data Type Declarations                                       }
{  ===========                                                }
{each model must define some data type                       }
{anEventType = <whatever event names are used by the model>   }

{***** EVENTs & EVENT NOTICEs for scheduling clusters of actions}

anEventPtrType = ↑ anEventNoticeType;
aTAPtrType     = ↑ aTArecordType;   {defined in QN_DCLS.LIB}

anEventNoticeType = record
{NOTE:anEventType must be provided by all programs using this}
                       eventTime: REAL;
                       eventClass: anEventType;
                       nextEvent: anEventPtrType;
                       thisTA    : aTAPtrType
                     end;

{***** CLOCK & EXECUTION MONITOR for time and event management}

anAsynchClockType = record
                        NOW            : REAL;
                        timeOfNextEvent: REAL;
                      end;

anEventListType = record
                    firstEvent: anEventPtrType;
                    length    : 0..MAXINT
                  end;

anEventSchedulingMonitorType = record
                                 agenda       : anEventListType;
                                 currentEvent : anEventPtrType;
                                 termCond     : record
                                                  simlength  : REAL;
                                                  sampleSize : 0..MAXINT;
                                                end;
                                 samplesTake  : 0..MAXINT;
                                 traceFlag    : BOOLEAN
                               end;
```

```
{ ========================================= }

{(A) Additional Utilities for Process-Oriented Modelling        }
{       +++++++++++++++++++++++++                               }

{(1)  CONSTants FirstPhase = <the TAs' initial state> ; and     }
{                LastPhase  = <the TAs' final state>  ;         }
{     must be defined by all programs using these utilities     }

{(2)  anEventType = <sequence of all of TAprocess'states>;      }
{     must be declared by all programs using a TAprocessType    }

{(3)  the TAprocesses' life cycles must be represented by a     }
{     procedure called TAprocessClass,in which a CASE-          }
{     simulated entry point (all values of "anEventType")       }
{     sequences its state transitions. Blocks of code           }
{     associated with case lables implement all actions         }
{     performed by these events. Procedure "TAprocessClass"     }
{     accepts a TAPtrType reference as a parameter. The data    }
{     structure associated with this pointer is a record of     }
{     the process' attributes                                   }

{***** Additional Data Types *****}

{!!!!! replace the old TAPtrType definitions !!!!!}
{ aTAPtrType = ↑ aTAprocessType;                  }

aTAprocessType = record
                   phase      : anEventType;
                   birthTime  : REAL;
                   nextInLine : aTAPtrTyp ;
                 end;
```

```
{(B) Additional Utilities for Activity-Oriented Modelling                          }
{      +++++++++++++++++++++++++                                                   }

{(1)  anActivityType  = <list of activityclasses>; and                             }
{     anActivityPhase = (start,finish); must be declared in any program using these }
{                                utilities.                                        }

{(2)  Each activity-oriented program using these utilities must provide a procedure }
{     for each class of activity; with a CASE statement defining the relevant actions }
{     associated with its start and finish entry points. The start actions must contain at }
{     least one call to the SetFinishTime procedure !                              }

{(3)  Procedures StartActivities and FinishActivities must also be provided, to define }
{     the conditions under which activities of each class may be triggered and when }
{     they may finish. The predefined function InspectNext can be used to step     }
{     through a list of activities kept by the monitor. This is a generator function, }
{     producing a pointer to an activity instance as a side effect. It also returns }
{     TRUE if the list is not yet empty.                                           }
{     If any activities are started by StartActivities, this must be signalled to the }
{     monitor using procedure ActivityStarted. Both StartActivities and            }
{     FinishActivities must advance the phase of the activities handled by them    }
{     (start to finish; finish to start)                                           }
```

```
{***** Additional Data Types *****}

anActivityPhaseType = (start, finish);

{***** ACTIVITIES & ACTIVITY NOTICES for clustering state-changing}
{        actions                                                  }

{!!!!! replace the old anEventPtrType and anEventNoticeType definitions !!!!!}

anActivityPtrType = ↑ anActivityNoticeType;

anActivityNoticeType = record
                          nextPhase    : anActivityPhaseType;
          {Unfortunately, start conditions can"t be put in here.They must}
          {be implemented "in line" in procedure StartActivities.        }
                          finishTime   : REAL;
                          activityClass : anActivityType;
                          nextActivity : anActivityPtrType
                          thisTA       : aTAPtrType;
                       end;

{***** EXECUTION MONITOR for time and activity management          }

{!!!!! replace the old anEventListType and anEventSchedulingMonitorType  }
{   definitions !!!!!                                                    }

anActivityListType = record
                        firstActivity : anActivityPtrType;
                        length        : 0..MAXINT;
                     end;

anActivityScanMonitorType = record
                               noActivityStarted : BOOLEAN;
                               agenda            : anActivityListType;
                               currentActivity   : anActivityPtrType;
                               termCond : record
                                             simLength : REAL;
                                             sampleSize : 0..MAXINT
                                          end;
                               samplesTaken      : 0..MAXINT
                            end;
```

```
{Utility Functions and Procedures for Event-Oriented Simulations     }
{=================================                                   }

{*****  Methods to manipulate events (place on and remove EVENT NOTICES }
{       from an event list)                                            }

procedure ScheduleEvent (var monitor      : anEventSchedulingMonitorType;
                             delay        : REAL;
                             kindOfEvent  : anEventType;
                             thisTA       : aTAPtrType);

{NOTE:this procedure assumes that data type "anEventType" has been}
{defined elsewhere !!!                                            }
```

```
{It creates an event notice of the appropriate type and files it into the     }
{global agenda at the appropriate position. It assumes that the agenda        }
{is sorted on ascending time values                                           }

var
  this, prev, next : anEventPtrType;

    procedure insert (x, before, after : anEventPtrType);
    begin
      before↑.nextEvent := x;
      x↑.nextEvent      := after;
    end; {of local insert procedure}

begin {body of procedure ScheduleEvent}
  if delay < 0.0 then
    begin
    WRITELN ('Time can''t run backwards ! Delay of 0 units assumed.');
    delay := 0.0;
    end;
    {create and initialize a new event notice}
  NEW(this);
  if monitor.traceFlag then
    begin
    WRITE   ('Event notice for event # ', ORD (kindOfEvent):2);
    WRITELN (' created at ', TheClock.NOW:7:2);
    end;
  this↑.eventClass:= kindOfEvent;
  this↑.eventTime:= TheClock.NOW + delay;
  this↑.thisTA    := thisTA;
  this↑.nextEvent:= nil;
  monitor.agenda.length := monitor.agenda.length + 1;
  {check for empty event list}
  if monitor.agenda.firstEvent = nil then
    begin
      monitor.agenda.firstEvent := this;
      if monitor.traceFlag then
        WRITELN ('Agenda empty ! New event filed at head of list.');
    end
  else
    begin
    {event list is not empty}
    {check whether notice goes at head of event list}
    if this↑.eventTime < monitor.agenda.firstEvent.eventTime then
      begin
      {this becomes the most imminent event. Insert at head.}
      if monitor.traceFlag then
        begin
        WRITE   ('Event filed at head of list, before event scheduled at ');
        WRITELN (monitor.agenda.firstEvent↑.eventTime:7:2);
        end;
      this↑.nextEvent := monitor.agenda.firstEvent;
      monitor.agenda.firstEvent   := this;
      TheClock.timeOfNextEvent := this↑.eventTime;
      end
```

```
    else
       begin
       {search for the appropriate position and insert the notice there}
       next := monitor.agenda.firstEvent;
       prev := nil;
       {search for position}
       while (this ↑ .eventTime >= next ↑ .eventTime) and
             (next ↑ .nextEvent <> nil) DO
          begin
          prev := next;
          next := next ↑ .nextEvent;
          end; {while}
       {insert notice}
       if (next ↑ .nextEvent = nil) and (this ↑ .eventTime >= next ↑ .eventTime) then
          begin
          insert (this, next, nil);
          if monitor.traceFlag then
             begin
             WRITE    ('File new event at tail of agenda, immediately following
                       event');
             WRITELN (' scheduled at ', next ↑ .eventTime:7:2);
             end;
          end
    else
          begin
          insert (this, prev, next);
          if monitor.traceFlag then
             begin
             WRITE    ('File new event following event scheduled at ');
             WRITE    (next ↑ .eventTime:7:2, ' and before event scheduled at ');
             WRITELN (prev ↑ .eventTime:7:2);
             end;
          end;
       end; {of search for insertion}
    monitor.agenda.length := monitor.agenda.length + 1;
    end; {of "event list is not empty" case}
end; {of procedure ScheduleEvent}

procedure GetNextEventNotice (var monitor : anEventSchedulingMonitorType);

{NOTE:this procedure resets the monitor's <currentEvent> and the Clock's }
{ <timeOfNextEvent> fields. It also changes the agenda by removing the     }
{ first event notice                                                       }
const
  aLongNumber = MAXINT;
var
  current : anEventPtrType;
begin
  current := nil;
  if monitor.agenda.firstEvent = nil then
    WRITELN ('ERROR! Can''t remove any item from empty event list.')
  else
    begin
      current := monitor.agenda.firstEvent;
```

```
        monitor.agenda.firstEvent := monitor.agenda.firstEvent ↑ .nextEvent;
        monitor.agenda.length     := monitor.agenda.length - 1;
        TheClock.NOW := current ↑ .eventTime;
        if monitor.traceFlag then
          begin
          WRITE    ('Event of type # ', ORD (current ↑ .eventClass):2);
          WRITELN (' triggered at ', TheClock.NOW:7:2);
          end;
        if monitor.agenda.firstEvent = nil then
          TheClock.timeOfNextEvent := aLongNumber
        else
          TheClock.timeOfNextEvent := monitor.agenda.firstEvent ↑ .eventTime;
        monitor.currentEvent := current;
      end; {else branch}
end; {of procedure GetNextEventNotice}


function KindOfEventQ (eventNotice : anEventPtrType) : anEventType;
begin
  KindOfEventQ := eventNotice ↑ .eventClass;
end;

function TAreferenceQ (eventNotice : anEventPtrType) : aTAPtrtype;
begin
  TAreferenceQ := eventNotice ↑ .thisTA;
end;

procedure FlushOutEvent (event : anEventPtrType);
begin
  DISPOSE(event);
end;

{***** Messages to ASYNCHCLOCKs & EVENTSCHEDULINGMONITORs}

procedure InitAsynchClock (var clock : anAsynchClockType);
const
  aLongNumber = MAXINT;
begin
  clock.NOW              := 0.0;
  clock.timeOfNextEvent := aLongNumber;
end; {of InitAsynchClock procedure}

procedure InitEventSchedulingMonitor
(var monitor     : anEventSchedulingMonitorType;
     simLengthP : REAL;
     samplesP    : INTEGER);
begin
  if(simLengthP < 0) or (samplesP < 0)then
    WRITELN ('Parameter Error: both length and sample size must be positive !')
  else
    with monitor do
      begin
        agenda.firstEvent     := nil;
        agenda.length         := 0;
        currentEvent          := nil;
```

```
        termCond.simLength := simLengthP;
        termCond.sampleSize:= samplesP;
        samplesTaken       := 0;
        traceFlag          := FALSE;
      end;{else branch}
end; {of InitEventSchedulingMonitor procedure}

procedure TraceMonitor (var monitor : anEventSchedulingMonitorType);
begin
  WRITE    ('***** Trace of Event scheduling & execution switched ON   ');
  WRITELN ('at ', TheClock.NOW:7:2);
  monitor.traceFlag := TRUE;
end;

procedure UntraceMonitor (var monitor : anEventSchedulingMonitorType);
begin
  WRITE    ('***** Trace of Event scheduling & execution switched OFF ');
  WRITELN ('at ', TheClock.NOW:7:2);
  monitor.traceFlag := FALSE;
end;

function MoreTimeToSimulateP (monitor : anEventSchedulingMonitorType) :
                                  BOOLEAN;
var
  nextEvent : anEventPtrType;
begin
  nextEvent := monitor.agenda.firstEvent;
  with monitor do
  if nextEvent = nil then
    begin
    WRITELN ('At ',TheClock.NOW:8:2,' the event list is empty!');
    MoreTimeToSimulateP := FALSE;
    end
  else
    MoreTimeToSimulateP := nextEvent ↑ .eventTime <= termCond.simLength;
end; {of MoreTimeToSimulate function}

function MoreEventsP (monitor : anEventSchedulingMonitorType) : BOOLEAN;
begin
  MoreEventsP := monitor.agenda.length > 0;
end;

function MoreSamplesToTakeP (monitor : anEventSchedulingMonitorType) :
                                  BOOLEAN;
begin
  with monitor do
  if agenda.length <= 0 then
    WRITELN('At ',TheClock.NOW:8:2,' the event list is empty!')
  else
    MoreSamplesToTakeP := samplesTaken < termCond.sampleSize;
end; {of MoreSamplesToTake function}

procedure RecordAnotherSample;
begin
  TheMonitor.samplesTaken := TheMonitor.samplesTaken + 1;
end;
```

```
function HowManySamplesQ (monitor : anEventSchedulingMonitorType) :INTEGER;
begin
  HowManySamplesQ := monitor.samplesTaken;
end;

procedure ResetMonitorStats (var monitor     : anEventSchedulingMonitorType;
                                 simLengthP : REAL;
                                 sampleSizeP: INTEGER );
begin
  if (simLengthP < 0) or (sampleSizeP < 0) then
    WRITELN ('*** ERROR in parameter to "ResetMonitor" procedure !')
  else
    begin
      WRITELN ('Monitor statistics reset at ', TheClock.NOW:7:2);
      monitor.termCond.simLength  := simLengthP;
      monitor.tcrmCond.sampleSize := sampleSizeP;
      monitor.samplesTaken        := 0;
    end;
end;

procedure ReportOnMonitor (monitor : anEventSchedulingMonitorType);
begin
  WRITE   (HowManySamplesQ (monitor):7, ' samples were taken since this ');
  WRITELN ('monitor was last started .');
  WRITELN ('The current clock time is: ', TheClock.NOW:7:2);
end;

procedure ShowAgenda (monitor : anEventSchedulingMonitorType);
var
  event : anEventPtrType;
begin
  WRITELN ('Agenda of pending events at ', TheClock.NOW:8:2);
  WRITELN ('====================  ');
  WRITELN;
  WRITELN ('Events are planned at:');
  WRITELN;
  event := monitor.agenda.firstEvent;
  if event = nil then
    WRITELN ('The agenda is empty!')
  else
    begin
    while event <> nil do
      begin
        WRITE   ('event of class # ', ORD (event↑.eventClass):2);
        WRITELN (' is scheduled at ', event↑.eventTime:8:2);
        event := event↑.nextEvent;
      end; {while}
    WRITELN;
  end; {if}
end; {of ShowAgenda procedure}
```

```
{***** Additional Procedures Understood by TAPROCESS Objects *****}

{necessary to allow model declarations in a "natural" sequence         }
```

```
procedure TAprocessClass (self : aTAPtrType); FORWARD;

procedure CreateAndStartTAprocess (initPhase : anEventType;
                                   delay     : REAL);
var
  thisOne : aTAPtrType;
begin
  NEW(thisOne);
  thisOne ↑ .phase      := initPhase;
  thisOne ↑ .birthTime  := TheClock.NOW;
  thisOne ↑ .NextinLine := nil;
  ScheduleEvent (TheMonitor, delay, thisOne ↑ .phase, thisOne);
end; {of CreateAndStartTAprocess procedure}

procedure FlowToNextEvent (thisTA : aTAPtrtype;
                           delay  : REAL);
begin
  if thisTA ↑ .phase = LastPhase then
    WRITELN('Process is already in its last phase!')
  else
    begin
    thisTA ↑ .phase := SUCC (thisTa ↑ .phase);
    ScheduleEvent (TheMonitor, delay, thisTA ↑ .phase, thisTA);
    end; {if}
end; {of FlowToNextEvent procedure}

procedure ResumeNextTAprocess;
var
  ta : aTAPtrType;
begin
  GetNextEventNotice (TheMonitor);
  ta := TheMonitor.currentEvent ↑ .thisTA;
  TAprocessClass (ta);
end; {of ResumeNextTAprocess procedure}
```

{***** **Additional Functions and Procedures for Activity-Based Models** *****}

{***** Messages to ACTIVITIES *****}

```
procedure InitAndStartActivity (actClass : anActivityType);
const
  aLongNumber = MAXINT;
var
  thisActivity : anActivityPtrType;
begin
  NEW(thisActivity);
  with thisActivity ↑ do
    begin
      nextPhase     := start;
      finishtime    := aLongNumber;
      activityClass := actClass;
      thisTA        := nil;
    end; {with}
  {insert in monitor's list of activities!}
```

```
  {check for empty list}
  if TheMonitor.agenda.firstActivity = nil then
    begin
      TheMonitor.agenda.firstActivity := thisActivity;
      TheMonitor.currentActivity      := thisActivity;
      thisActivity ↑ .nextactivity    := nil;
    end
  else
    begin
      {list of activities is not empty! Insert new activity at tail of list.}
      TheMonitor.agenda.firstActivity ↑ .nextActivity := thisActivity;
      thisActivity ↑ .nextActivity     := nil;
    end; {inner if}
  TheMonitor.agenda.length := TheMonitor.agenda.length + 1;
end; {of initAndStartActivity procedure}

{***** Messages to ACTIVITYSCANMONITORS *****}

procedure InitActivityScanMonitor(var monitor     : anActivityScanMonitorType;
                                      simLengthP : REAL;
                                      samples    : INTEGER );
begin
  if (simLengthP < 0) or (samples < 0) then
    WRITELN ('Sim. length and samplesize must have positive values!')
  else
    with monitor do
      begin
        noActivityStarted     := TRUE;
        agenda.firstActivity  := nil;
        agenda.length         := 0;
        currentActivity       := nil;
        termCond.simlength    := simLengthP;
        termCond.sampleSize   := samples;
        samplesTaken          := 0;
      end; {with}
end; {of InitActivityScan procedure}

procedure AdvanceActivityScan;
begin
  if TheMonitor.currentActivity ↑ .nextActivity = nil then
    TheMonitor.currentActivity := nil
  else
    TheMonitor.currentActivity := TheMonitor.currentActivity ↑ .nextActivity;
end;

procedure ResetActivityScan;
begin
  TheMonitor.currentActivity := TheMonitor.agenda.firstActivity;
end;

function EndOfAgendaP : BOOLEAN;
begin
  EndOfAgendaP := TheMonitor.currentActivity = nil;
end;
```

```
{!!!!! Activities are assumed to become a permanent part of the model      }
{      once they are created. A procedure which parallels FlushOutEvents }
{      is therefore not provided, although it can easily be implemented.   }

procedure SetFinishTime (thisActivity: anActivityPtrType;
                         delay       : REAL;
                         thisTAP     : aTAPtrType );
begin
  if delay < 0 then
    WRITELN ('Time can''t run backwards!')
  else
    begin
      thisActivity↑.finishTime := TheClock.NOW + delay;
      thisActivity↑.thisTA     := thisTAP;
    end; {if}
end; {of SetFinishTime procedure}

{Procedures MoreSamplesToTakeP, RecordAnotherSample and}
{HowManySamplesQ do not change !!!                      }

procedure ActivityStarted;
begin
  TheMonitor.noActivityStarted := FALSE;
end;

procedure StartNewScan;
begin
  TheMonitor.noActivityStarted := TRUE;
  ResetActivityScan;
end;

function ScanWithoutAnyStartsP : BOOLEAN;
begin
  ScanWithoutAnyStartsP := TheMonitor.noActivityStarted;
end;

procedure SetClock (var clock : anAsynchClockType);
const
  aLongNumber = MAXINT;
var
  nextActivity    : anActivityPtrType;
  minTimeValue : REAL;
begin
  minTimeValue := aLongNumber;
  ResetActivityScan;
  while not EndOfAgendaP do
    begin
      nextActivity := TheMonitor.currentActivity;
      if (nextActivity↑.nextPhase = finish) and
        (nextActivity↑.finishTime < minTimeValue) then
          minTimeValue := nextActivity↑.finishTime;
      AdvanceActivityScan;
    end; {while}
  clock.NOW := minTimeValue;
end; {of SetClock procedure}
```

```
function StillMoreSamplesP (monitor : anActivityScanMonitorType) : BOOLEAN;
begin
  StillMoreSamplesP := monitor.samplesTaken < monitor.termCond.sampleSize;
end;

function GetNextActivityQ (monitor : anActivityScanMonitorType)
                                    : anActivityPtrType;
begin
  GetNextActivityQ := monitor.currentActivity;
end;

function PhaseOfActivityQ (act : anActivityPtrType) : anActivityPhaseType;
begin
  PhaseOfActivityQ := act ↑ .nextPhase;
end;

function ClassOfActivityQ (act : anActivityPtrType) : anActivityType;
begin
  ClassOfActivityQ := act ↑ .activityClass;
end;

function FinishTimeOfActivityQ (act : an ActivityPtrType) : REAL;
begin
  FinishTimeOfActivityQ := act ↑ .finishTime;
end;

procedure AdvanceActivityPhase (act : anActivityPtrType);
begin
  if act ↑ .nextPhase = start then
    act ↑ .nextPhase := finish
  else
    act ↑ .nextPhase := start;
end;
```

1.6 Queueing network scenarios

The queueing network library

Object	: *aSourceType*;
Methods	: **private** *HowManyCreationsQ* (*aSourceType*); **public** *InitSource* (*aSourceType*); *ResetSourceStats* (*aSourceType*); *CreateNewTA* (*aSourceType*); *TraceTAgeneration* (*aSourceType*); *UntraceTAgeneration* (*aSourceType*); *ReportOnSource* (*aSourceType*);

Object	: *aSinkType*;
Methods	: **private** *How ManyDeparturesQ* (*aSinkType*): *INTEGER*; **public** *InitSink* (*aSinkType*); *ResetSinkStats* (*aSinkType*); *RemoveTA* (*aSinkType*); *TraceTAdepartures* (*aSinkType*); *UntraceTAdepartures* (*aSinkType*); *ReportOnSink* (*aSinkType*);

Object	: *aTArecordType*;
Methods	: **private** *FlowTimeQ* (*aTAPtrType*): *REAL*;

(cont.)

Object : *aServerType*;

Accessible Instance Variables : *capacity*, *state* {*idle*/*busy*};

Methods : **private**

ServerUtilizationQ (*aServerType*): *BOOLEAN*;

public

InitServer (*aServerType*, *capacity*);
ResetServerStats (*aServerType*);
SeizeServer (*aServerType*, *howMuch*);
ReleaseServer (*aServerType*, *howMuch*);
EnoughServerUnitsAvailableP (*aServerType*, *units*): *BOOLEAN*;
TraceService (*aServerType*);
UntraceService (*aServerType*);
ReportOnServer (*aServerType*);

Object : *aQueueType*;

Accessible Instance Variables : *length*;

Methods : **private**

MeanQlengthQ (*aQueueType*): *REAL*;
MaxQlengthQ (*aQueueType*): *INTEGER*;
MeanQdelayQ (*aQueueType*): *REAL*;

public

InitQueue (*aQueueType*);
ResetQueueStats (*aQueueType*);
FileIntoFIFOq (*aQueueType*, *aTAPtrType*);
TakeFirstFromQ (*aQueueType*): *aTAPtrType*;
TraceQops (*aQueueType*);
UntraceQops (*aQueueType*);
ReportOnQueue (*aQueueType*);

```
{UTILITIES FOR SIMULATING QUEUEING NETWORK SCENARIOS }
{**************************************************************}

{**************************************************************}
{* These utilities should be used together with the          *}
{* DES or COMBINED simulation utility packages               *}
{**************************************************************}

{Data Type Declarations}
{ =========== }

{***** SOURCEs & SINKs used to record generation}
{       and destructions of workload items       }

aSourceType = record
                startTime     : REAL;
                creationCount : 0..MAXINT;
                traceFlag     : BOOLEAN
              end;

aSinkType = record
              startTime        : REAL;
              destructionCount : 0..MAXINT;
              traceFlag        : BOOLEAN
            end;

{***** TRANSACTIONs used to represent workload items}

{ "aTAPtrType" is declared in DES-DCLS.LIB !}

aTArecordType = record
                  birthTime  : REAL;
                  nextInLine : aTAPtrType;
                 {these TAs can only be in at most one queue at a time !}
                 end;

{***** Resources used to represent SERVER units}

aServerType = record
                startTime        : REAL;
                capacity         : 0..MAXINT;
                inUse            : 0..MAXINT;
                state            : (idle, busy);
                timeOfLastChange : REAL;
                useIntegral      : REAL;
                traceFlag        : BOOLEAN;
              end;

{***** QUEUEs to represent waiting lines of transactions}
```

```
aQueueType = record
               startTime          : REAL;
               first, last        : aTAPtrType;
               length, maxlength  : 0..MAXINT;
               timeOfLastChange   : REAL;
               lengthIntegral     : REAL;
               noOfEntries        : 0..MAXINT;
               noOfDepartures     : 0..MAXINT;
               traceFlag          : BOOLEAN
             end;
```

```
{Utility Functions and Procedures for Building Queueing Scenarios}
{ ================================= }

{***** Messages to SOURCEs}

procedure InitSource (var source : aSourceType);
begin
  source.startTime     := TheClock.NOW;
  source.creationCount := 0;
  source.traceFlag     := FALSE;
end;

procedure ResetSourceStats (var source : aSourceType);
begin
  WRITELN ('Source statistics reset at ', TheClock.NOW:7:2);
  InitSource (source);
end;

function CreateNewTA (var source : aSourceType) : aTAPtrType;
  var
  thisOne : aTAPtrType;
begin
  source.creationCount := source.creationCount + 1;
  if source.traceFlag then
    WRITELN ('New TA created at ',TheClock.NOW:7:2);
  NEW(thisOne);
  thisOne ↑ .birthTime  := TheClock.NOW;
  thisOne ↑ .nextinLine := nil;
  CreateNewTA           := thisOne;
end;

procedure TraceTAGeneration (var source : aSourceType);
begin
  WRITELN;
  WRITE    ('***** Trace of TA generations switched ON at ');
  WRITELN (TheClock.NOW:7:2); WRITELN;
  source.traceFlag := TRUE;
end;

procedure UntraceTAGeneration (var source : aSourcetype);
begin
  WRITELN;
  WRITELN (' *****Trace of TA generations switched OFF at ' );
  WRITELN (TheClock.NOW:7:2); WRITELN;
  source.traceFlag := FALSE;
end;
```

```
function HowManyCreationsQ (source : aSourceType) :INTEGER;
begin
  HowManyCreationsQ := source.creationCount;
end;

procedure ReportOnSource (source : aSourceType);
begin
  WRITE   ('Since last reset ', HowManyCreationsQ (source):5,' transactions ');
  WRITELN ('were created by this source.');
end;

{***** Messages to SINKs}

procedure InitSink (var sink : aSinkType);
begin
  sink.startTime        := TheClock.NOW;
  sink.destructionCount:= 0;
  sink.traceFlag        := FALSE;
end;

procedure ResetSinkStats (var sink : aSinkType);
begin
  WRITELN ('Sink statistics reset at ', TheClock.NOW:7:2);
  InitSink (sink);
end;

procedure RemoveTA (var sink : aSinkType; ta : aTAPtrType);
begin
  if sink.traceFlag then
    WRITELN('TA departs at ',TheClock.NOW:8:2);
  DISPOSE(ta);
  sink.destructionCount := sink.destructionCount + 1;
end;

procedure TraceTADepartures (var sink : aSinkType);
begin
  WRITELN;
  WRITE   (' *****Trace of TA departures switched ON  at ');
  WRITELN (TheClock.NOW:7:2); WRITELN;
  sink.traceFlag := FALSE;
end;

procedure UntraceTADepartures (var sink : aSinkType);
begin
  WRITELN;
  WRITE   (' *****Trace of TA departures switched OFF at ');
  WRITELN (TheClock.NOW:7:2); WRITELN;
end;

function HowManyDeparturesQ (sink : aSinkType) : INTEGER;
begin
  HowManyDeparturesQ := sink.destructionCount;
end;

procedure ReportOnSink (sink : aSinkType);
begin
  WRITE   ('Since last reset ', HowManyDeparturesQ (sink):5);
```

```
  WRITELN (' transactions were removed through this sink.');
end;

{***** Messages to TRANSACTIONs}

function FlowTimeQ (ta : aTAPtrType) : REAL;
begin
  FlowTimeQ := TheClock.NOW - ta↑.birthTime;
end;

{***** Messages to SERVERs}

procedure InitServer (var server : aServerType; capacityP : INTEGER);
begin
  if capacityP <= 0 then
    WRITELN ('Capacity must be > 0!')
  else
    with server do
      begin
        startTime        := TheClock.NOW;
        capacity         := capacityP;
        inUse            := 0;
        state            := idle;
        timeOfLastChange := 0.0;
        useIntegral      := 0.0;
        traceFlag        := FALSE;
      end;
end; {of InitServer procedure}

procedure ResetServerStats (var server : aServerType);
begin
  WRITELN ('Server statistics reset at ', TheClock.NOW:7:2);
  server.startTime   := TheClock.NOW;
  server.useIntegral := 0.0;
end;

procedure SeizeServer (var server : aServerType; howMuch : INTEGER);
begin
  if ((howMuch < 0) and (server.capacity < (server.inUse + howMuch))) then
    begin
      WRITE    ('The number of server units requested must be > 0');
      WRITELN ('and <= the available capacity!');
    end
  else
    with server do
      begin
        if traceFlag then
          WRITELN ('A service starts at',TheClock.NOW:8:2);
        useIntegral := useIntegral + (inUse * (TheClock.NOW -
                       timeOfLastChange));
        inUse       := inUse + howMuch;
        state       := busy;
        timeOfLastChange := TheClock.NOW;
      end; {with}
end; {of SeizeServer procedure}
```

```
procedure ReleaseServer (var server : aServerType; howMuch : INTEGER);
begin
  if howMuch > server.inUse then
    WRITELN ('Can''t release more than you have!')
  else
    with server do
      begin
        if traceFlag then
          WRITELN ('A service finishes at', TheClock.NOW:8:2);
        useIntegral := useIntegral + (inUse * (TheClock.NOW -
                       timeOfLastChange));
        inUse       := inUse - howMuch;
        if inUse <= 0 then
          state := idle;
        timeOfLastChange := TheClock.NOW;
      end;{with}
end; {of ReleaseServer procedure}

function EnoughServerUnitsAvailableP (server : aServerType;
                                      units  : INTEGER   ) : BOOLEAN;
begin
  EnoughServerUnitsAvailableP := (server.inUse + units) <= server.capacity;
end;

procedure TraceService (var server : aServerType);
begin
  WRITELN;
  WRITE    (' ***** Trace of service actions switched ON at ');
  WRITELN (TheClock.NOW:7:2); WRITELN;
  server.traceFlag := TRUE;
end;

procedure UntraceService (var server : aServerType);
begin
  WRITELN;
  WRITE    (' ***** Trace of service actions switched OFF at ');
  WRITELN (TheClock.NOW:7:2); WRITELN;
  server.traceFlag := FALSE;
end;

function ServerUtilizationQ (server : aServerType) : REAL;
begin
  if TheClock.NOW > 0 then
    ServerUtilizationQ := server.useIntegral / (TheClock.NOW - server.startTime)
  else
    ServerUtilizationQ := 0;
end;

procedure ReportOnServer (server : aServerType);
begin
  WRITE    ('The capacity of this server is ', server.capacity:3  );
  WRITELN (' units.');
  WRITELN (server.inUse:3, ' of these are currently in use .');
  WRITE    ('Utilization  of this server was ');
  WRITELN ((ServerUtilizationQ (server) * 100):5:2, ' %, since last reset.');
end;
```

```
{***** Messages to (Fifo) QUEUEs}

procedure InitQueue (var q : aQueueType);
begin
  with q do
    begin
      startTime        := TheClock.NOW;
      first            := nil;
      last             := nil;
      length           :=   0;
      maxLength        :=   0;
      timeOfLastChange := 0.0;
      lengthIntegral   := 0.0;
      noOfEntries      :=   0;
      noOfDepartures   :=   0;
      traceFlag        := FALSE;
    end; {with}
end; {of InitQueue procedure}

procedure ResetQueueStats (var q : aQueueType);
begin
  WRITELN ('Queue statistics reset at ', TheClock.NOW:7:2);
  with q do
    begin
      startTime      := TheClock.NOW;
      maxLength      :=   0;
      lengthIntegral := 0.0;
      noOfEntries    :=   0;
      noOfDepartures :=   0;
    end;
end;

procedure FileIntoFIFOq (var q : aQueueType; ta : aTAPtrType);
begin
  with q do
    begin
      {FIFO queue! Append transactions to tail of queue.}
      noOfEntries := noOfEntries + 1;
      length := length + 1;
      if traceFlag then
        begin
          WRITELN ('A TA enters queue at ',TheClock.NOW:8:2);
          WRITELN ('Its length is now       ', q.length:5);
        end;
      lengthIntegral := lengthIntegral + length * (TheClock.NOW -
                        timeOfLastChange);
      timeOfLastChange := TheClock.NOW;
      if length > maxLength then
        maxLength := length;
      if (length - 1) = 0 then
      {length has already been incremented to new value}
        first := ta
      else
        last ↑ .nextinLine := ta;
      ta ↑ .nextInLine := nil;
```

```
      last    := ta;
    end; {with}
end; {of FileIntoFIFOQ procedure}

function TakeFirstFromQ (var q : aQueueType) : aTAPtrType;
var
  ta : aTAPtrType;
{FIFO queue! Remove transactions from head of queue.}
begin
  if q.length <= 0 then
    WRITELN ('Cant't remove items from an empty queue!')
  else
    with q do
      begin
        noOfDepartures := noOfDepartures + 1;
        length := length - 1;
        if traceFlag then
          begin
            WRITELN ('A TA leaves queue at',TheClock.NOW:8:2);
            WRITELN ('Its length is now      ', q.length:5);
          end;
        lengthIntegral := lengthIntegral + length * (TheClock.NOW -
                                   timeOfLastChange);
        timeOfLastChange := TheClock.NOW;
        ta := first;
        first := first ↑ .nextinLine;
        ta ↑ .nextInLine := nil;
        if length = 0 then
        q.last := nil;
        TakeFirstFromQ := ta;
      end; {with}
end; {of TakeFirstFromFIFOQ function}

procedure TraceQops (var q : aQueueType);
begin
  WRITELN;
  WRITE    (' ***** Trace of queue movements switched ON at ');
  WRITELN (TheClock.NOW:7:2); WRITELN;
  q.traceFlag := TRUE;
end;

procedure UntraceQops (var q : aQueueType);
begin
  WRITELN;
  WRITE    (' ***** Trace of queue movements switched OFF at ');
  WRITELN (TheClock.NOW:7:2); WRITELN;
  q.traceFlag := FALSE;
end;

function MeanQlengthQ (q : aQueueType) : REAL;
begin
  if TheClock.NOW > 0 then
    MeanQlengthQ := q.lengthIntegral / (TheClock.NOW - q.startTime)
  else
    MeanQlengthQ := 0;
end;
```

```
function MaxQlengthQ (q : aQueueType) : INTEGER;
begin
  MaxQlengthQ := q.maxLength;
end;

function MeanQdelayQ (q : aQueueType) : REAL;
begin
  if TheClock.NOW > 0 then
    MeanQdelayQ := q.lengthIntegral / q.noOfDepartures
  else
    MeanQdelayQ := 0;
end;

procedure ReportOnQueue (q : aQueueType);
begin
  WRITE    ('Sincc its statistics were last reset ', q.noOfEntries:5, ' items');
  WRITELN (' entered and ');
  WRITELN (q.noOfDepartures:5, ' left the queue.');
  WRITELN ('Its current length is ', q.length:5);
  WRITELN;
  WRITELN ('The queue''s main characteristics were: ');
  WRITELN (' ------------------- ');
  WRITELN;
  WRITELN ('Mean length = ', MeanQlengthQ (q):5:2);
  WRITELN ('Max. length = ', MaxQlengthQ  (q):5  );
  WRITELN ('Mean delay  = ', MeanQdelayQ  (q):5:2);
end;
```

1.7 Combined simulation

The combined library

Object	: {various previously declared classes}
Methods	: **public** *ChangeState (aContVarType, someRealQtty)*; *WaitUnitlP (condition): BOOLEAN*; *TimeEventNOWp (anEventSchedulingMonitorType): BOOLEAN*;

```
{UTILITIES FOR COMBINED DISCRETE/CONTINUOUS SIMULATION}
{**************************************************************}

{************************************************************}
{*ChangesState & WaitUntil should be used together with both the CS   *}
{* and the DES utility libraries (add the QN utilities if required).   *}
{************************************************************}

{*****       No additional Data Types       *****}

{***** Utility Procedures and Functions *****}

procedure ChangeState (var variable : aContVarType;
                           by       : REAL );
{can be used by a TIME EVENT to change the state of a quasi-continuously}
{varying variable                                                        }
begin
  variable.level := variable.level + by;
end;

function WaitUntilP (condition : BOOLEAN) : BOOLEAN;
{can be used to implement a STATE EVENT through a loop monitoring}
{changes in the state of the model.The condition will be re-evaluated   }
{each time the procedure is invoked                                      }
begin
  WaitUntilP := condition;
end;

function TimeEventNOWp (monitor : anEventSchedulingMonitorType) : BOOLEAN;
begin
  TimeEventNOWp := monitor.agenda.firstEvent ↑ .eventTime <= TheClock.NOW;
end;
```

1.8 A SIMULA context for queueing network simulations: Class MiniQNcontext

```
EXTERNAL CLASS simulation;

simulation CLASS miniQNcontext;
begin
process CLASS aProcessObject (aName); VALUE aName; TEXT aName;
VIRTUAL  : PROCEDURE reset, report;
BEGIN

  TEXT PROCEDURE hasName;
    hasName :- aName;

  PROCEDURE reset;
  BEGIN
  OUTTEXT ("I "); OUTTEXT (hasName); OUTTEXT (" have nothing to reset!");
  OUTIMAGE;
  END of reset procedure;

  PROCEDURE report;
  BEGIN
  OUTTEXT ("I "); OUTTEXT (hasName); OUTTEXT (" have nothing to report!");
  OUTIMAGE;
  END of report procedure;

END of aProcessObject;

link CLASS aStructureObject (aName); VALUE aName; TEXT aName;
VIRTUAL  : PROCEDURE reset, report;
BEGIN

  TEXT PROCEDURE hasName;
    hasName :- aName;

  PROCEDURE reset;
  BEGIN
  OUTTEXT ("I "); OUTTEXT (hasName); OUTTEXT (" have nothing to reset!");
  OUTIMAGE;
  END of reset procedure;

  PROCEDURE report;
  BEGIN
  OUTTEXT ("I "); OUTTEXT (hasName); OUTTEXT (" have nothing to report!");
  OUTIMAGE;
  END of report procedure;

END of aStructureObject;

  aStructureObject CLASS aDist (aSeed); INTEGER aSeed;
  VIRTUAL  : REAL PROCEDURE sample;
  BEGIN

  COMMENT: private data;
```

```
  INTEGER startingSeed;

COMMENT: public routines;

  INTEGER PROCEDURE hasSeed;
    hasSeed := startingSeed;

  PROCEDURE reset;
    aSeed := StartingSeed;

  PROCEDURE report;
  BEGIN
  OUTTEXT ("Seed value for this ");OUTTEXT (aName);
  OUTTEXT (" object = ");          OUTINT   (StartingSeed, 5);
  OUTIMAGE; OUTIMAGE;
  END of report procedure;

initialize:
  startingSeed := aSeed;
END of aDist;

  aDist CLASS anExpoDist (aRate); REAL aRate;
  BEGIN

    REAL PROCEDURE sample;
      sample := NEGEXP (aRate, aSeed);

  END of ExpoDist;

  aDist CLASS aNormalDist (aMean, aStDev); REAL aMean, aStDev;
  BEGIN

    REAL PROCEDURE sample;
      sample := NORMAL (aMean, aStDev, aSeed);

  END of NormalDist;

  aDist CLASS aUnifDist (aMin, aMax); REAL aMin, aMax;
  BEGIN

    REAL PROCEDURE sample;
      sample := UNIFORM (aMin, aMax, aSeed);

  END of UnifDist;

aProcessObject CLASS aCustomer (anEntryTime); VALUE anEntryTime;
                                              REAL  anEntryTime;
BEGIN

  REAL PROCEDURE hasEntryTime;
    hasEntryTime := anEntryTime;

END of aCustomer;
```

```
aProcessObject CLASS aSource (anArrivalDist, anOutputQueue);
                              REF (aDist   ) anArrivalDist;
                              REF (aQueue) anOutputQueue;
VIRTUAL  : REF (aCustomer) PROCEDURE createaCustomer;
BEGIN

COMMENT: private data;

  INTEGER          creationCount;
  REF (aCustomer) Fred;

COMMENT: public procedures;

  REF (aCustomer) PROCEDURE createaCustomer;
    createaCustomer :- NEW aCustomer ("Customer", TIME);

  PROCEDURE recordCreation;
    creationCount := creationCount + 1;

  INTEGER PROCEDURE hasCreated;
    hasCreated := creationCount;

  REAL PROCEDURE sampleTimeToNextArrival;
    sampleTimeToNextArrival := anArrivalDist.sample;

  PROCEDURE reset;
    creationCount := 0;

  PROCEDURE report;
  BEGIN
  OUTINT (hasCreated,5); OUTTEXT (" customers were created by this ");
  OUTTEXT (hasName); OUTIMAGE; OUTIMAGE;
  END of report procedure;

initialize:
  creationCount :- 0;
END of Source;

aProcessObject CLASS aSink (aSampleSize, anInputQueue);
                            INTEGER      aSampleSize;
                            REF (aQueue) anInputQueue;
VIRTUAL  : PROCEDURE eataCustomer;
BEGIN

COMMENT: private data;

  INTEGER samples;

COMMENT: public procedures;

  INTEGER PROCEDURE hasSampleSize;
    hasSampleSize := aSampleSize;

  INTEGER PROCEDURE hasEaten;
    hasEaten := samples;
```

```
  PROCEDURE recordDeparture;
    samples := samples + 1;

  PROCEDURE reset;
    samples := 0;

  PROCEDURE report;
  BEGIN
  OUTTEXT (hasName); OUTTEXT(" ate "); OUTINT (hasEaten, 5);
  OUTTEXT (" customers."); OUTIMAGE; OUTIMAGE;
  END of report procedure;

END of Sink;

aStructureObject CLASS aQueue;
VIRTUAL  : PROCEDURE insertaCustomer, removeaCustomer,
           wakeUpNextInLine; REF (aCustomer) PROCEDURE isFirstItem;
BEGIN

COMMENT: private data;

  REF (HEAD) items;
  INTEGER    max;
  REAL       TimeOfLastEntry, ResetTime;
  REAL       LengthIntegral;

COMMENT: private procedure;

  PROCEDURE updateLengthIntegral;
  BEGIN
  LengthIntegral    := LengthIntegral + hasLength * (TIME - TimeOfLastEntry);
  TimeOfLastEntry := TIME;
  END of updateLengthIntegral procedure;

COMMENT: public procedures;

  REAL PROCEDURE hasLength;
    hasLength := items.CARDINAL;

  BOOLEAN PROCEDURE isEmpty;
    isEmpty := IF items.CARDINAL <= 0 THEN TRUE ELSE FALSE;

  REAL PROCEDURE hadMeanLength;
    hadMeanLength := LengthIntegral / (TIME - ResetTime);

  REAL PROCEDURE hadMaxLength;
    hadMaxLength := max;

  PROCEDURE reset;
  BEGIN
  ResetTime      := TIME;
  max            := 0;
  LengthIntegral := 0.0;
  END of reset procedure;
```

```
  PROCEDURE report;
  BEGIN
  OUTTEXT ("Reporting on queue: "); OUTTEXT (hasName); OUTIMAGE;
                                                       OUTIMAGE;
  OUTTEXT ("Current   queuelength is   : "); OUTINT(hasLength, 5); OUTIMAGE;
  OUTTEXT ("The Max. queuelength was : "); OUTINT(hadMaxLength, 5); OUTIMAGE
  OUTTEXT ("The Mean queuelength was : "); OUTFIX(hadMeanLength, 2, 5);
  OUTIMAGE;
  OUTIMAGE;
  END of report procedure;

initialize:
  items             :- NEW HEAD;
  max               := 0;
  TimeOfLastEntry   := TIME;
  LengthIntegral    := 0.0;
  ResetTime         := 0;
END of Queue;

  aQueue CLASS aFIFOqueue;
  BEGIN

    PROCEDURE insertaCustomer (Fred); REF (aCustomer) Fred;
    BEGIN
    Fred.INTO (items);
    UpdateLengthIntegral;
    IF hasLength > hadMaxLength THEN
      max := hasLength;
    END of insertaCustomer procedure;

    PROCEDURE removeaCustomer (Fred); REF (aCustomer) Fred;
    IF fred =/= NONE THEN
      BEGIN
      Fred.OUT;
      UpdateLengthIntegral;
      END of rcmovcaCustomer procedure;

    REF (aCustomer) PROCEDURE isFirstItem;
      isFirstItem :- items.FIRST;

    PROCEDURE wakeUpNextInLine;
    BEGIN
      REF (aCustomer) Fred;
    IF NOT isEmpty THEN
      BEGIN
      Fred :- isFirstItem;
      ACTIVATE Fred;
      END if;
    END of wakeUpNextInLine procedure;

  END of aFIFOqueue;

aProcessObject CLASS aServer (aServiceDist, anInputQueue, anOutputQueue);
                             REF (aDist   ) aServiceDist;
                             REF (aQueue) anInputQueue , anOutputQueue ;
```

```
BEGIN

COMMENT: private data;

   INTEGER Capacity;
   REAL    UseIntegral;
   REAL    ResetTime;

COMMENT: public procedures;

   INTEGER PROCEDURE hasCapacity;
     hasCapacity := Capacity;

   REAL PROCEDURE sampleServiceTime;
   BEGIN
     REAL st;
   st                := aServiceDist.sample;
   UseIntegral       := UseIntegral + st;
   sampleServiceTime := st;
   END of sampleServiceTime procedure;

   REAL PROCEDURE hasUtilization;
   BEGIN
     REAL util;
   util := UseIntegral / (TIME - ResetTime) * Capacity;
   hasUtilization := util;
   END hasUtilization procedure;

   PROCEDURE reset;
   BEGIN
   UseIntegral := 0.0;
   ResetTime   := TIME;
   END of reset procedure;

   PROCEDURE report;
   BEGIN
   OUTTEXT (hasName); OUTTEXT ("'s Utilization was : "); OUTFIX
              (hasUtilization,2,5);}
   OUTIMAGE; OUTIMAGE;
   END of report procedure;

initialize:
   capacity    := 1;
   UseIntegral := 0.0;
   ResetTime   := TIME;
END of Server;

   aServer CLASS aSingleServer;
   BEGIN
     BOOLEAN busy;

   BOOLEAN PROCEDURE isIdle;
     isIdle := IF busy THEN FALSE ELSE TRUE;
```

```
    BOOLEAN PROCEDURE isBusy;
      isBusy := IF busy THEN TRUE ELSE FALSE;

    PROCEDURE makeIdle;
      busy := FALSE;

    PROCEDURE makeBusy;
      busy := TRUE;

    initialize:
      Capacity := 1;
      busy     := FALSE;
    END of SingleServer;

  aStructureObject CLASS anObserver;
  VIRTUAL  : PROCEDURE update;
  BEGIN

  COMMENT: private data;

    INTEGER obs;

  COMMENT: public procedure;

    INTEGER PROCEDURE hasObserved;
      hasObserved := obs;

  initialize:
    obs := 0;
  END of Observer;

    anObserver CLASS aCounter;
    BEGIN

    COMMENT: private data;

      REAL count;

    COMMENT: public procedures;

      PROCEDURE update (number); INTEGER number;
        count := count + number;

      PROCEDURE reset;
      BEGIN
      obs   := 0;
      count := 0;
      END of reset procedure;

      PROCEDURE report;
      BEGIN
      OUTTEXT (hasName); OUTTEXT (" counted "); OUTINT (count, 5);
      OUTTEXT (" occurrences."); OUTIMAGE;
      END of report procedure;
```

```
initialize:
  count := 0;
END of Counter;

anObserver CLASS aTally;
BEGIN

COMMENT: private data;

  REAL       SumX, SumXSq;

COMMENT: private procedure;

  REAL PROCEDURE sigma;
    sigma := (SumXSq - (sumx * sumx) / obs) / (obs - 1);

COMMENT: public procedures;

PROCEDURE update (aValue); REAL aValue;
BEGIN
obs := obs + 1;
Sum     := SumX + aValue;
SumXSq := SumXSq + aValue * aValue;
END of update procedure;

REAL PROCEDURE hadMean;
BEGIN
  REAL mean;
mean := SumX / obs;
hadMean := mean;
END of hadMean procedure;

REAL PROCEDURE hadStDev;
  hadStDev := SQRT (sigma);

PROCEDURE reset;
BEGIN
obs      := 0;
SumX   := SumXSq := 0.0;
END of reset procedure;

PROCEDURE report;
BEGIN
OUTTEXT ("Reporting on Tally: "); OUTTEXT (hasName); OUTIMAGE;
OUTIMAGE;
OUTINT (obs, 5); OUTTEXT (" customers were observed leaving the shop");
OUTIMAGE; OUTIMAGE;
OUTTEXT ("The distribution of times the customers spent in the shop");
OUTIMAGE;
OUTTEXT ("  ------------------------------------------- ");
OUTIMAGE; OUTIMAGE;
OUTTEXT ("       Mean = "); OUTFIX (hadMean, 2, 10);
OUTTEXT ("       Standard Deviation = "); OUTFIX (hadStDev, 2, 10);
OUTIMAGE; OUTIMAGE;
END of report procedure;
```

```
    initalize:
      obs     := 0;
      SumX  := SumXSq := 0.0;
    END of Tally;

end of context;
```

Appendix 2
A Survey of SIMULA Implementations

(From *SIMULA Newsletter* 1986. Prices are subject to change and should serve as guidelines only.)

Computer System	*Academic Price* (P: purchase; L: leasing; S: single release)	*Commercial Price*
DEC 10/20	P: 3.600 \$ L: 1.296 \$	P: 10.800 \$ L: 3.888 \$ year
Cyber 170s NOS and NOS/BE NDRE-SIMULA	P: 16.000 \$ L: 400 \$ month 3.500 \$ init. fee	same as for universities
Data General Eclipse, Desktop and MV Series	L: 200 \$ year	L: 1.000 \$ year reducing to 500 \$ after first year (for desktop)
IBM SIMULA (OS, VS, VM, MVS, . . .)	P: 2.000 \$ L: 150 \$ month S: 1.000 \$	P: 7.000 \$ L: 550 \$ month S: 3.500 \$
ICL 2900	P: 1.400 \$	P: 7.000 \$
NORSK DATA 100	P: 9.000 \$ L: 225 \$ month	same as for universities
NORSK DATA 500	P: 10.000 \$ L: 250 \$ month	same as for universities
CROMEMCO	P: 2.000 \$	P: 2.000 \$
PRIME	P: 5.000 \$	P: 10.000 \$
VAX/UNIX (Simprog)	P: 2.000 \$	P: 8.000 \$
VAX/VMS	P: 4.800 \$	P: 12.000 \$
VAX BERKELEY/ UNIX 4.2	P: 2.000 \$	P: 8.000 \$

The following list gives distributors' addresses:

DEC 10/20
VAX VMS & VAX/UNIX
Stockholms Datorcentral (QZ)
Box 27322
S-102 54 Stockholm
SWEDEN

DG Eclipse/Desktop/MV Series
Department of Computer Science
Lund Institute of Technology
Box 725
S-220 07 Lund
SWEDEN

CDC 3000
Computer & Automation Institute
Hungarian Academy of Sciences
PO Box 63
H-1502 Budapest 112
HUNGARY

CDC 6000/CYBER
Control Data Norway A/S
PO Box 112, Refstad
0513 Oslo 5
NORWAY

CROMEMCO
A/S Omnibus
Wdm. Thranes gate 75
0171 Oslo 1
NORWAY

VAX/UNIX
SIMPROG AB
øѕtermalmsgatan 21
S-114 26 Stockholm
SWEDEN

VAX BERKELEY/UNIX 4.2
Unger Huntsinger & Assoc. Ltd.
5001 13th Ave. N.W.
Calgary, Alberta
CANADA T3B 1J1

IBM SIMULA
SIMULA A/S
Norwegian Computing Center
Postbox 335, Blindern
0314 Oslo 3
NORWAY

S-2000
Simulation Technology A/S
PO Box 15, Refstad
0513 Oslo 5
NORWAY

BESM 6
ADW der DDR
Zl für Mathematik & Mechanik
Mohrenstrasse 39
108 Berlin
GDR

NORD 100/500
Norsk Data A/S
PO Box 25 – Bogerud
0621 Oslo 6
NORWAY

PRIME
PRIME Computer Scandinavia AB
Box 18
S-163 93 Spånga
SWEDEN

ICL 2900 & PERQ
ERCC
University of Edinburgh
59 George Square
Edinburgh EH8 9JU
SCOTLAND

HB DPS8
CH Honeywell Bull
94 Avenue Gambetta
75960 Paris Cedex 20
FRANCE

SPERRY
Sperry Corp.
PO Box 500
PA 19424-0001
USA

Appendix 3

Textbooks on Simulation Methodology: An Annotated Bibliography

Banks, J. and Carson J.S., II. (1984). *Discrete Event System Simulation*. Englewood Cliffs, NJ: Prentice-Hall.

Discusses Monte Carlo and discrete-event simulation and emphasizes statistical methodology: Brief examples of SIMSCRIPT, GPSS and SLAM.

Bratley, P., Fox, B.C. and Schrage, L.E. (1983). *A Guide to Simulation*. New York/Heidelberg: Springer-Verlag.

Monte Carlo and discrete-event simulation. Not predominantly a programming-oriented text. Good coverage of statistical methodology. Uses mainly FORTRAN, but has a 'languages' chapter with examples in PASCAL, SIMSCRIPT, GPSS, SIMULA and DEMOS.

Bulgren, W.G. (1980). *Discrete Event Simulation*. Englewood Cliffs, NJ: Prentice-Hall.

Discrete-event simulation. Oriented towards simulation programming. Uses SIMSCRIPT and GPSS.

Deo, N. (1979). *System Simulation with Digital Computers*. Englewood Cliffs, NJ: Prentice-Hall (Indian edition).

Monte Carlo, discrete-event and continuous-system simulation. Emphasizes methodology over programming. Uses mainly FORTRAN. Mentions SIMSCRIPT, GPSS and SIMULA. Very cheap!

Emshoff, J.R. and Sisson, C.R. (1970). *Design and Use of Computer Simulation Models*. London: Macmillan.

Mainly discrete-event simulation. Old classic. Emphasizes methodology. Uses FORTRAN and has a few examples in GPSS and DYNAMO.

Fishman, G.S. (1978). *Principles of Discrete Event Simulation*. New York: John Wiley.

Monte Carlo and discrete-event simulation. Good coverage of random-number generation and statistical aspects of simulation. Uses FORTRAN throughout. Has brief summaries of SIMSCRIPT, GPSS; examples in SIMPL/1, SIMULA.

Franta, W.R. (1977). *The Process View of Simulation*. New York/Amsterdam: Elsevier.

Discrete-event simulation. A programming-oriented text on aspects of simulation methodology. Also a good, brief introduction to SIMULA. Uses SIMULA throughout.

Gordon, G. (1978). *System Simulation* (2nd edn.). Englewood Cliffs, NJ: Prentice-Hall.

Monte Carlo, discrete-event and continuous-system simulation. A classic, which covers simulation methodology and simulation programming. Has a bias towards FORTRAN, DYNAMO and GPSS, although CSMP and SIMSCRIPT are also quite well described.

Graybeal, W.R. and Pooch, U.W. (1980). *Simulation: Principles and Methods*. Cambridge, MA: Winthrop.

Monte Carlo, discrete-event and continuous-system simulation. A blend of simulation programming and statistical methodology. Uses FORTRAN predominantly. Mentions GPSS, SIMSCRIPT and GASP.

Law, A. and Kelton, W.D. (1982). *Simulation Modelling and Analysis*. New York: McGraw-Hill.

Monte Carlo and discrete-event simulation. Good textbook on simulation methodology. Uses FORTRAN as its main programming tool.

Lewis, T.G. and Smith, B.J. (1979). *Computer Principles of Modelling and Simulation*. Boston: Houghton and Mifflin.

Monte Carlo and discrete-event simulation. Uses pseudocode. Has brief examples in GPSS, SIMSCRIPT and SIMPL/1.

Maisel, H. and Gnugnoli, C. (1972). *SIMULATION of Discrete Stochastic Systems*. Chicago: Science Research Associates.

Monte Carlo and discrete-event simulation. Good chapters on statistical methodology for simulation experiments. Uses FORTRAN and GPSS. SIMSCRIPT is briefly mentioned.

Martin, F.R. (1968). *Computer Modelling and Simulation*. New York: John Wiley.

Monte Carlo and discrete-event simulation. Now rather an old text on simulation techniques. Some programming in FORTRAN.

Maryanski, F. (1980). *Digital Computer Simulation*. Rochelle Park: Hayden.

Monte Carlo, discrete-event and continuous-system simulation. Covers DYNAMO, CSMP, GPSS, SIMSCRIPT and simulation methodology.

Meier, R.C., Newell, W.T. and Pazer, H.L. (1969). *Simulation in Business and Economics*. Englewood Cliffs, NJ: Prentice-Hall.

A somewhat outdated classic. FORTRAN based.

Mihram, A.G. (1972). *Simulation: Statistical Foundations and Methodology*. New York: Academic Press.

Monte Carlo and discrete-event simulation. Good text on statistical foundations of simulation and experimental design. Not programming oriented.

Mitrani, I. (1982). *Simulation Techniques for Discrete Event Systems*. Cambridge: Cambridge University Press.

Monte Carlo and discrete-event simulation. Good short text on simulation methodology. Although use of SIMULA is recommended (nice introduction), the treatment is largely language independent.

Naylor, T.H. (ed.) (1971). *Computer Simulation with Models of Economic Systems.* New York: John Wiley.

Monte Carlo, discrete-event and continuous-system simulation. Contains contributions to statistical methodology for simulation experiments, a number of case studies, and a classic paper (by Kiviat) on discrete-event simulation languages. Somewhat outdated.

Naylor, T.H., Balintfly, J.L., Burdick, D.S. and Chu, D.S. (1966). *Computer Simulation Techniques.* New York: John Wiley.

Monte Carlo, discrete-event and continuous-system simulation. Old classic. Uses FORTRAN as programming vehicle. Somewhat outdated.

Payne, J.A. (1982). *Introduction to Simulation.* New York: McGraw-Hill.

Monte Carlo, discrete-event and continuous-system simulation. Nice introduction to simulation methodology. Uses PL/1 as its main programming vehicle. Gives examples in SIMSCRIPT and GPSS.

Reitman, J. (1971). *Computer Simulation Applications.* New York: John Wiley.

Monte Carlo and discrete-event simulation. Programming-oriented textbook of simulation methodology. GPSS is the main programming tool. Example in SIMSCRIPT. Some exemplary case studies (not going into any depth). Well written.

Rubinstein, R.J. (1981). *Simulation and the Monte Carlo Method.* New York: John Wiley.

Monte Carlo and discrete-event simulation.

Schmidt, J.W. and Taylor, R.E. (1970). *Simulation and Analysis of Industrial Systems.* Homewood: Irwin.

Monte Carlo and discrete-event simulation. Somewhat outdated classic on simulation methodology. Uses FORTRAN, but GPSS and SIMSCRIPT are mentioned.

Shannon, R.E. (1975). *System Simulation – The Art and Science.* Englewood Cliffs, NJ: Prentice-Hall.

Monte Carlo, discrete-event and continuous-system simulation. A nice textbook on simulation methodology. Not language oriented, although FORTRAN is used for examples and utilities. A number of complete case studies are given in an appendix.

Solomon, S.L. (1983). *Simulation of Waiting Line Systems.* Englewood Cliffs, NJ: Prentice-Hall.

Monte Carlo and discrete-event simulation. Book on simulation methodology. Lists and exemplifies a number of FORTRAN programs for random sampling, various statistical tests, and experimental analysis and design. FORTRAN is the main programming vehicle. GPSS is also covered in some depth. A number of complete case studies are given (not too well integrated with the rest of the text). Good bibliography.

Tocher, K.D. (1963). *The Art of Simulation.* New York: Van Nostrand.

Old and outdated classic.

Watson, H.J. (1981). *Computer Simulation in Business.* New York: John Wiley.

Good coverage of basic simulation techniques and their application to business problems. Nice exercises and examples.

Zeigler, B.P. (1976). *Theory of Modelling and Simulation.* New York: John Wiley.

Monte Carlo, discrete-event and continuous-system simulation. Still the classical text on simulation theory (or what theory there is). Not a programming-oriented text.

Zeigler, B.P., Elzas, M.S., Klir, G.J. and Oren, T.I. (eds.) (1979). *Methodology in Systems Modelling and Simulation.* Amsterdam/New York: North-Holland.

Monte Carlo, discrete-event and continuous-system simulation. A good up-to-date summary of simulation methodology. Not a programming-oriented textbook.

Zeigler, B.P. (1984). *Multifacetted Modelling and Discrete Event Simulation.* London/ Orlando: Academic Press.

Covers discrete-event simulation from a system–theoretic viewpoint. Uses pseudo codes. Not primarily a programming oriented text.

Appendix 4
Solutions to Simulation Problems

program *LouisDestiny (INPUT, OUTPUT);*
{Solution to 5.2.2 in PASCAL (& toolboxes)}

```
const
  MaxHistoClasses = 50;
  MaxPlotPts      = 50;
  LongNumber      = MAXINT;
type
  aCreatureType   = (xxy, yyy);
  aPlaceNameType  = (home, work);
  aPlacesType     = set of aCreatureType;
  aWeightType     = (light, heavy);

{insert declaration part of Monte Carlo toolbox}
{insert declaration part of statistics  toolbox}

var
  TheClock   : aStaticClockType;
  TheMonitor : aStateSamplingMonitorType;
  Louis      : record
                 where : aPlaceNameType;
                 load  : aWeightTyp;
               end;
  AtHome, AtWork                                  : aPlacesType;
  WendelinComes, StartWhenLight, StartWhenHeavy: aDrawGenType;
  FandCfill                                       : aNormalGenType;
  DayCount                                        : 0..MAXINT;
  StatusQuoCount, BlissCount, MorvenCount         : aCounterType;

{insert procedures from Monte Carlo toolbox}
{insert procedures from statistics  toolbox}

procedure LouisEats;
var
  foodIntake : 0..MAXINT;
begin
  foodIntake := ROUND (Normal (FandCfill));
  if foodIntake < 10 then
    Louis.load := light
  else
    Louis.load := heavy;
end;

procedure ResetUniverse;
begin
  DayCount    := 0;
  Louis.where := home;
  LouisEats;
  AtHome  := [xxy,yyy];
  AtWork  := [];
end;

procedure ReadAndInitialize;
begin
  InitStaticClock          (TheClock);
```

```
  InitStateSamplingMonitor  (TheMonitor, 150);
  InitDrawGen   (WendelinComes   , 12345, 0.3);
  InitDrawGen   (StartWhenLight  , 67891, 0.7);
  InitDrawGen   (StartWhenHeavy  , 23457, 0.5);
  InitNormalGen (FandCfill, 89123, 9, 3);
  InitCounter   (StatusQuoCount);
  InitCounter   (BlissCount);
  InitCounter   (MorvenCount);
  ResetUniverse;
end;

function CentipedeStarts (where : aPlacesType; LouisWeight : aWeightType) :
                              boolean;
var
  startDist : aDrawGenType;
  startedP : boolean;

  procedure moveCentipede (who : aCreatureType);
  begin
  if Louis.where = home then
     begin
       AtHome := AtHome - [who];
       AtWork  := AtWork  + [who];
     end;
  if Louis.where = work then
     begin
       AtWork  := AtWork  - [who];
       AtHome := AtHome + [who];
     end;
  end,

begin
  startedP := false;
  if (Louis.load = heavy) then
     startDist := StartWhenHeavy
  else
     startDist := StartWhenLight;
  if (xxy in where) and (Draw (startDist)) then
     begin
       startedP := true;
       moveCentipede (xxy);
     end
  else
     if (yyy in where) and (Draw (startDist)) then
        begin
          startedP := true;
          moveCentipede (yyy);
        end;
  CentipedeStarts := startedP;
end;

procedure WriteReport;
begin
  WRITELN ('Loui''s Destiny');
  WRITELN ('*************'); WRITELN;
  WRITELN ('150 samples were taken'); WRITELN;
```

```
  WRITELN ('Louis continues to mine icycles ');
  ReportOnCounter (StatusQuoCount); WRITELN;
  WRITELN ('Louis & Wendelin live in Eternal Bliss ');
  ReportOnCounter (BlissCount); WRITELN;
  WRITELN ('Louis & Harold elope to Morven ');
  ReportOnCounter (MorvenCount); WRITELN;
end;

begin {Main body}

ReadAndInitialize;
while AnotherSampleP (TheMonitor) do
  begin
  while (DayCount < 3) do
    begin
    DayCount := DayCount + 1;
    if (not Draw (WendelinComes)) then
      begin
      if (not CentipedeStarts (AtHome, Louis.load)) then
        begin
          UpdateCounter (BlissCount);
          DayCount := LongNumber;
          {stop this sample}
        end
      else
        Louis.where := work;
      end
    else
      Louis.where := work;
    if (Louis.where = work) then
    if (not CentipedeStarts (AtWork, Louis.load)) then
      begin
        UpdateCounter (MorvenCount);
        DayCount := LongNumber;
      end
    else
      begin
        Louis.where := home;
        LouisEats;
        if DayCount = 3 then
          UpdateCounter (StatusQuoCount);
      end;
    end {while over 3 days};
  ResetUniverse;
  end {while over 150 samples};
WriteReport;
end.
```

INPUT: Loui's Destiny {program needs no input}

OUTPUT: Loui's Destiny

150 samples were taken

```
Louis continues to mine icycles
Value of this Counter at         1.00:       51
Louis & Wendelin live in Eternal Bliss
Value of this Counter at         1.00:        0
Louis & Harold elope to Morven
Value of this Counter at         1.00:       99
```

Begin
comment: Solution to 5.2.3 in SIMULA class MCcontext;

```
External Class MCcontext;

MCcontext begin
  ref (idist)  ClockSelection;
  ref (rdist)  ClockDeviation;
  ref (count) TickCount, TockCount, SampleCount;

  ref (Clock) array TickClocks  (1:100);
  ref (Clock) array TockClocks (1:100);

  real     IntervalBetweenMeetings;
  real     MinutesPerYear, MinutesPerDay, DaysPerMeeting;
  integer Meetings, FirstClock, SecondClock;

  procedure GetClocks (c1, c2); name c1, c2; integer c1, c2;
  begin
  comment: returns indices for two distinct (!) clocks;
  c1 := ClockSelection.sample;
  c2 := c1;
  while c1 = c2 do
    c2 := ClockSelection.sample;
  end of GetClocks procedure;

  procedure UpdateTime (thisClock, intervalNo); ref (Clock) thisClock;
                                                integer intervalNo;
  begin
  comment: updates a clock's time, ready for a master's meeting;
  thisClock.clockTime := thisClock.clockTime +
                         (intervalNo - thisClock.lastSample) *
                         (IntervalBetweenMeetings + DaysPerMeeting *
                         thisClock.deviation);
  thisClock.lastSample := intervalNo;
  end of UpdateTime procedure;

class Clock;
begin
  real deviation, clockTime;
  integer lastSample;
deviation   := 0.0;
clockTime := 0.0;
lastSample :=   0;
end of class Clock;
```

```
comment: *** Body of main program ***;

outf :- new outfile ("TickAndTock.results");
outf.open (blanks (80));
box ("The Clocks of Tick & Tock");

outtext ("Enter number of samples: "); outimage;
Meetings := inint;
ClockSelection  :- new randint ("Masters", 1, 100);
ClockDeviation :- new uniform("Clock Rate", -0.99, +0.99);
TickCount       :- new count   ("Tick Failures");
TockCount       :- new count   ("Tock Failures");
SampleCount     :- new count   ("No. of Meetings");

MinutesPerYear := 300 * 20 * 60;
MinutesPerDay  := 20 * 60;

begin
  integer index;
for index := 1 step 1 until 100 do
  begin
  comment: initialize all clocks;
  TickClocks (index) :- new Clock;
  TockClocks (index) :- new Clock;
  TickClocks (index).deviation := ClockDeviation.sample;
  TockClocks (index).deviation := TickClocks (index).deviation;
  comment: must be the same !;
  end;
end of initialization block;

IntervalBetweenMeetings := MinutesPerYear / Meetings;
DaysPerMeeting          := IntervalBetweenMeetings / MinutesPerDay;

Comment: ******* Sampling Loop *******;

begin
  integer index;
  real    temp;
for index := 1 step 1 until Meetings do
  begin
  GetClocks (FirstClock, SecondClock); comment: for TickClocks;
  UpdateTime (TickClocks (FirstClock)   , index);
  UpdateTime (TickClocks (SecondClock), index);
  temp := TickClocks (FirstClock)    .clockTime +
          TickClocks (SecondClock).clockTime;
  TickClocks (FirstClock)     .clockTime := temp;  comment: set both to mean !;
  TickClocks (SecondClock) .clockTime := temp;

  UpdateTime (TockClocks (FirstClock)     , index); comment: use the same clocks
                                                              for Tock !;
  UpdateTime (TockClocks (SecondClock) , index);
  temp := TockClocks (FirstClock).clockTime;      comment: swap times !;
  TockClocks (FirstClock)    .clockTime := TockClocks (SecondClock).clockTime;
  TockClocks (SecondClock).clockTime := temp;
  end;
end of sampling block;
```

```
SampleCount.update (Meetings);

comment: now determine the number of tragedies;
begin
  integer index;
for index := 1 step 1 until 100 do
  begin
  UpdateTime (TickClocks (index), Meetings); comment: synchronize all clocks;
  if abs (TickClocks (index).clockTime - MinutesPerYear) > 60 then
     TickCount.update (1);
  UpdateTime (TockClocks (index), Meetings);
  if abs (TockClocks (index).clockTime - MinutesPerYear) > 60 then
     TockCount.update (1);
  end;
end of clock matching block;

end of MCcontext;
end of model;
```

INPUT: MCcontext

either 150 (a) or 300 (b) or 600 (c) {number of meetings per year!}

OUTPUT: MCcontext

Case A: Sample size of 150.

```
*******************************************************************************
*                                                                             *
*                          The Clocks of Tick & Tock                          *
*                                                                             *
*******************************************************************************

               CLOCK TIME =                    0.000

*******************************************************************************
*                                                                             *
*                                 REPORT                                      *
*                                                                             *
*******************************************************************************

                             DISTRIBUTIONS
                             *************
```

TITLE	(RE)SET	OBS	TYPE	A	B	SEED
Masters	0.000	301	RANDINT	1	100	33427485
Clock Rate	0.000	100	UNIFORM	−0.990	0.990	22276755

```
                         C O U N T S
                         **********

          TITLE               (RE)SET      OBS
          Tick Failure         0.000        97
          Tock Failure         0.000        64
          No. of Meeti         0.000       150
```

Case B: Sample size of 300;

```
*****************************************************************************
*                                                                           *
*                       The Clocks of Tick & Tock                           *
*                                                                           *
*****************************************************************************

              CLOCK TIME =                      0.000
*****************************************************************************
*                                                                           *
*                              R E P O R T                                  *
*                                                                           *
*****************************************************************************

                       D I S T R I B U T I O N S
                       ********************

TITLE          (RE)SET     OBS      TYPE          A          B       SEED
Masters         0.000      601    RANDINT         1        100   33427485
Clock Rate      0.000      100    UNIFORM    -0.990      0.990   22276755

                         C O U N T S
                         **********

          TITLE               (RE)SET      OBS
          Tick Failure         0.000       100
          Tock Failure         0.000        50
          No. of Meeti         0.000       300
```

Case C: Sample size 600;

```
*****************************************************************************
*                                                                           *
*                       The Clocks of Tick & Tock                           *
*                                                                           *
*****************************************************************************

              CLOCK TIME =                      0.000
*****************************************************************************
*                                                                           *
*                              R E P O R T                                  *
*                                                                           *
*****************************************************************************
```

DISTRIBUTIONS
****

TITLE	(RE)SET	OBS	TYPE	A	B	SEED
Masters	0.000	1204	RANDINT	1	100	33427485
Clock Rate	0.000	100	UNIFORM	−0.990	0.990	22276755

COUNTS
****

TITLE	(RE)SET	OBS
Tick Failure	0.000	100
Tock Failure	0.000	43
No. of Meeti	0.000	600

Glowfish/Salmon Model – Solution to 5.3.2 (b) in DYNAMO

```
* ***** Glowfish section
*
R BGLOW.KL =GBRATE*GLOWFH.K                      Glowfish birth rate
R DGLOW.KL =FIFZE(0,GDRATE*SALMON.K,             Glowfish death rate
                  GLOWFH.K)
L GLOWFH.K =MIN(MXGLOW,MAX(0,GLOWFH.J+           Glowfish population
                 CHGLOW.J)
A CHGLOW.K=(DT)(BGLOW.JK−DGLOW.JK)               Change in population
*
* ***** SALMON section
*
R BSALM.KL =FIFGE(0,SBRATE*SALMON.K,             Salmon birth rate
                  PROP.K,SALMPC)
A PROP.K= FIFZE(1,SALMON.K/GLOWFH.K+1),          Proportion salmon/
                 GLOWFH.K)                       glowfish
R DSALM.KL =0                                    Salmon death rate
L SALMON.K=SALMON.J+(DT)(BSALM.JK−               Salmon population
              DSALM.JK)+INSALM.J
A INSALM.K =STEP(RLSALM,RLTIME)*FIFZE(1,0,       Initial salmon
              TTRELS.K)                          population
A TTRELS.K =RLTIME−TIME.K                        Time to release of
*                                                salmon
* ***** constant values
*
C MXGLOW=500000 Max. size of glowfish population
C BCGLOW =3     % growth of glowfish population/year
C DCGLOW =120   Number of yearly glowfish deaths
C BCSALM =1.5   % growth of salmon population/year
C RLYEAR =3     Number of years before salmons are released
C RLSALM =200   Initial number of released salmon
C SALMPC =.1    Ratio: glowfish/salmon (when salmon don't grow)
C SPYEAR =365   (365/DT -> samples per year)
*
* ***** initial values
```

```
*
N GLOWFH =5000                                 Glowfish population
N RLTIME  =SPYEAR*RLYEAR                       Salmon release time
                                               (in periods)
N SALMON =0                                    Salmon population
N GBRATE =EXP(LOGN(BCGLOW)/SPYEAR)-1           Glowfish birth rate
N GDRATE =DCGLOW/SPYEAR                        Glowfish death rate
N SBRATE  =EXP(LOGN(BCSALM)/SPYEAR)-1          Salmon birth rate
*
* ***** control section
*
SPEC DT=1,LENGTH=7300,PRTPER=73,PLTPER=73
PLOT GLOWFH=G/SALMON=S
*
* ***** run control
*
RUN 200 SALMON   set up experiment with 200 salmons;
...              further experiments ...
```

INPUT: A Glowfish/Salmon model {program needs no data}

OUTPUT: A Glowfish/Salmon Model

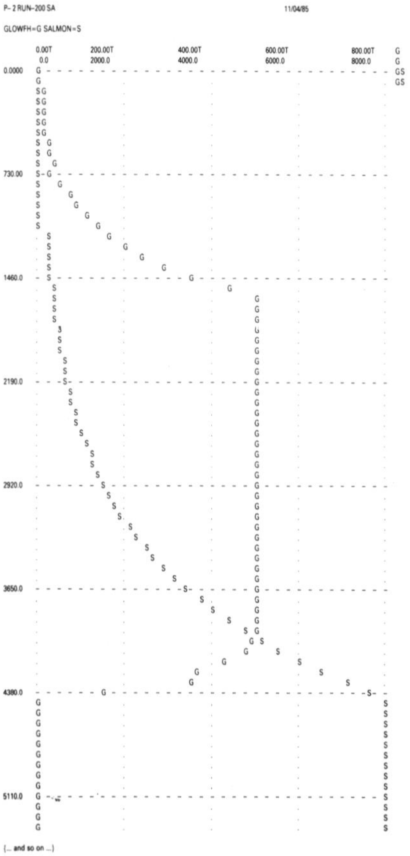

```
begin
comment: Solution to 5.4.1 in SIMULA;

external class simset;

simset begin

comment: Writes into an outfile referenced by 'outf',
         the file is opened and named "hereafter";

  ref (outfile) Outf;
  ref (infile  ) Inf ;
  ref (Universe) Giddeon;
  integer Size , Generations;
  integer RowRim , ColRim;
  integer Rows , Cols;

  link class Image;
  begin
    character array b (0:RowRim, 0:ColRim);

    procedure read;
    begin
      integer i, j;
    inspect inf do
      begin
      for i := 1 step 1 until Rows do
        begin
        inimage;
        for j := 1 step 1 until Cols do
          if inchar = '.'
            then b(i,j) := ' '
            else  b(i,j) := '*';
        end of row loop;
      end of "inspect inf ...";
    end of read procedure;

    procedure paintrim (marking); character marking;
    begin
      integer i;
    for i := 0 step 1 until ColRim  do
      b (0,i) := b (RowRim, i) := marking;
    for i := 0 step 1 until RowRim do
      b (i,0) := b (i, ColRim ) := marking;
    end of paintrim procedure;

    procedure print1row (row); integer row;
    begin
    inspect outf do
      begin
        integer j;
      for j := 0 step 1 until ColRim do
        outchar ( b( row, j));
      end of operation on outfile;
    end of print1row procedure;
```

```
    procedure printaboard;
    begin
      integer i;
    inspect outf do
      begin
      for i := 0 step 1 until RowRim do
        begin
        print1row (i);
        outimage;
        end;
      end;
    end of printaboard procedure;

  end of class Image *****;

  link class Board (rows, cols); integer rows, cols;
  begin
    ref (head ) memory;
    ref (image) oldboard, newboard;
    integer     width, imagesperline, line, boardno, boardrow;

    class boardprinter;
    begin
      ref (image) it;
    end of class boardprinter *****;

    procedure makenewimage;
    begin
    oldboard :- newboard;
    oldboard.into (memory);
    newboard :- new image;
    newboard .paintrim ('+');
    end of makenewimage procedure;

    procedure recallboards;
    begin
    tryagain:
      outtext ("What shall be the width of the printfile ?");
      outimage;
      inimage; width:= inint;
      if width < 0 or width > 130 then
        begin
        outtext ("No way"); outimage;
        goto tryagain;
        end;

    outf :- new outfile ("Life.results");
    outf .OPEN (blanks (width));
    imagesperline := (width - 10) / (colrim + 2);

    while not memory.empty do
      begin
        ref (boardprinter) array elves (1:imagesperline);
      for boardno := 1 step 1 until imagesperline do
        begin
        elves (boardno) :- new boardprinter;
```

```
        if not memory.empty then
           begin
           elves (boardno).it :- memory.first;
           elves (boardno).it .out;
           end
        else
           elves (boardno).it :- none;
        end of "for boardno ...";
      for line := 0 step 1 until rowrim do
        begin
        for boardno := 1 step 1 until imagesperline do
           begin
           if elves (boardno).it =/= none then
              elves (boardno).it.print1row (line);
           outf.outtext ("   ");
           end of "for boardno ...";
        outf.outimage;
        end of "for line ...";
      outf.outimage;
      end of "while not memory.empty ...";
   end of procedure recallboards;

comment: ***** Main Body of class Board *****;

memory :- new head;
oldboard :- new image;
oldboard .read;
oldboard .paintrim ('#');
newboard :- new image;
newboard .paintrim ('+');

end of class Board *****;

Board class universe (generations); integer generations;
begin
   integer    genno, kirlians;
   character dead, alive;
   integer procedure liveneighbours (b,i,j); integer i, j;
                               character array b;
   begin
      integer t;
   t := 0;
   if b (i-1, j-1) = alive then t := t+1;
   if b (i-1, j   ) = alive then t := t+1;
   if b (i-1, j+1) = alive then t := t+1;
   if b (i   , j-1) = alive then t := t+1;
   if b (i   , j+1) = alive then t := t+1;
   if b (i+1, j-1) = alive then t := t+1;
   if b (i+1, j   ) = alive then t := t+1;
   if b (i+1, j+1) = alive then t := t+1;
   liveneighbours := t;
   end of liveneighbours procedure;

comment: ***** Main Body of class universe *****;

dead := ' ';
```

```
alive := '*';

inspect outf do
  begin
    integer i;
  outtext ("First Generation"); outimage;
  outtext ("**************"); outimage; outimage;
  oldboard.printaboard;
  outimage;
  end of print of the seeding generation;

  for genno := 0 step 1 until generations do
    begin
      integer   i, j;
      character ch;
    inspect oldboard do
      for i := 1 step 1 until rows do
        for j := 1 step 1 until cols do
          begin
          kirlians    := liveneighbours (b,i,j);
          ch          := b (i,j);
          newboard.b (i,j) := if (kirlians = 3) or
                                (kirlians = 2 and ch = alive) then
                                alive
                              else
                                dead;
          end of for nest;
      makenewimage;
    end of simulation loop;

  recallboards;
  end of class universe *****;

comment: ***** Body of main PROGRAM *****;

outtext ("Results will be provided in file Life.results"); outimage; outimage;
outimage;
outtext ("What is the size of the universe ?"); outimage;
outtext ("(always a rectangle)"); outimage;
inimage; size:= inint; Rows := Cols := size;
RowRim := Rows + 1;
ColRim  := Cols  + 1;
outtext ("How many generations do you wish to study ?"); outimage;
inimage; generations:= inint;
outimage;
inf :- new infile ("Life.data");
inf.open (blanks (80));
outimage;
outtext ("Thank You"); outimage;
giddeon :- new universe (size, size, generations);
inf.close;
outf.close;

end of simset context;

end of program;
```

INPUT: Life Game

```
7   {number of row in col. cells in our "universe"}
7   {number of new generations to be simulated   }
40  {number of columns in printfile               }

***** this program reads data from file "Life.data" (see below),
      and prints its results into file "Life.results"    *****

. . . . . . .        {configuration of "seedling" generation in file "Life.data"}
. . . . . . .
. . . * . . .
. . * * * . .
. . . * . . .
. . . . . . .
. . . . . . .
```

OUTPUT: Life Game

```
+++++++++   +++++++++   +++++++++
+       +   +       +   +       +
+       +   +   *   +   +   *   +
+  ***  +   +  * *  +   +  ***  +
+  * *  +   + *   * +   + ** ** +
+  ***  +   +  * *  +   +  ***  +
+       +   +   *   +   +   *   +
+       +   +       +   +       +
+++++++++   +++++++++   +++++++++

+++++++++   +++++++++   +++++++++
+       +   |   *   +   +  ***  +
+  ***  +   +  ***  +   +       +
+ *   * +   + * * * +   +*     *+
+ *   * +   +*** ***+   +*     *+
+ *   * +   + * * * +   +*     *+
+  ***  +   +  ***  +   +       +
+       +   +   *   +   +  ***  +
+++++++++   +++++++++   +++++++++

+++++++++   +++++++++
+   *   +   +       +
+   *   +   +       +
+       +   +       +
+**   **+   +       +
+       +   +       +
+   *   +   +       +
+   *   +   +       +
+++++++++   +++++++++
```

```
program DiningPhilosophers (INPUT, OUTPUT);
{Solution to 5.4.2 - an event-oriented version - in PASCAL (& toolboxes)}

const
  MaxHistoClasses =       20;
  MaxPlotPts      =       50;
  NoDelay         =      0.0;
  AlongNumber     =   MAXINT;
  PhilNumber      =        5;
  ForkNumber      =        5;
type
  anEventType    = (startThinking , finishThinking, startSleeping,
                    finishSleeping, startEating   , finishEating );
  aForkIndexType = 1 .. ForkNumber;
  aPhilstateType = (think, sleep, eat, inBetween);

{insert declarations for  Monte Carlo       toolbox}
{insert declarations for  statistics        toolbox}
{insert declarations for  discrete-event    toolbox}
{insert declarations for  queueing network  toolbox}
{it must be modified to include philosophers'
 attributes : leftFork , rightFork : aForkIndexType;
              state                : aPhilStateType ;
              startWaiting         : real;
              id                   : 1 .. PhilNumber ;
 as part of the "TArecordType" declaration !!!}

var
  TheClock    : anAsynchClockType;
  TheMonitor  : anEventSchedulingMonitorType;
  Source      : aSourceType;
  Forks       : array [1 .. ForkNumber] of aServerType;
  EatDelay    : array [1 .. PhilNumber ] of aTallyType;
  Thinking    : aUniformGenType;
  Sleeping    : aNegExpGenType;
  Eating      : aNormalGenType ;
  ForkQ       : aQueueType;
  SimTime     : real;

{insert procedures from the  Monte Carlo       toolbox}
{insert procedures from the  statistics        toolbox}
{insert procedures from the  discrete-event    toolbox}
{insert procedures from the  queueing network toolbox }

procedure ReadAndInitialize;
var
  simTime : real;
  index   : 0 .. MAXINT;
  him     : aTAPtrType;
begin
  READ (SimTime);            {30 hours}
  InitAsynchClock (TheClock);
  InitEventSchedulingMonitor (TheMonitor, simTime * 60, AlongNumber);
  InitUniformGen          (Thinking, 12345, 20,  30);
  InitNegExpGen           (Sleeping , 67891,     180);
```

```
  InitNormalgen              (Eating   , 23457, 20,   5);
  InitSource                 (Source);
  InitQueue                  (ForkQ);
  for index := 1 to ForkNumber do
    InitServer (Forks [index], 1);
  for index := 1 to PhilNumber  do
    begin
    him              := CreateNewTA (Source);
    him ↑ .id        := index;
    him ↑ .leftFork  := index;
    if index = 5 then
      him ↑ .rightFork := 1        {it is a circular (!!!) table}
    else
      him ↑ .rightFork := index + 1 ;
    him ↑ .state     := inBetween;
    ScheduleEvent (TheMonitor, NoDelay, startThinking, him);
    InitTally (EatDelay [index]);
    end;
end;

procedure WriteReport;
var
  index : 1..MAXINT;
begin
WRITE    ('After ', SimTime, ' hours the following waiting times were');
WRITELN (' encountered by each philosopher:'); WRITELN;
for index := 1 to PhilNumber do
  begin
  WRITELN ('Phil. Number ', index:2);
  ReportOnTally (EatDelay [index]);
  WRITELN;
  end;
WRITELN; WRITELN ('The Fork utilizations were:'); WRITELN;
for index := 1 to ForkNumber do
  begin
  WRITELN ('Fork Number ', index:2);
  ReportOnServer (Forks [index]);
  WRITELN;
  end;
end;

procedure StartThinkingEvent   (who : aTAPtrType);
begin
who ↑ .state := think;
ScheduleEvent (TheMonitor, Uniform (Thinking), FinishThinking, who);
end;

procedure FinishThinkingEvent (who : aTAPtrType);
begin
who ↑ .state := inBetween;
ScheduleEvent (TheMonitor, NoDelay, startSleeping, who);
end;

procedure StartSleepingEvent   (who : aTAPtrType);
begin
```

```
who ↑ .state := sleep;
ScheduleEvent (TheMonitor, NegExp (Sleeping), finishSleeping, who);
end;

procedure FinishSleepingEvent  (who : aTAPtrType);
begin
who ↑ .state          := inBetween;
who ↑ .startWaiting := TheClock.NOW;
FileIntoFIFOq (ForkQ, who);
ScheduleEvent (TheMonitor, NoDelay, startEating, nil);
end;

procedure StartEatingEvent;
var
  who : aTAPtrType;
begin
who := TakeFirstFromQ (ForkQ);
if (EnoughServerUnitsAvailableP (Forks [who ↑ .leftFork  ], 1)) and
  (EnoughServerUnitsAvailableP (Forks [who ↑ .rightFork], 1)) then
  begin
  SeizeServer (Forks [who ↑ .leftFork  ], 1);
  SeizeServer (Forks [who ↑ .rightFork], 1);
  who ↑ .state:= eat;
  ScheduleEvent (TheMonitor, Normal (Eating), FinishEating, who);
  end
else
  begin
  {back to the end of the waiting line}
  FileIntoFIFOq (ForkQ, who);
  end;
end;

procedure FinishEatingEvent    (who : aTAPtrType);
begin
who ↑ .state := inBetween;
ReleaseServer (Forks [who ↑ .leftFork  ], 1);
ReleaseServer (Forks [who ↑ .rightFork], 1);
UpdateTally (EatDelay [who ↑ .id], TheClock.NOW - who ↑ .startWaiting);
ScheduleEvent (TheMonitor, NoDelay, startThinking, who);
end;

begin {*** main Body ***}

ReadAndInitialize;
while  MoreTimeToSimulateP (TheMonitor) and MoreEventsP (TheMonitor) do
  with TheMonitor do
    begin
    GetNextEventNotice (TheMonitor);
    case KindOfEventQ (currentEvent) of
      startThinking   : StartThinkingEvent    (currentEvent ↑ .thisTA);
      finishThinking : FinishThinkingEvent (currentEvent ↑ .thisTA);
      startSleeping   : StartSleepingEvent     (currentEvent ↑ .thisTA);
      finishSleeping  : FinishSleepingEvent   (currentEvent ↑ .thisTA);
      startEating      : StartEatingEvent;
      finishEating     : FinishEatingEvent      (currentEvent ↑ .thisTA);
    end {case};
```

```
    FlushOutEvent (currentEvent);
    end {simulation loop};
WriteReport;
end.
```

INPUT: Dining Philosophers

30 {length of simulation (in minutes) }

OUTPUT: Dining Philosophers

After 30 hours the following waiting times were encountered by each customer:

Phil. Number 1
Until 1791.52 , 8 observations were made.

This tally's characteristics were as follows:

Mean : 182.01
Sigma : 53.04
Min. obs.: 63.45
Max. obs.: 243.67

Phil. Number 2
Until 1791.52 , 9 observations were made.

This tally's characteristics were as follows:

Mean : 163.76
Sigma : 60.06
Min. obs.: 22.34
Max. obs.: 203.83

Phil. Number 3
Until 1791.52 , 9 observations were made.

This tally's characteristics were as follows:

Mean : 172.35
Sigma : 38.77
Min. obs.: 101.35
Max. obs.: 201.95

Phil. Number 4
Until 1791.52 , 9 observations were made.

This tally's characteristics were as follows:

Mean : 167.42
Sigma : 49.13
Min. obs.: 58.82
Max. obs.: 199.66

Phil. Number 5
Until 1791.52 , 9 observations were made.

```
This tally's characteristics were as follows:

Mean     : 155.28
Sigma    :  67.04
Min. obs.:  19.18
Max. obs.: 209.90


The Fork utilizations were:

Fork Number 1
The capacity of this server is    1 units.
1 of these are currently in use.
Utilization of this server was 18.42 %, since last reset.

Fork Number 2
The capacity of this server is    1 units.
0 of these are currently in use.
Utilization of this server was 18.86 %, since last reset.

Fork Number 3
The capacity of this server is    1 units.
0 of these are currently in use.
Utilization of this server was 20.52 %, since last reset.

Fork Number 4
The capacity of this server is    1 units.
0 of these are currently in use.
Utilization of this server was 19.56 %, since last reset.

Fork Number 5
The capacity of this server is    1 units.
1 of these are currently in use.
Utilization of this server was 19.03 %, since last reset.
```

```
begin
comment: Solution to 5.4.4 in SIMULA class DEMOS;

External class demos;

demos begin
   ref (waitq) array SpellQ (1:3);
   ref (tally)  SpellTransits;
   ref (count)  BoatCount;
   real         TotalBoatCapacity, UsedBoatCapacity;
   integer      Kadatheron, Morven, SonaNyl;
   real         array MeanVoyageFromCity       (1:3);
   integer      array NextCity                 (1:3);
   ref (rdist)  array LengthOfVoyageFromCity   (1:3);
   ref (rdist)  array TimeToSinkFromCity       (1:3);
   ref (rdist)  BoatBuildingTime;
   ref (bdist)  SinkProb;
```

```
entity class aSource (city); integer city;
begin
  real lambda;
  ref (rdist) timeBetweenSpells;
  ref (idist) destinationDist;
  ref (aSpell) curse;
lambda := inreal;
timeBetweenSpells :- new negexp ("between spells", lambda);
destinationDist   :- new randint ("destinations", 1, 3);
loop:
  curse :- new aSpell ("spell", city);
  curse.dest := destinationDist.sample;
  while curse.dest = city do
    curse.dest := destinationDist.sample;
  curse.schedule (0.0);
  hold (timeBetweenSpells.sample);
repeat;
end;

entity class aSpell (city);integer city;
begin
  boolean reachedP;
  real    creationTime;
  integer dest;
creationTime := time;
reachedP     := false;
SpellQ (city) .wait;
if reachedP then
  SpellTransits.update (time - creationTime);
end;

entity class aBoat (city); integer city;
begin
  real                 tripLength;
  ref (aSpell) array spellLoad (1:3);
  integer              spellsCarried;
  boolean              sunkP;
spellsCarried  := 0;
sunkP := false;
BoatCount.update (1);

while not sunkP do
  begin
    integer i;
  for i := 1 step 1 until 3 do
    begin     comment: unload and load spells in slot i;
    if spellLoad (i) =/= none then
      begin
      if spellLoad (i).dest = city then
        begin
        spellLoad(i).reachedP := true;
        spellLoad(i).schedule (now);
        spellLoad(i):- none;
        spellsCarried := spellsCarried - 1;
        end of inner if;
      end of outer if;
    end for;
```

```
    for i := 1 step 1 until 3 do
      if (spellLoad (i) == none) and (spellQ (city).length > 0) then
        begin        comment: load waiting spell into empty slot;
        spellLoad (i) :- spellQ (city).coopt;
        spellsCarried := spellsCarried + 1;
        end of inner if & for;

    TripLength := LengthOfVoyageFromCity (city).sample;
    comment: days the next leg of the voyage is going to take;

    if SinkProb.sample then
      begin        comment: boats sink during voyage;
      sunkP := true;
      hold (TimeToSinkFromCity (city).sample);
      spellsCarried := 0;
      comment: order a new boat to be built at SonaNyl !;
      new aBoat ("Boat", SonaNyl).schedule (BoatBuildingTime.sample);
      for i := 1 step 1 until 3 do
        begin        comment: lay all spells to rest;
        if spellLoad (i) =/= none then
          begin
          spellLoad (i).reachedP := false;
          spellLoad (i).schedule (now);
          spellLoad (i)  :- none;
          end;
        end;
      end
    else
      begin        comment: boat will arrive safely at next destination;
      hold (TripLength);
      TotalBoatCapacity := TotalBoatCapacity + 3 * TripLength;
      UsedBoatCapacity := UsedBoatCapacity + spellsCarried * TripLength;
      city := NextCity (city);
      end of outer if;
    end of while;
end of class Boat;

comment: ********* main program *********;

outf :- new outfile ("SpellNet.results");
outf.open (blanks (80));

comment: initialize town indices;
Kadatheron := 1;
Morven     := 2;
SonaNyl    := 3;

trace;
comment : initialize arrays of "next cities";
NextCity (Kadatheron) := Morven;
NextCity (Morven     ) := SonaNyl;
NextCity (SonaNyl    ) := Kadatheron;

comment: read array of mean lengths of voyages from each city to the next;
begin
  integer i;
```

```
for i := 1 step 1 until 3 do
  MeanVoyageFromCity (i) := inreal;
end;

comment: instantiate the probability distributions;
begin
  integer i;
for i := 1 step 1 until 3 do
  LengthOfVoyageFromCity (i) :- new normal ("Voy. Lengths",
                                    MeanVoyageFromCity (i),
                              0.1 * MeanVoyageFromCity (i));
for i := 1 step 1 until 3 do
  TimeToSinkFromCity (i) :- new uniform ("Time to sink", 0,
                                  MeanVoyageFromCity (i) );
SinkProb          :- new draw ("Prob. of sinking", inreal);
BoatBuildingTime :- new uniform ("to build a boat" , inreal, inreal);
end of instantiation block for global distributions;

comment: create the three dungeons;
new aSource ("Kadatheron", Kadatheron).schedule (0.0);
new aSource ("Morven     ", Morven    ).schedule (0.0);
new aSource ("Sona Nyl   ", SonaNyl   ).schedule (0.0);

comment: create the first three boats and position them;
new aBoat ("Boat", Kadatheron).schedule (0.0);
new aBoat ("Boat", Morven    ).schedule (0.0);
new aBoat ("Boat", SonaNyl   ).schedule (0.0);

comment: instantiate and initialize everything else;
TotalBoatCapacity    :-  UsedBoatCapacity  :=  0;
BoatCount            :- new count("Boats used");
SpellTransits        :- new tally  ("Spell's transit times");
SpellQ (Kadatheron):- new waitq("Kadatheron Q ");
SpellQ (Morven     ):- new waitq("Morven Q.");
SpellQ (SonaNyl   ) :- new waitq("SonaNyl Q.");

box ("A Simulation-Analysis of SpellNet");
hold (10);
notrace;
hold (inreal);        comment: presimulation period;
TotalBoatCapacity := UsedBoatCapacity := 0;
reset;
hold (inreal);        comment: simulation period;

report;
if TotalBoatCapacity > 0 then
  begin
  outf.outimage;
  outf.outtext ("The Boats' total utilization was: ");
  outf.outfix (UsedBoatCapacity / TotalBoatCapacity, 3, 6);
  outf.outimage;         outf.close;
  end;
noreport;

end of demos context;
end of program;
```

INPUT: SpellNet Simulation

```
3       {travelling time (in days) from Kadatheron to Morven     }
4       {travelling time           from Morven     to Sona Nyl   }
2       {travelling time           from Sona Nyl   to Kadatheron}
0.3     {probability of sinking on each journey                  }
2 5     {min. & max. time it takes to build a new boat           }
1/7     {spell generation rate at Kadatheron                     }
1/3     {spell generation rate at Morven                         }
1/5     {spell generation rate at Sona Nyl                       }
15 60   {duration of presimulation & simulation (in days)        }
```

OUTPUT: SpellNet Simulation

```
                              CLOCK TIME =          0.000
*******************************************************************************
*                                                                             *
*                           TRACING COMMENCES                                 *
*                                                                             *
*******************************************************************************

TIME   CURRENT       AND ITS ACTION(S)

0.000  DEMOS         SCHEDULES Kadatheron1 NOW
                     SCHEDULES Morven    1 NOW
                     SCHEDULES Sona Nyl  1 NOW
                     SCHEDULES Boat 1 NOW
                     SCHEDULES Boat 2 NOW
                     SCHEDULES Boat 3 NOW

*******************************************************************************
*                                                                             *
*                    A Simulation Analysis of SpellNet                        *
*                                                                             *
*******************************************************************************

                     HOLDS FOR 10.000,  UNTIL 10.000
       Kadatheron 1  SCHEDULES spell  1 NOW
                     HOLDS FOR 19.953,  UNTIL 19.953
       Morven     1  SCHEDULES spell  2 NOW
                     HOLDS FOR 6.670,   UNTIL 6.670
       Sona Nyl   1  SCHEDULES spell  3 NOW
                     HOLDS FOR 0.612,   UNTIL 0.612
       Boat 1        HOLDS FOR 1.891,   UNTIL 1.891
       Boat 2        HOLDS FOR 4.318,   UNTIL 4.318
       Boat 3        HOLDS FOR 2.002,   UNTIL 2.002
       spell  1      WAITS IN Kadatheron Q
       spell  2      WAITS IN Morven Q.
       spell  3      WAITS IN SonaNyl Q.
0.612  Sona Nyl   1  SCHEDULES spell  4 NOW
                     HOLDS FOR 1.777,   UNTIL 2.389
       spell  4      WAITS IN SonaNyl Q.
1.891  Boat 1        SCHEDULES Boat 4 AT 4.302
                     ***TERMINATES
```

```
2.002  Boat 3          COOPTS spell  1 FROM Kadatheron Q
                       HOLDS FOR 2.314,    UNTIL 4.315
2.389  Sona Nyl     1  SCHEDULES spell  5 NOW
                       HOLDS FOR 2.645,    UNTIL 5.034
       spell  5        WAITS IN SonaNyl Q.
4.302  Boat 4          COOPTS spell  3 FROM SonaNyl Q.
                       COOPTS spell  4 FROM SonaNyl Q.
                       COOPTS spell  5 FROM SonaNyl Q.
                       HOLDS FOR 2.221,    UNTIL 6.523
4.315  Boat 3          SCHEDULES spell  1 NOW
       spell  1        ***TERMINATES
       Boat 3          COOPTS spell  2 FROM Morven Q.
                       HOLDS FOR 2.890,    UNTIL 7.205
4.318  Boat 2          HOLDS FOR 1.141,    UNTIL 5.459
5.034  Sona Nyl     1  SCHEDULES spell  6 NOW
                       HOLDS FOR 0.870,    UNTIL 5.905
       spell  6        WAITS IN SonaNyl Q.
5.459  Boat 2          SCHEDULES Boat 5 AT 10.292
                       ***TERMINATES
5.905  Sona Nyl     1  SCHEDULES spell  7 NOW
                       HOLDS FOR 1.330,    UNTIL 7.234
       spell  7        WAITS IN SonaNyl Q.
6.523  Boat 4          SCHEDULES spell  4 NOW
       spell  4        ***TERMINATES
       Boat 4          SCHEDULES spell  5 NOW
       spell  5        ***TERMINATES
       Boat 4          HOLDS FOR 3.390,    UNTIL 9.914
6.670  Morven       1  SCHEDULES spell  8 NOW
                       HOLDS FOR 0.605,    UNTIL 7.275
       spell  8        WAITS IN Morven Q.
7.205  Boat 3          SCHEDULES Boat 6 AT 11.853
                       SCHEDULES spell  2 NOW
       spell  2        ***TERMINATES
       Boat 3          ***TERMINATES
7.234  Sona Nyl     1  SCHEDULES spell  9 NOW
                       HOLDS FOR 0.740,    UNTIL 7.974
       spell  9        WAITS IN SonaNyl Q.
7.275  Morven       1  SCHEDULES spell 10 NOW
                       HOLDS FOR 0.438,    UNTIL 7.713
       spell 10        WAITS IN Morven Q.
7.713  Morven       1  SCHEDULES spell 11 NOW
                       HOLDS FOR 3.584,    UNTIL 11.298
       spell 11        WAITS IN Morven Q.
7.974  Sona Nyl     1  SCHEDULES spell 12 NOW
                       HOLDS FOR 2.004,    UNTIL 9.979
       spell 12        WAITS IN SonaNyl Q.
9.914  Boat 4          SCHEDULES spell  3 NOW
       spell  3        ***TERMINATES
       Boat 4          COOPTS spell  8 FROM Morven Q.
                       COOPTS spell 10 FROM Morven Q.
                       COOPTS spell 11 FROM Morven Q.
                       HOLDS FOR 4.305,    UNTIL 14.219
9.979  Sona Nyl     1  SCHEDULES spell 13 NOW
                       HOLDS FOR 0.180,    UNTIL 10.159
       spell 13        WAITS IN SonaNyl Q.
```

```
                         CLOCK TIME =            10.000
*****************************************************************************
*                                                                           *
*                        TRACING SWITCHED OFF                               *
*                                                                           *
*****************************************************************************

                         CLOCK TIME =           430.000
*****************************************************************************
*                                                                           *
*                              REPORT                                       *
*                                                                           *
*****************************************************************************
```

D I S T R I B U T I O N S
****

TITLE	(RE)SET	OBS	TYPE	A	B	SEED
Voy. Lengths	70.000	101	NORMAL	3.000	0.300	33427485
Voy. Lengths	70.000	78	NORMAL	4.000	0.400	22276755
Voy. Lengths	70.000	143	NORMAL	2.000	0.200	46847980
Time to sink	70.000	23	UNIFORM	0.000	3.000	43859043
Time to sink	70.000	19	UNIFORM	0.000	4.000	64042082
Time to sink	70.000	41	UNIFORM	0.000	2.000	44366385
Prob. of sin	70.000	322	DRAW	0.300		41357879
to build a b	70.000	83	UNIFORM	2.000	5.000	11320893
between spel	70.000	64	NEGEXP	0.200		6528269
destinations	70.000	93	RANDINT	1	3	47478000
between spel	70.000	180	NEGEXP	0.500		46802881
destinations	70.000	279	RANDINT	1	3	59224073
between spel	70.000	357	NEGEXP	1.000		22046052
destinations	70.000	533	RANDINT	1	3	65931384

C O U N T S
****

TITLE	(RE)SET	OBS
Boats used	70.000	84

T A L L I E S
****

TITLE	(RE)SET	OBS	AVERAGE	EST.ST.DV	MIN	MAX
Spell's tran	70.000	379	14.493	6.360	2.525	31.501

W A I T Q U E U E S
****

TITLE	(RE)SET	OBS	QMAX	QNOW	Q AVERAGE	ZEROS	AV. WAIT
Kadatheron Q	70.000	64	1	0	0.000	64	0.000
Kadatheron Q*	70.000	64	5	0	0.548	0	3.080
Morven Q.	70.000	182	1	0	0.000	182	0.000
Morven Q. *	70.000	182	13	2	4.805	0	9.620

SonaNyl Q.		70.000	342	1	0	0.000	342	0.000
SonaNyl Q.	*	70.000	342	27	23	11.550	0	11.733

The Boats' total utilization was: 0.815

```
begin
comment: Solution to 5.5.1 in SIMULA class DISCO;

external class disco;

disco begin
ref (head)      FireStation, AlarmQ;
ref (histogram) Fires, Damages, Remains;
ref (count)     Burnt, Saved;
integer         I, Seed;
integer         NoOfDragons;
real            MinMat,       MaxMat;
real            MinTravel,    MaxTravel;
real            MinFireRate, MaxFireRate;
real            MinFireSize, MaxFireSize;
real            MinsBetweenFires;
real            SimLength;

link class Alarm (House); ref (HouseOnFire) House; ;

process class HouseOnFire;
begin
  real material, extinguishrate, traveltime, intensity;
  ref (statvar) size, damage;
  ref (Fire   ) aFire;
extinguishrate := 0.0;
material      := uniform (MinMat, MaxMat, Seed)                ;
traveltime    := uniform (MinTravel, MaxTravel, Seed)          ;
intensity     := uniform (MinFireRate, MaxFireRate, Seed)      ;
size     :- new statvar(uniform (MinFireSize, MaxFireSize, Seed));
damage   :- new statvar(size.state)                            ;
aFire    :- new Fire (this HouseOnFire)                        ;

new Alarm (this HouseOnFire).into (AlarmQ);
aFire.start; damage.start; size.start;
outtext ("call a dragon at: "); outfix (time, 2, 10); outimage;
activate FireStation.first;
waituntil (size.state <= 0 or damage.state >= material);
if size.state <= 0 then
  begin
  outtext ("Fire out at: "); outfix (time, 2, 10);
  Saved.update (1);
  Remains.update (material - damage.state);
  end
```

```
else
  begin
  outtext ("House burnt at: "); outfix (time, 2, 10);
  Burnt.update (1);
  end;
outimage; outimage;
aFire.stop; damage.stop; size.stop;
Fires     .update (size     .max);
Damages .update (damage.state);
end;

continuous class Fire (House); ref (HouseOnFire) House;
inspect House do
  begin
  size.rate     := intensity * size.state - extinguishRate;
  damage.rate := size.state;
  end;

process class FireDragon (Capacity); real Capacity;
while true do
  inspect AlarmQ.first when Alarm do
    begin
    outtext ("Attend to alarm at: "); outfix (time, 2, 10); outimage;
    this Alarm.out;
    if House.aFire.active then
      begin
      this FireDragon.out;
      hold (House.travelTime);
      outtext ("Arrive at fire at: "); outfix (time, 2, 10); outimage;
      if House.size.rate > capacity then
        begin
        this Alarm.into (AlarmQ);
        outtext ("Call another dragon at: "); outfix (time, 2, 10); outimage;
        House.aFire.stop; House.damage.stop; House.size.stop;
        activate FireStation.first;
        House.aFire.start; House.damage.start; House.size.start;
        end;
      if House.aFire.active then
        begin
        outtext ("Start fighting at: "); outfix (time, 2, 10); outimage;
        House.extinguishRate := House.extinguishRate + Capacity;
        waituntil (not House.aFire.active);
        end;
      outtext ("rush home at: "); outfix (time, 2, 10); outimage;
      hold (House.travelTime);
      this FireDragon.into (FireStation);
      end;
    end
  otherwise passivate;

process class FireBug;
begin
while true do
  begin
    real timeToNextFire;
```

```
  timeToNextFire := negexp (1 / MinsBetweenFires, Seed);
  outimage;
  outtext ("Plan the next fire at: "); outfix (time + timeToNextFire, 2, 10);
  outimage;
  hold (timeToNextFire);
  outtext ("Start a fire at: "); outfix (time, 2, 10); outimage;
  activate new HouseOnFire;
  end;
end;

comment: ***** main Body *****;

outtext ("Enter the model's parameters: "); outimage;
dtmax            := 10000;
maxrelerror      := 0.00001;
Seed             := 12345;
MinMat           := inreal; MaxMat      := inreal;
MinTravel        := inreal; MaxTravel   := inreal;
MinFireRate      := inreal; MaxFireRate := inreal;
MinFireSize      := inreal; MaxFireSize := inreal;
MinsBetweenFires := inreal; outimage;

FireStation :- new head;
AlarmQ      :- new head;
Burnt       :- new count ("Burned Houses");
Saved       :- new count ("Saved   Houses");
outtext ("Now enter parameters of histograms: "); outimage;
Fires       :- new histogram   ("Fire Sizes", inreal, inreal, inint);
Damages     :- new histogram   ("Damages", inreal, inreal, inint);
Remains     :- new histogram   ("Remains ", inreal, inreal, inint);
outimage;

outtext ("How many dragons and what is each of their capacity ?");
outimage;
NoOfDragons := inint;
for i := 1 step 1 until NoOfDragons do
  new FireDragon (inint).into (FireStation);
activate new FireBug;

outtext ("Length of simulation (in mins.) ?"); outimage;
SimLength := inreal;
hold (SimLength / 2);
dump;
hold (SimLength / 2);

outimage; outtext ("Statistics about the fires: "); outimage; outimage;

end;
end;
```

INPUT: FireDragons

```
100 600   {Min. & Max. material                  }
5 15      {Min. & Max. travel time (in mins.)    }
0.05 0.3  {Min. & Max. fire intensity            }
0 5       {Min. & Max. size of initial fire      }
3600      {Mean mins. between fires              }
0 200 10  {lb, ub and number of classes for      }
          {"FireSize"  histogram                 }
0 600 10  {lb, ub and number of classes for      }
          {"Damage" histogram                    }
0 600 10  {lb, ub and number of classes for      }
          {"Remains" histogram                   }
3         {number of fire-fighting dragons       }
3 *       {Extinguishing capacity for each       }
          {dragon   (*times)                     }
86400     {length of simulation (in mins.)       }
          { (= 60 days !)                        }
```

OUTPUT: FireDragons

```
Plan the next fire at:      1522.39
Length of simulation (in mins.) ?
Start a fire at:           1522.39
Call a dragon at:           1522.39
Attend to alarm at:           1522.39

Plan the next fire at:             4092.90

Arrive at fire at:            1534.13
Start fighting at:            1534.13
Fire out at:      1536.45

Rush home at:     1536.45

Start a fire at:           4092.90
Call a dragon at:           4092.90
Attend to alarm at:           4092.90

{ ... and so on ... }

Arrive at fire at:            4790.86
Call another dragon at:            4790.86
Attend to alarm at:           4790.86
Start fighting at:            4790.86
House burnt at:    4794.85

{ ... and so on ...}

********************************************************************************
*                                                                              *
*                                DISCO-DUMP                                    *
*                                                                              *
********************************************************************************
```

```
                    TIME
              43200.0000

       DTMIN           DTMAX       MAXRELERROR      MAXABSERROR
       0.000      10000.0000        1.000&-05            0.000

          DT        LASTTIME     NEXTEVENTTIME   NEXTREPORTTIME
   4146.5039      39053.4961        43200.0000       43200.0000

                WAITPRIORITY         WAITPRIOR
                       0.000             FALSE

              INTEGRATION METHOD: RKE

******************************************************************************
*                                                                            *
*                     ACTIVE VARIABLE OBJECTS                                *
*                                                                            *
******************************************************************************

                          - NONE -

******************************************************************************
*                                                                            *
*                    ACTIVE CONTINUOUS OBJECTS                               *
*                                                                            *
******************************************************************************

                          - NONE -

******************************************************************************
*                                                                            *
*                     ACTIVE REPORTER OBJECTS                                *
*                                                                            *
******************************************************************************

                          - NONE -

******************************************************************************
*                                                                            *
*                           TIME EVENTS                                      *
*                                                                            *
******************************************************************************

      EVTIME
1:    43200.0000 - CURRENT - MAIN
2:    48496.3516

******************************************************************************
*                                                                            *
*                           STATE EVENTS                                     *
*                                                                            *
******************************************************************************

                          - NONE -
```

```
{ ... and so on ...}
                              CLOCK TIME = 86400.0000
*****************************************************************************
*                                                                           *
*                                  REPORT                                   *
*                                                                           *
*****************************************************************************

                                  COUNTS
                                **********

                    TITLE                  (RE)SET     OBS
                    Burned House             0.000      10
                    Saved  House             0.000      13

                                HISTOGRAMS
                             ****************

                                 SUMMARY

 TITLE          (RE)SET     OBS      AVERAGE      EST.ST.DV       MIN       MAX
 Fire Sizes       0.000      23      27.2166        31.2565    0.2494  125.7445

 CELL     LOWER LIM      N      FREQ    CUM %
                                              | ----------------------
   0      -INFINITY      0      0.00     0.00 |
   1        0.000       13      0.57    56.52 |************************************
   2       20.0000       3      0.13    69.57 |*******
   3       40.0000       5      0.22    91.30 |*************
   4       60.0000       0      0.00    91.30 |
   5       80.0000       1      0.04    95.65 |**
   6      100.0000       0      0.00    95.65 |
   7      120.0000       1      0.04   100.00 |**
                                                  **REST OF TABLE EMPTY**
                                              | ----------------------

                                 SUMMARY

 TITLE          (RE)SET     OBS      AVERAGE      EST.ST.DV       MIN       MAX
 Damages          0.000      23     170.4196       173.0197    1.9223  585.7393

 CELL     LOWER LIM      N      FREQ    CUM %
                                              | ----------------------
   0      -INFINITY      0      0.00     0.00 |
   1        0.000        8      0.35    34.78 |************************************
   2       60.0000       4      0.17    52.17 |*****************
   3      120.0000       3      0.13    65.22 |*************
   4      180.0000       2      0.09    73.91 |*********
   5      240.0000       1      0.04    78.26 |****
   6      300.0000       1      0.04    82.61 |****
   7      360.0000       1      0.04    86.96 |****
   8      420.0000       1      0.04    91.30 |****
   9      480.0000       1      0.04    95.65 |****
  10      540.0000       1      0.04   100.00 |****
  11      600.0000       0      0.00   100.00 |
                                              | ----------------------
```

```
                              SUMMARY

TITLE       (RE)SET     OBS      AVERAGE      EST.ST.DV      MIN       MAX
Remains      0.000      13      266.2441      168.7674     38.3181   543.9957

CELL     LOWER LIM      N       FREQ     CUM %
                                               | ----------------------
  0      -INFINITY      0       0.00      0.00 |
  1        0.000        2       0.15     15.38 |**********************************
  2       60.0000       2       0.15     30.77 |**********************************
  3      120.0000       0       0.00     30.77 |
  4      180.0000       1       0.08     38.46 |***************
  5      240.0000       2       0.15     53.85 |**********************************
  6      300.0000       2       0.15     69.23 |**********************************
  7      360.0000       1       0.08     76.92 |***************
  8      420.0000       1       0.08     84.62 |***************
  9      480.0000       1       0.08     92.31 |***************
 10      540.0000       1       0.08    100.00 |***************
 11      600.0000       0       0.00    100.00 |
                                               | ----------------------
```

References

Abelson, H., Sussman, J. and Sussman, J. (1985). *Structure and Interpretation of Computer Programs*. Cambridge, MA: MIT Press.

Ackoff, R.L. (1962). *Scientific Method: Optimizing Applied Research Decisions*. New York: John Wiley.

Ackoff, R.L. (1970). *A Concept of Corporate Planning*. New York: John Wiley.

Adams, N.R. and Dogramaci, A. (eds.) (1979). *Current Issues in Computer Simulation*. New York: John Wiley.

Adelsberger, H. (1982). 'ASSE – Ada Simulation Support Environment'. *Proceedings Winter Simulation Conference*, ACM, pp. 89–101.

Adelsberger. H. (1984). 'Transaction Flow Models in Ada: Technical Background'. In *Proceedings of an SCS Conference*, La Jolla, 1984, R. Bryant and B.W. Unger (eds.), pp. 110–118.

Alemparte, M., Cheda, D., Seeley, D. and Walker, W. (1975). 'Interaction with Discrete Simulation Using On-Line Graphic Animation', *Computers and Graphics* **1**, pp. 309–318.

Allen, J.R. (1978). *Anatomy of Lisp*. New York: McGraw-Hill.

Allison, L. (1983). 'Stable Marriages by Coroutines', *Inf. Proc. Letters* **16**, pp. 61–65.

Amarel, S. (1968). 'On Representation of Problems of Reasoning about Actions'. In *Machine Intelligence* **3**, D. Michie (ed.), Edinburgh University Press.

Amarel, S. (1971). 'Representations and Modelling in Problems of Program Formation'. In *Machine Intelligence* **6**, B. Meltzer and D. Michie (eds.), Edinburgh University Press.

Andersen, N.E. and Kyng, M. (1978). 'Construction of System Descriptions in SIMULA and DELTA'. Report DAIMI PB-81, Department of Computer Science, University of Aarhus, Denmark.

Anderson, J.R. (1980). *Cognitive Psychology and its Implications*. San Francisco: W.H. Freeman.

Andrews, G.R. and Schneider, F.B. (1983). 'Concepts and Notations for Concurrent Programming', *Computing Surveys* **15**(1), pp. 3–44.

Atkinson, L.V. and Harley, P.J. (1983). *An Introduction to Numerical Methods with PASCAL*. Wokingham: Addison-Wesley.

Bachman, C.W. (1969). 'Data Structure Diagrams', *DATABASE (ACM SIGMOD)* **1**(2).

Banks, J. and Malave, C.O. (1984). 'The Simulation of Inventory Systems: An Overview', *SIMULATION* (June), pp. 283–290.

Barnden, J.A. (1981). 'Nonsequentiality and Concrete Activity Phases in Discrete Event Simulation Languages', *ACM TOPLAS* **3**(3), pp. 293–317.

Barr, A., Cohen, P.R. and Feigenbaum, E.A. (eds.) (1981). *The Handbook of AI, Vol. 1: Knowledge Representation*. Los Altos, CA: W. Kaufman.

Barron, D.W. (1977). *An Introduction to the Study of Programming Languages*. Cambridge: Cambridge University Press.

Barto, A.G. (1978). 'Discrete and Continuous Models', *Int. Journal of General Systems* **4**, pp. 163–177.

Birtwistle, G.M. (1979). *Discrete Event Modelling on SIMULA*. London/Basingstoke: Macmillan.

Birtwistle, G.M. (1984). 'Future Directions in Simulation Software', Panel Discussion. In *Proceedings of an SCS Conference*, La Jolla, 1984, R. Bryant and B.W. Unger (eds.), pp. 120–121.

Birtwistle, G.M. (ed.) (1985). 'AI, Graphics and Simulation'. *Proceedings of Multi-Conference of Society for Computer Simulation*, San Diego, 1985.

Birtwistle, G.M., Dahl, O.J., Myhrhaug, B. and Nygaard, K. (1973). *SIMULA BEGIN*. Philadelphia: Auerbach/Studentliteratur.

Birtwistle, G., Liblong, B., Unger, B. and Witten, I. (1982). 'Simulation Environments'. *Proceedings of a Conference on Simulation: A Research Focus*, Rutgers University.

Birtwistle, G.M., Liblong, B. and Wyvill, B. (1984). 'SIMULA Bibliography'. Report 84/159/17, Computer Science Department, University of Calgary.

Birtwistle, G.M., Lomov, G., Unger, B. and Luker, P. (1984). 'Process Style Packages for Discrete Event Modelling: Data Structures and Packages in SIMULA', *Transactions of the SCS* **1**(1), pp. 61–82.

Birtwistle, G. and Luker, P. (1984). 'Dialogs for Simulation'. In *Proceedings of an SCS Conference*, La Jolla, 1984, R. Bryant and B.W. Unger (eds.), pp. 90–97.

Birtwistle, G.M., Luker, P., Lomov, G. and Unger, B. (1985). 'An Approach to Discrete Event Modelling'. Part 1: SIMULA; Part 2: The Process Approach to Discrete Event Modelling; Part 3: The Transaction, Activity and Events Approaches; Part 4: DEMOS: A Process Based Package. Submitted for publication to Transactions of the SCS.

Birtwistle, G.M., Wyvill, B., Lewinson, D. and Radford, N. (1984). 'Visualizing a Simulation with Animated Pictures'. In *Proceedings of an SCS Conference*, La Jolla, 1984, R. Bryant and B.W. Unger (eds.), pp. 57–61.

Bobillier, P.A., Kahan, B.C. and Probst, A.R. (1976). *Simulation with GPSS and GPSS V*. Englewood Cliffs, NJ: Prentice-Hall.

Bobrow, D.G. and Stefik, M. (1983). *The LOOPS Manual – Preliminary Version*. Xerox PARC.

Bobrow, D.G. and Winograd, T. (1977). 'An Overview of KRL, A Knowledge Representation Language', *Cognitive Science* **1**(1), pp. 3–46.

Borning, A.H. (1979). 'THINGLAB – A Constraint-oriented Simulation Laboratory. Xerox PARC, Report SSL-79-3, Palo Alto.

Borning, A.H. and Ingalls, D.H.H. (1982). 'Multiple Inheritance in SMALLTALK-80'. *Proceedings AAAI-82*, Carnegie-Mellon University, pp. 234–237.

Brachman, R.J. and Smith, B.C. (eds.) (1980). 'Special Issue: Knowledge Representation', *SIGART Newsletter (ACM SIGART)* **70**.

Bratley, P., Fox, D.C. and Schrage, L.E. (1983). *A Guide to Simulation*. New York/Heidelberg: Springer-Verlag.

Brooks, F.P. (1975). *The Mythical Man Month*. Reading, MA: Addison-Wesley.

Brown, H.I. (1977). *Perception, Theory and Commitment. The New Philosophy of Science*. Chicago/London: University of Chicago Press.

Brown, M.H. and Sedgewick, R. (1985). 'Techniques for Algorithm Animation', *IEEE Software* (January), pp. 28–39.

Bruno, G. (1982). 'Discrete Event Simulation in ADA'. *Proceedings Summer Computer Simulation Conference*, SCS, pp. 181–185.

Bruno, G. (1984). 'Rationale for the Introduction of Discrete Simulation Primitives in ADA'. In *Proceedings of an SCS Conference*, La Jolla, 1984, R. Bryant and B.W. Unger (eds.), pp. 10–15.

Bryant, R.M. (1980). 'SIMPAS – A Simulation Language Based on PASCAL'. *Proceedings Winter Simulation Conference*, ACM, pp. 25–40.

Bryant, R.M. (1981a). 'A Tutorial for PASCAL Users on Simulation Programming with SIMPAS'. Computer Sciences Technical Report #454, University of Wisconsin, Madison.

Bryant, R.M. (1981b). 'SIMPAS 5.0 User Manual'. Technical Report #456, Computer Science Department, University of Wisconsin, Madison.

Bryant, R.M. (1981c). 'Discrete Event Simulation with ADA'. Technical Report #458, Computer Science Department, University of Wisconsin, Madison.

Bryant, R. and Unger, B.W. (eds.) (1984). 'Simulation in Strongly Typed Languages: Ada, Pascal, Simula...'. *Proceedings of an SCS Conference*, La Jolla, 1984.

Bundy, A. (ed.) (1978). *Artificial Intelligence – An Introductory Course*. New York: North-Holland.

Bundy, A. (ed.) (1984). *Catalogue of Artificial Intelligence Tools*. Berlin/New York: Springer-Verlag.

Buxton, J.N. (ed.) (1968). 'Simulation Programming Languages'. *Proceedings of the IFIP Working Conference on Simulation Programming Languages*, Amsterdam/New York: North-Holland.

Cajori, F. (1974). *History of Mathematical Notations* (Vol. I and II). La Salle, IL: Open Court Publ. Reprinted 1974.

Cavouras, J.C. (1983). 'Implementing Simulation in a High Level Language with no Multi-tasking Facilities', *SOFTWARE Practice and Experience* **13**, pp. 809–815.

Cellier, F. (1979a). 'Combined Continuous Discrete Simulation by Use of Digital Computers: Techniques and Tools'. Ph.D. thesis, Swiss Federal Institute of Technology, Zurich.

Cellier, F. (1979b). 'Combined Continuous/Discrete System Simulation Languages: Usefulness, Experience and Future Development'. In *Methodology in Systems Modelling and Simulation*, B.P. Zeigler *et al.* (eds.), Amsterdam/ New York: North-Holland, pp. 201–210.

Cellier, F. (ed.) (1982). *Progress in Modelling and Simulation*. London/New York: John Wiley.

Chandy, K.M. and Misra, J. (1984). 'The Drinking Philosophers Problem', *ACM Transactions on Programming Languages and Systems* **6**(4), pp. 632–646.

Charniak, E., Riesbeck, C.K. and McDermott, D.V. (1980). *Artificial Intelligence Programming*. Hillsdale, NJ: L. Erlbaum Assoc.

Cleary, J.G. and Dewar, A. (1984). 'Interpreters in Logic Programming – A Powerful Tool for Simulation'. In *Proceedings of an SCS Conference*, La Jolla, 1984, R. Bryant and B.W. Unger (eds.), pp. 84–89.

Clementson, A.T. (undated). 'ECSL/CAPS Detailed Reference Manual', Department of Engineering Production, University of Birmingham, Birmingham, UK.

Clementson, A.T. (1966). 'Extended Control and Simulation Language', *The Computer Journal* **9**, pp. 215–220.

Clocksin, W.F. and Mellish, C.S. (1981). *Programming in PROLOG*. Berlin/New York: Springer-Verlag.

Coffman, E.G., Jr. (ed.) (1976). *Computer and Job Shop Scheduling Theory*. London/ New York: John Wiley.

COMTEN Inc. (1976). *SCERT 76 – System Guide*. Washington.

Coombs, M.J and Alty, J.L. (eds.) (1981). *Computer Skills and the User Interface*. London/New York/Toronto: John Wiley.

Crane, A. and Lemoine, L. (1977). *An Introduction to the Regenerative Method for Simulation Analysis*. Berlin/New York: Springer-Verlag.

Crosbie, R.E., Javey, S., Hay, J.L. and Pearce, J.G. (1985). 'ESL – A New Continuous System Simulation Language', *SIMULATION* (May), pp. 242–246.

Dahl, O.J. (1968). 'Discrete Event Simulation Languages'. In *Programming Languages*, F. Genuys (ed.), New York/London: Academic Press, pp. 349–395.

Dahl, O.J., Dijkstra, E.W. and Hoare, C.A.R. (1972). *Structured Programming*. New York/London/San Francisco: Academic Press.

Dahl, O.J. and Nygaard, K. (1966). 'SIMULA – An ALGOL-based Simulation Language', *CACM* **9**(9), pp. 671–678.

Dambolena, I.G. (1983). 'The Length of Simulation Runs'. *Proceedings Winter Simulation Conference*, SCS, pp. 152–155.

Davies, N.R. (1979). 'Interactive Simulation Program Generation'. In *Methodology in Systems Modelling and Simulation*, B.P. Zeigler *et al.* (eds.), North-Holland, Amsterdam/New York, pp. 179–200.

Davis, M.W. (1980). 'A Comparison of ADA and SIMULA'. *Proceedings of the 7th SIMULA Users Conference*, NCC, Oslo.

Davis, P.J. and Hersch, R. (1981). *The Mathematical Experience*. Brighton: Harvester Press.

Davis, P. and King, J. (1977). 'An Overview of Production Systems'. In *Machine Intelligence* **8**, E.W. Elcock and D. Michie (eds.) Edinburgh University Press, pp. 300–334.

Davis, P.J. and Rabinowitz, P. (1984). *Methods of Numerical Integration* (2nd edn.). New York/London: Academic Press.

Decker, H. and Maierhofer, J. (1984). 'Very High Level Model Description and Simulation'. In *Proceedings of an SCS Conference*, La Jolla, 1984, R. Bryant and B.W. Unger (eds.), pp. 44–48.

Deering, M., Faleti, J. and Wielensky, R. (1981). 'PEARL – A Package for Efficient Access to Representations in LISP'. *Proceedings International Joint Conf. on AI*, pp. 930–932.

Delfosse, C.M. (1976). 'Continuous and Combined Simulation in SIMSCRIPT II.5'. Technical Report CACI Inc. (L.A.).

Denning, P.J. (ed.) (1974). 'Special Issue: Programming', *Computing Surveys* **6**(4).

DeSwaan A.H. (1983). 'Expert Systems in the Simulation Domain', *Math. and Computers in Simulation* **25**, pp. 10–16.

Dewdney, A.K. (1985). 'Five Easy Pieces for a DO Loop and a Random Number Generator', *Scientific American* (April), pp. 12–16.

Dijkstra, E.W. (1976). *A Discipline of Programming*. Englewood Cliffs, NJ: Prentice-Hall.

Dowham, P.Y. and Roberts, F.D.K. (1967). 'Multiplicative Congruential Pseudo Random Number Generators', *The Computer Journal* **11**(1), pp. 74–77.

Doyle, P. (1976). *Every Object is a System: A Cognitive Basis for Systems Description*. Duncannon: Doyle Publ.

Dunning, K.A. (1981). *Getting Started in GPSS*. San Jose: Engineering Press.

Dyck, V.A., Lawson, J.D., Smith, J.A. and Beach, R.J. (1982). *Computing: An Introduction to Structured Problem Solving Using PASCAL*. Reston: Reston.

Enderle, G., Kansy, K. and Pfaff, G. (1984). *Computer Graphics Programming: GKS – The Graphics Standard*. Berlin/New York: Springer-Verlag.

Engelbrecht-Wiggans, R. and Maxwell, W.L. (1978). 'Analysis of the Time Indexed List Procedure for Synchronization of Discrete Event Simulations', *Management Science* **24**(13), pp. 1417–1427.

Evans, J.B. (1984). 'Simulation and Intelligence'. Technical Report TR-A5-84, University of Hong Kong.

Evans, J.B. (1984). 'A Characterisation of Discrete Simulation Complexity'. Technical Report TR-A6-84, University of Hong Kong.

Fahrland, D.A. (1970). 'Combined Discrete-Event Continuous System Simulation', *SIMULATION* **14**(2), pp. 61–72.

Fishman, G.S. (1977). 'Achieving Specific Accuracy in Simulation Output Analysis', *CACM* **20**(5), pp. 310–315.

Fishman, G.S. (1978). *Principles of Discrete Event Simulation*. London/New York/ Sydney: John Wiley.

Floyd, R. (1979). 'The Paradigms of Programming', *Comm. of the ACM* **22**, pp. 455–460.

Forrester, J.W. (1968). *Principles of Systems*. Cambridge, MA: MIT Press.

Frankowski, E. and Franta, W.R. (1980). 'A Process Oriented Simulation Model Specification and Documentation Language', *SOFTWARE Practice and Experience* **10**, pp. 721–742.

Franta, W.R. (1977). *The Process View of Simulation*. New York: Elsevier.

Franta, W.R. and Maly, K. (1973). 'The Suitability of a Very High Level Language (SETL) for Simulation Structuring and Control'. Technical Report TR 73–4, Department of Computer Science, University of Minnesota.

Franta, W.R. and Maly, K. (1977). 'An Efficient Data Structure for the Simulation Event Set', *CACM* **20**(8), pp. 596–602.

Franta, W.R. and Maly, K. (1978). 'A Comparison of Heaps and the TL Structure for the Simulation Event Set', *CACM* **21**(10), pp. 873–875.

Frantz, F.K. and Trott, K.C. (1984). 'Extensions to PASCAL for Discrete Event Simulations'. *In Proceedings of an SCS Conference*, La Jolla, 1984, R. Bryant and B.W. Unger (eds.), pp. 32–36.

Futo, I. and Szeredi, J. (1982). 'A Very High Level Discrete Simulation System', *T-PROLOG Computational Linguistics and Computer Languages* **25**, pp. 111–131.

Gaines, B.R. (1976). 'System Identification, Approximation, and Complexity', *Int. Journal of General Systems* **3**, pp. 145–174.

Gale, D. and Shapley, L.S. (1962). 'College Admissions and the Stability of Marriage', *Am. Math. Monthly* **69**, pp. 9–15.

Gardner, M. (1984). *Wheels, Life and other Mathematical Amusements*. San Francisco: W.H. Freeman.

Goldberg, A. (1984). *SMALLTALK-80 – The Interactive Programming Environment*. Reading, MA: Addison-Wesley.

Goldberg, A. and Robson, D. (1979). 'A Metaphor for User Interface Design'. *Proceedings 12th Hawaii International Conference on System Sciences* **6**(1), pp. 148–157.

Goldberg, A. and Robson, D. (1983). *SMALLTALK-80. The Language and its Implementation*. Reading, MA: Addison-Wesley.

Goodman, S.E. and Hedctniemi, S.T. (1977). *Introduction to the Design & Analysis of Algorithms*. New York/St. Louis: McGraw-Hill.

Gordon, G. (1975). *The Application of GPSS V to Discrete Event Simulation*. Englewood Cliffs, NJ: Prentice-Hall.

Gordon, G. (1978a). 'The Development of the General Purpose Simulation System (GPSS)'. In *Preprints ACM SIGPLAN Conference on History of Programming Languages*, *SIGPLAN Notices* **13**(8), pp. 183–198.

Gordon, G. (1978b). *System Simulation* (2nd edn.). Englewood Cliffs, NJ: Prentice-Hall.

Gordon, G. (1979). 'The Design of the GPSS Language'. In *Current Issues in Computer Simulation*, N.R. Adams and A. Dogramaci (eds), New York: John Wiley, pp. 15–26.

Green, T. (1980). 'Programming as a Cognitive Activity'. In *Human Interaction with Computers*, H.T. Smith and T.R.G. Green (eds.), New York/London: Academic Press, pp. 271–320.

Green, T.R., Sime, M.E. and Fitter, M.J. (1980). 'The Art of Notation'. In *Human Interaction with Computers*, H.T. Smith and T.R.G. Green (eds.), New York/London: Academic Press, pp. 221–251.

Greenberg, S. (1972). *GPSS Primer*. London/New York/Sydney: John Wiley.

Guedji, R.A., ten Hagen, P.W., Hopgood, F.R.A., Tucker, H.A. and Duce D.D. (eds.) (1980). *Methodology of Interaction*. Amsterdam/New York: North-Holland.

Haberman, A.N. and Dewayne, E.P. (1983). *ADA for Experienced Programmers*. Reading, MA: Addison-Wesley.

Hac, A. (1982). 'Computer System Simulation in PASCAL', *SOFTWARE Practice and Experience* **12**, pp. 777–784.

Hac, A. (1984). 'PL/1 as a Discrete Event Simulation Tool', *SOFTWARE Practice and Experience* **14**(7), pp. 697–702.

Hammersley, J.M. and Handscombe, D.C. (1964). *Monte Carlo Methods*. London: Methuen.

Hamming, R.W. (1962). *Numerical Methods for Scientists and Engineers*. New York: McGraw-Hill.

Hasemer, T. (1984). *A Beginner's Guide To LISP*. Reading, MA: Addison-Wesley.

Hayes, J.R. and Simon, H.A. (1976). 'The Understanding Process: Problem Isomorphs', *Cognitive Psychology* **8**.

Heidorn, G.E. (1974). 'English as a Very High-Level Language for Simulation Programming'. *Proceedings Symposium on Very High-Level Languages, SIGPLAN Notices* **9**, pp. 91–100.

Helsgaun, K. (1980). 'DISCO – A SIMULA-based Language for Continuous, Combined, and Discrete Simulation', *SIMULATION* (July), pp. 1–12.

Henriksen, J.O. (1975). 'Building a Better GPSS: A 3:1 Performance Improvement'. *Proceedings Winter Simulation Conference*, SCS, pp. 465–469.

Henriksen, J.O. (1983a). 'State of the Art: GPSS'. *Proceedings Winter Simulation Conference*, SCS, pp. 918–933.

Henriksen, J.O. (1983b). 'The Integrated Simulation Environment (Simulation Software of the 1990s)', *Operations Research* **31**(6), pp. 1053–1073.

Hewitt, C. (1977). 'Viewing Control Structures as Patterns of Passing Messages', *Artificial Intelligence* **8**, pp. 323–384.

Hill, I.D. (1983). 'Natural Language versus Computer Language'. In *Designing for Human Computer Communication*, M.E. Sime and M.J. Coombs (eds.), London: Academic Press, pp. 55–72.

Hills, P.R. (1965). 'SIMON – A Simulation Language in ALGOL'. In *Simulation in O.R.*, S.M. Hollingdale (ed.).

Hills, P.R. (1971). *HOCUS – Basic and Advanced Manuals.* P.E. Group, Surrey (UK).

Hills, P.R. (1975). *SIMON 75 Reference Manual.* P.R. Hills Consultants.

Hoffman, D.B. and Carlson, C.M. (1981). 'Using Computer Graphics to Interpret Simulation Output', *SIMULATION* (August), pp. 59–64.

Hofstadter, D.R. (1979). *Goedel, Escher, Bach: The Eternal Golden Braid.* New York: Basic Books.

Hofstadter, D.R. (1985). *Metamagical Themas: Questing for the Essence of Mind and Pattern.* Harmondsworth: Penguin.

Holt, R.C. (1972). 'Some Deadlock Properties of Computer Systems', *Computing Surveys* **4**(3), pp. 179–196.

Hooper, J.W. (1983). 'The Effects of Strategy on Simulation Language Characteristics'. *Proceedings of the Summer Simulation Conference* (SSC), SCS, pp. 48–53.

Hooper, J.W. and Reilly, K.D. (1982). 'An Algorithmic Analysis of Simulation Strategies', *Int. Journal of Computer and Information Sciences* **11**(2), pp. 101–122.

Hooper, J.W. and Reilly, K.D. (1983). 'The GPSS-GASP Combined (GGC) System', *Int. Journal of Computer and Information Sciences* **12**(2), pp. 111–136.

Hoover, S.V. and Perry, R.F. (1983). 'Concurrent Graphical Display of GPSS Simulations'. *Proceedings Winter Simulation Conference*, SCS, pp. 132–135.

Horton, I.A. and Turner, J. (1986). 'Using Coroutines in Pascal', *SOFTWARE Practice and Experience* **16**(1), pp. 45–61.

Huber, O. (1983). 'Zur Psychologie des Problemsloesens'. In *Psychologie des Programmierens*, H. Schauer and M. Tauber (eds.), Wien/Muenchen: Oldenbourg, pp. 59–78.

Hull, T.E. and Dobell, A.R. (1964). 'Mixed Congruential Random Number Generators for Binary Machines', *Journal ACM* **11**, pp. 31–40.

Hutchinson, G.K. (1975a). 'Introduction to the Use of Activity Cycles as a Basis for Systems Decomposition and Simulation', *SIMULETTER (ACM SIGSIM)* **7**(1), pp. 15–23.

Hutchinson, G.K. (1975b). 'An Introduction to CAPS – Computer Aided Programming for Simulation', *SIMULETTER (ACM SIGSIM)* **7**(1), pp. 35–51.

IBM (UK) (1972). 'SIMPL/1 General Information Manual'. IBM-Form GH-19-5035-0, Croydon (UK).

Iglehart, D.L. (1978). 'The Regenerative Method for Simulation Analysis'. In *Current Trends in Programming Methodology – Software Modelling*, K.M. Chandy and R.T. Yen (eds.), Englewood Cliffs, NJ: Prentice-Hall.

Inkster, J., Lomov, G. and Unger, B. (1984). 'Combined Discrete and Continuous Simulation in Ada'. In *Proceedings of an SCS Conference*, La Jolla, 1984, R. Bryant and B.W. Unger (eds.), pp. 16–21.

Intellicorp (1984). *Knowledge Engineering Environment (KEE) – User Manual.* Menlo Park, CA.

Itoga, S. (1978). 'The Upper Bound for the Stable Marriage Problem', *Journal Operations Res. Society* **29**, pp. 811–814.

Itoga, S. (1981). 'A Generalization of the Stable Marriage Problem', *Journal Operations Res. Society* **32**, pp. 1069–1074.

Itoga, S. (1983). 'A Probabilistic Version of the Stable Marriage Problem', *BIT* **23** (1983), pp. 161–169.

Jansson, B. (1966). *Random Number Generators*. Stockholm: Almquist and Wicksell.

Karplus, W.J. (1977). 'The Spectrum of Mathematical Modelling and Systems Simulation', *Mathematics and Computers in Simulation* **19**, pp. 3–10.

Karplus, W.J. (1982). 'Support Languages for Model Simulation'. In *Computer-aided Modelling and Simulation*, J.A. Spriet and G.C. Vansteenkiste (eds.), New York/London: Academic Press, pp. 337–387.

Kaubisch, W.H., Perrot, R.H. and Hoare, C.A.R. (1976). 'Quasiparallel Programming', *SOFTWARE Practice and Experience* **6**, pp. 341–356.

Kelley, D.H. and Buxton, J.N. (1962). 'MONTECODE – An Interpretative Program for Monte Carlo Simulations', *The Computer Journal* **5**(2), pp. 88–93.

Kerckhoffs, E.J.H. and Vansteenkiste, G.C. (1986). 'The Impact of Advanced Information Processing on Simulation – An Illustrative Review', *SIMULATION* **46**(1), pp. 17–26.

Kieburtz, R.B. (1978). *Structured Programming and Problem Solving with PASCAL.* Englewood Cliffs, NJ: Prentice-Hall.

Kindler, E. (1979). 'Dynamic Systems and Theory of Simulation', *Kybernetika* **15**(2), pp. 77–87.

Kiviat, P.J. (1971). 'Simulation Languages'. In *Computer Simulation Experiments with Models of Economic Systems*, T.H. Naylor (ed.), New York: John Wiley, pp. 406–489.

Kiviat, P.J., Villanueva, R. and Markowitz, H.M. (1968). *The SIMSCRIPT II Programming Language*. Englewood Cliffs, NJ: Prentice-Hall.

Klahr, P. and Faught, W.S. (1980). 'Knowledge Based Simulation'. *Proceedings AAAI*, pp. 181–183.

Kleine, H. (1970). 'A Survey of Users' Views of Discrete Simulation Languages', *SIMULATION* (May), pp. 225–229.

Kleine, H. (1971). 'A Second Survey of Users' Views of Discrete Simulation Languages', *SIMULATION* (August), pp. 89–93.

Kleinjen, J.P.C. (1974, 1975). *Statistical Techniques in Simulation* (2 volumes). New York: M. Dekker.

Kleinjen, J.P.C. (1979). 'The Role of Statistical Methodology in Simulation'. In *Methodology in Systems Modelling and Simulation*, B.P. Zeigler *et al.* (eds.), Amsterdam/New York: North-Holland, pp. 425–445.

Kleinjen, J.P.C. (1982). 'Experimentation with Models: Statistical Design and Analysis Techniques'. In *Progress in Modelling and Simulation*, F. Cellier (ed.), London/New York: John Wiley, pp. 173–188.

Kleinrock, L. (1976). *Queueing Systems. Volume I: Theory*. London/New York/Sydney: John Wiley.

Klir, G.J. (ed.) (1972). *Trends in General Systems Theory*. New York/Cincinnatti: John Wiley.

Knuth, D.E. (1969). *The Art of Computer Programming. Vol. II: Seminumerical Algorithms*. Reading, MA: Addison-Wesley.

Knuth, D.E. and McNeley, J.E. (1966). 'SOL – A Symbolic Language for General Purpose System Simulation', *IEEE Transactions on Electronic Computers* (August), pp. 401–408.

Koopman, B.O. (1977). 'Intuition in Mathematical Operations Research', *Operations Res.* **25**, pp. 186–206.

Koopman, B.O. (1979). 'An Operational Critique of Deduction', *Operations Res.* **27**, pp. 115–133.

Korf, R.E. (1980). 'Towards a Model of Representation Changes', *Artificial Intelligence* **14**, pp. 41–78.

Korn, G.A. and Wait, J.V. (1978). *Digital Continuous System Simulation (DARE)*. Englewood Cliffs, NJ: Prentice-Hall.

Kowalski, R. (1979). *Logic for Problem Solving*. New York/Oxford: North-Holland.

Krasner, G. (ed.) (1983). *SMALLTALK-80: Bits of History, Words of Advice*. Reading, MA: Addison-Wesley.

Kreutzer, W. (1977). 'Survey of Simulation Packages'. In *INFOTECH Report on Performance Modelling and Prediction*, London, pp. 239–276.

Kreutzer, W. (1983). 'Of Clouds and Clocks: A Guided Walk through a Simulator's Toolbox', *NZ Operations Research* **11**(1), pp. 51–120.

Kriz, J. and Sandmayr, H. (1980). 'Extension of PASCAL by Coroutines and its Application to Quasi-parallel Programming and Simulation', *SOFTWARE Practice and Experience* **11**, pp. 773–789.

Kuhn, T.S. (1962). *The Structure of Scientific Revolutions* (2nd edn). Chicago: University of Chicago Press.

Kupka, I. (1983). 'Paradigmen des Programmierens'. *In Psychologie des Programmierens*, H. Schauer and M. Tauber (eds.), Wien/Muenchen: Oldenbourg, pp. 11–34.

Lamprecht, G. (1982). *Einfuehrung in die Programmiersprache SIMULA*. Braunschweig: Vieweg (English publication pending).

Langlois, L. (1984). 'Simulation Visualization with SIMSEA, a General Purpose Animation Language'. In *Proceedings of an SCS Conference*, La Jolla, 1984, R. Bryant and B.W. Unger (eds.), pp. 62–67.

Lavery, R.G. (ed.) (1985). *Modelling and Simulation on Microcomputers*. San Diego: Society for Computer Simulation.

Law, A.M. and Kelton, D. (1982). *Simulation Modelling and Analysis*. New York/London/Sydney: McGraw-Hill.

Lebel, J.D. (1982). 'System Dynamics'. In *Progress in Modelling and Simulation*, F. Cellier (ed.), London/New York: John Wiley, pp. 119–158.

Lehmer, D.H. (1951). 'Mathematical Methods in Large Scale Computing Units', *Ann. Comp. Lab. Harvard University* **26**, pp. 141–146.

Lemon, M.J. and Lindstrom, G. (1983). 'Structuring Simulation through Coordinated Local Viewpoints and Roles'. *Proceedings Summer Simulation Conference*, SCS, pp. 79–84.

Lindstrom, H. and Skansholm, J. (1981). 'How to Make Your Own Simulation System', *SOFTWARE Practice and Experience* **11**, pp. 629–637.

Lomov, G. and Unger, B. (1982). 'The Process View of Simulation in ADA'. *Proceedings Winter Simulation Conference*, SCS, pp. 77–86.

Lomov, G., Unger, B. and Inkster, J. (1984). 'Combined Discrete and Continuous Simulation in Ada'. In *Proceedings of an SCS Conference*, La Jolla, 1984, R. Bryant and B.W. Unger (eds.), pp. 16–21.

Lorin, H. (1972). *Parallelism in Hardware and Software: Real and Apparent Concurrency*. Englewood Cliffs, NJ: Prentice-Hall.

Loyer, M. (1983). 'Modularity Mechanisms in SIMULA and ADA'. *Proceedings of the 11th SIMULA Users' Conference*, Norwegian Computing Center, Oslo.

Luker, P.A. (1984). 'MODELLER: Computer-assisted Modelling of Continuous Systems', *SIMULATION* (May), pp. 205–214.

Magnenat-Thalmann, N. and Thalmann, D. (1984). 'Animated Types and Actor Types in Computer Simulation and Animation'. In *Proceedings of an SCS Conference*, La Jolla, 1984, R. Bryant and B.W. Unger (eds.), pp. 51–56.

Maisel, H. and Gnugnoli, C. (1972). *SIMULATION of Discrete Stochastic Systems.* Chicago: Science Research Associates.

Manguel, A. and Gualdalupi, G. (1980). *The Dictionary of Imaginary Places.* London/Toronto/Sydney/New York: Macmillan.

Markowitz, H.M. (1979a). 'SIMSCRIPT: Past, Present, and some Thoughts about the Future'. In *Current Issues in Computer Simulation*, N.R. Adams and A. Dogramaci (eds.), New York: John Wiley, pp. 27–60.

Markowitz, H.M. (1979b). 'SIMSCRIPT'. In *Encyclopedia of Computer Science & Technology, Volume 13*, J. Belzer, A.G. Holzman and A. Kent (eds.), New York/ Basel: M. Dekker.

Marsaglia, G. (1972). 'The Structure of Linear Congruential Sequences'. In *Applications of Number Theory to Numerical Analysis*, S.K. Zaremba (ed.), New York/ London: Academic Press, pp. 249–285.

Marsden, B.W. (1984). 'A Standard Pascal Event Simulation Package', *SOFTWARE Practice and Experience* **14**(7), pp. 659–684.

Masterman, M. (1970). 'The Nature of a Paradigm'. In *Criticism and the Growth of Knowledge*, I. Lakatos and A. Musgrave (eds.), Cambridge: Cambridge University Press, pp. 59–90.

Mathewson, S.C. (1974). 'Simulation Program Generators', *SIMULATION* (June), pp. 181–189.

Mathewson, S.C. (1983). 'User Acceptance: Design Considerations for a Program Generator', *SOFTWARE Practice and Experience* **13**, pp. 101–117.

Milner, R., Sabudo, J.D. and Wortman, D.B. (1982). 'Graphics and Simulation: Tools and Applications'. *Proceedings Winter Simulation Conference*, SCS, pp. 161–174.

Minsky, M. (1975). 'A Framework for Representing Knowledge'. In *The Psychology of Computer Vision*, P. Winston (ed.), New York: McGraw-Hill.

Mitchell, E.E.L. and Gauthier, J.S. (1976). 'Advanced Continuous Simulation Language (ACSL)', *SIMULATION* **26**(3), pp. 72–78.

Mitchell, E.E.L. and Gauthier J.S. (1981). *ACSL User Guide/Reference Manual.*

Mitrani, I. (1982). *Simulation Techniques for Discrete Event Systems.* Cambridge/ London/New York: Macmillan.

Moore, L.J. and Clayton, E.R. (1976). *GERTS Modelling and Simulation: Fundamentals and Applications.* New York: Petrocelli/Charter.

MacLaren, M.D. and Marsaglia, G. (1965). Uniform Random Number Generators', *Journal ACM* **12**, pp. 83–89.

MacLennan, B.J. (1982). 'Values and Objects in Programming Languages', *ACM SIGPLAN Notices* **17**(12), pp. 70–79.

MacLennan, B.J. (1983). *Principles of Programming Languages: Design, Evaluation and Implementation*. New York: Holt/Rhinehart/Winston.

McArthur, D. and Sowizral, H. (1981). 'An Object Oriented Language for Constructing Simulations'. *Proceedings of the 7th Int. Joint Conf. on AI.*

McCormack, W.M. and Sargent, R.G. (1979). 'Comparison of Event Set Algorithms for Simulations of Closed Queueing Systems'. In *Current Issues in Computer Simulation*, N.R. Adams and A. Dogramaci (eds.), New York: John Wiley, pp. 70–82.

McDougall, M.H. (1975). 'Process and Event Control in ASPOL'. *Proceedings of a Symposium on the Simulation of Computer Systems, ACM SIGSIM*, pp. 187–195.

McVitie, D.G. and Wilson, L.B. (1969). 'Three Procedures for the Stable Marriage Problem Algorithm 411'. *Collected Algorithms (ACM).*

McVitie, D.G. and Wilson, L.B. (1970). 'Stable Marriage Assignment for Unequal Sets', *BIT* **10**, pp. 295–309.

Nance, R.E. (1979). 'Model Representation in Discrete Event Simulation: Prospects for Developing Documentation Standards'. In *Current Issues in Computer Simulation*, N.R. Adams and A. Dogramaci (eds.), New York: John Wiley, pp. 83–97.

Nance, R.E. (1981). 'The Time and State Relationships in Simulation Modelling', *CACM* **24**(4), pp. 173–179.

Nance, R. and Overstreet, C. (1972). 'Bibliography on Random Number Generation', *Computing Reviews* **13**, pp. 495–508.

Nelson, S.S. (1979). 'CONSIM: A Study of Control Issues in Conversational Simulation', *The Computer Journal* **22**(2), pp. 119–126.

Neumann, von J. (1951). 'Various Techniques Used in Connection with Random Digits', *US Nat. Bureau of Standards Appl. Math. Series* **12**, pp. 36–38.

Newell, A. and Simon, H.A. (1972). *Human Problem Solving*. Englewood Cliffs, NJ: Prentice-Hall.

Nilsson, N.J. (1980). *Principles of Artificial Intelligence*. Palo Alto: Tioga Press.

Novak, G.S., Jr. (1983). 'Data Abstraction in GLISP'. *Proceedings ACM Conference on Principles of Programming Languages*, ACM, pp. 170–177.

Nygaard, K. and Dahl, O.J. (1978). 'The Development of the SIMULA Languages'. In *Preprints of the Conference on History of Programming Languages, SIGPLAN Notices* **13**(8).

O'Donovan, T.M. (1979). *GPSS Simulation Made Simple*. London/New York/ Sydney: John Wiley.

O'Keefe, R.M. (1985). 'Expert Systems and Operational Research – Mutual Benefits', *Journal Opl. Res. Soc.* **36**(2), pp. 125–129.

O'Keefe, R.M. (1986). 'Simulation and Expert Systems – A Taxonomy and some Examples', *SIMULATION* **46**(1), pp. 10–16.

Olsson, G. (1980). *Birds in Egg – driB ni sggE*. London: Pion Ltd.

Oren, T. (1973). 'Digital Simulation Languages for Combined Systems – An Overview'. *Proceedings Summer Computer Simulation Conference*, SCS.

Oren, T.I. (1977). 'Software for Simulation of Combined Continuous and Discrete Systems: A State of the Art Review', *SIMULATION* **28**(2), pp. 33–45.

Oren, T.I. and Zeigler, B.P. (1979). 'Concepts for Advanced Simulation Methodologies', *SIMULATION* **30**(3), pp. 70–82.

Overstreet, C.M. and Nance, R.E. (1983). 'A Specification Language to Assist in Analysis of Discrete Event Simulation Models'. Technical Report 83026-R, Virginia Polytechnic and State University, Blacksburg, VA.

Panel Discussions (1984). '(a) Simulation Workstations of the Future; (b) Future Directions in Simulation Software'. In *Proceedings of an SCS Conference*, La Jolla, 1984, R. Bryant and B.W. Unger (eds.), pp. 119–125.

Papazoglou, M.R., Georgiadis, P.I. and Maritsas, D.G. (1984a). 'Architectural Considerations of the Parallel SIMULA Machine', *The Computer Journal* **27**(3), pp. 254–259.

Papazoglou, M.R., Georgiadis, P.I. and Maritsas, D.G. (1984b). 'An Outline of the Programming Language SIMULA', *Computer Languages* **9**(2), pp. 107–131.

Pattee, H. (ed.) (1973). *Hierarchy Theory*. New York: Brazillier.

Payne, J.A. (1982). *Introduction to Simulation*. New York/St. Louis: McGraw-Hill.

Perrot, R.H., Raja, A.K. and Kane, P.C.O. (1980). A Simulation Experiment Using Two Languages', *The Computer Journal* **23**(2), pp. 142–146.

Peterson, J.L. (1977). 'Petri Nets', *Computing Surveys* **9**(3), pp. 223–252.

Piatelli-Palmarini, M. (ed.) (1980). *Language and Learning. The Debate between Jean Piaget and Noam Chomsky*. Cambridge, MA: Harvard University Press.

Polya, G. (1957). *How to Solve It*. New York: Doubleday.

Polya, G. (1962). *Mathematical Discovery* (Volumes 1 and 2). New York: John Wiley.

Poole, T. and Szymankiewicz, J. (1977). *Using Simulation to Solve Problems*. New York/Sydney: John Wiley.

Pritsker, A.A.B. (1974). *The GASP IV Simulation Language*. London/New York/Sydney: John Wiley.

Pritsker, A.A.B. (1979). 'GASP: Present Status and Future Aspects'. In *Current Issues in Computer Simulation*, N.R. Adams and A. Dogramaci (eds.), New York: John Wiley, pp. 61–69.

Pritsker, A.A.B. (1983). 'SLAM II and Applications'. *Proceedings Winter Simulation Conference*, SCS, pp. 911–917.

Pritsker, A.A.B. and Pegden, P. (1979). *Introduction to Simulation and SLAM*. New York: Halsted Press.

Pritsker, A.A.B. and Young, R.E. (1975). *Simulation Using GASP-PL/1: A PL/1 based Continuous/Discrete Simulation Language*. London/New York/Sydney: John Wiley.

Pugn, A., III. (1973). *DYNAMO II Users' Manual*. Cambridge, MA: MIT Press.

Pylyshyn, Z. (1984). *Computation and Cognition*. Cambridge, MA: MIT Press.

RAND Corporation (1955). *A Million Random Digits with 100,000 Normal Deviates*. Glencoe, IL: Rand Corporation.

Randers, J. (ed.) (1980). *Elements of the System Dynamics Method*. Cambridge, MA: MIT Press.

Reddy, Y.V. and Fox, M.S. (1982). 'KBS: An Artificial Intelligence Approach to Flexible Simulation'. Technical Report CMU-RI-TR 82-1, Carnegie Mellon University.

Reitman, J. (1971). *Computer Simulation Applications*. London/New York/Sydney/Toronto: John Wiley.

Reitman, J. (1982). 'The Use of Enhanced GPSS (GPSS/H)', *SIMULATION* (January), pp. 23–25.

Rentsch, T. (1982). 'Object Oriented Programming', *SIGPLAN Notices* **17**(9), pp. 51–57.

Rich, E. (1983). *Artificial Intelligence*. New York: McGraw-Hill.

Richmond, B. (1985). *A User's Guide to Stella*. Lyme, NH: High Performance Systems Inc.

Rivett, P. (1972). *Principles of Model Building*. London/New York/Sydney/Toronto: John Wiley.

Roberts, N., Andersen, D., Deal, R., Garet, M. and Shafter, W. (1983). *Introduction to Computer Simulation. A System Dynamics Modelling Approach*. Reading, MA: Addison-Wesley.

Roberts, I. and Goldstein, R. (1977). 'The FRL Primer'. AI Memo #408, Cambridge, MA: MIT.

Roberts, S.D. and Sadlowski, T.E. (1975). 'INS: Integrated Network Simulator'. *Proceedings Winter Simulation Conference*, ACM, pp. 575–586.

Robson, D. (1981). 'Object Oriented Software Systems', *BYTE* (August), pp. 75–86.

Rubinstein, M.F. (1975). *Patterns of Problem Solving*. Englewood Cliffs, NJ: Prentice-Hall.

Rubinstein, R.Y. (1981). *Simulation and the Monte Carlo Method*. New York: John Wiley.

Russel, E.C. (1976). *Simulation with Processes and Resources in SIMSCRIPT II.5*. CACI Inc., Los Angeles.

Sandewall, E. (1978). 'Programming in the Interactive Environment: The LISP Experience', *Computing Surveys* **10**(1), pp. 35–72.

Sargent, R.G. (1977). 'Statistical Analysis of Simulation Output Data', *SIMULETTER (ACM SIGSIM)* **8**(3), pp. 21–31.

Savage, C.W. (ed.) (1978). 'Perception and Cognition Issues in the Foundation of Psychology'. *Minnesota Studies in the Philosophy of Science Vol. 9*, Minneapolis: University of Minnesota Press.

Schauer, H. and Tauber, M. (eds.) (1983). *Psychologie des Programmierens*. Wien/ Muenchen: Oldenbourg.

Schmidt, B. (1980). *GPSS – FORTRAN*. Chichester/New York/Brisbane/Toronto: John Wiley.

Schriber, T.J. (1974). *Simulation Using GPSS*. London/New York/Sydney: John Wiley.

Schwetman, H.D. and Bruell, S.C. (1976). 'When to Stop a Simulation Run: A Case Study', *SIMULETTER (ACM SIGSIM)* **7**(4), pp. 131–137.

SCS (Simulation Council Inc.) (1984, 1985). 'Catalogue of Simulation Software', *SIMULATION* (October 1984), pp. 180–192; *SIMULATION* (October 1985), pp. 196–211.

SCS (Simulation Council Inc.) (1985a). 'Artificial Intelligence and Simulation'. *Proceedings 1985 Eastern Multi-conference*.

Sedgewick, R. (1983). *Algorithms*. Reading, MA: Addison-Wesley.

Shearn, D.C.S. (1975). 'Discrete Event Simulation in ALGOL 68', *SOFTWARE Practice and Experience* **5**(4), pp. 279–293.

Sheil, B.A. (1981). 'The Psychological Study of Programming', *ACM Computing Surveys* (Special Issue) **13**, pp. 101–120.

Sheil, B.A. (1983). 'Power Tools for Programmers', *Datamation* (February), pp. 131–144.

Sheppard, S. (1984). 'Simulation Workstations of the Future', Panel Discussion. In *Proceedings of an SCS Conference*, La Jolla, 1984, R. Bryant and B.W. Unger (eds.), p. 119.

Sheppard, S., Friel, P. and Reese, D. (1984). 'Simulation in Ada: An Implementation of Two World Views'. In *Proceedings of an SCS Conference*, La Jolla, 1984, R. Bryant and B.W. Unger (eds.), pp. 3–9.

Shneiderman, B. (1980). *Software Psychology*. Cambridge, MA: Winthrop.

Sim, R. (1975). *CADSIM Users' Guide and Reference Manual*. London: Imperial College Publ. #75/23.

Simon, H.A. (1969). *The Sciences of the Artificial*. Cambridge, MA: MIT Press.

Sloman, A. (1978). *The Computer Revolution in Philosophy*. Hassocks: Harvester Press.

Smith, B.J. and Smith, M.Z. (1984). 'The PASCAL/VS Simulation Tool: Overview and Examples'. In *Proceedings of an SCS Conference*, La Jolla, 1984, R. Bryant and B.W. Unger (eds.), pp. 27–31.

Smith, H.T. and Green, T.R.G. (eds.) (1980). *Human Interaction with Computers*. New York/London: Academic Press.

Sol. H. (1984). 'A SIMULA Problem Solving Environment'. In *Proceedings of an SCS Conference*, La Jolla, 1984, R. Bryant and B.W. Unger (eds.), pp. 99–104.

Solomon, S.L. (1982). 'SIMSTAT – A Simple Statistical Package to Support Simulation'. SCS, pp. 307–311.

Solomon, S.L. (1983). *Simulation of Waiting Line Systems.* Englewood Cliffs, NJ: Prentice-Hall.

Speckhart, H. and Green, W.H. (1976). *A Guide to Using CSMP.* Englewood Cliffs, NJ: Prentice-Hall.

Stanridge, C.R., Pritsker, A.A.B. and O'Reilly, J. (1984). 'Integrated Simulation Support Systems. Concepts and Examples'. *Proceedings 1984 Annual Simulation Symposium*, IEEE, pp. 141–151.

Stirling, W.D. (1978). 'The Use of a Procedure Oriented Language for Process Oriented Simulations', *SOFTWARE Practice and Experience* **8**, pp. 137–148.

Strauss, J.C. *et al.* (1967). 'The SCI Continuous System Simulation Language (CSSL)', *SIMULATION*, pp. 281–303.

Subrahmanian, E. and Cannon, R.L. (1981). 'A Computer Program for Models of Discrete Systems', *SIMULATION* (March), pp. 93–101.

Tennent, R.D. (1981). *Principles of Programming Languages.* New York/London: Prentice-Hall.

Theriault, D. (1982). 'A Primer for the ACT-1 Language'. AI-Memo No. 672, Cambridge, MA: MIT.

Thomas, J.C. and Carroll, J.M. (1981). 'Human Factors in Communication', *IBM Syst. Journal* **20**(2), pp. 237–263.

Tocher, K.D. (1964). 'Some Techniques of Model Building', *Proceedings of an IBM Symposium on Simulation Models and Planning*, New York, pp. 119–155.

Tocher, K.D. (1965). 'Review of Simulation Languages', *Operations Res. Quarterly* **16**(2), pp. 189–217.

Ulrich, E.G. (1978) 'Event Manipulation for Discrete Simulations Requiring Large Numbers of Events', *CACM* **21**(9), pp. 777–785.

Unger, B. (1978). 'Programming Languages for Computer System Simulation', *SIMULATION* (April), pp. 101–111.

Unger, B., Birtwistle, G., Cleary, J., Hill, D., Lomov, G., Neale, R., Peterson, M., Witten, I. and Wyvill, B. (1984). 'JADE: A Simulation and Software Prototyping Environment'. In *Proceedings of an SCS Conference*, La Jolla, 1984, R. Bryant and B.W. Unger (eds.), pp. 77–83.

Uyeno, D.H. (1980). 'PASSIM: A Discrete Event Simulation Package for PASCAL', *SIMULATION* (December), pp. 183–190.

Vaucher, J. (1974). *GPSS User Notes.* Department d'Informatique, Universite de Montreal.

Vaucher, J. (1984). 'Future Directions in Simulation Software', Panel Discussion. In *Proceedings of an SCS Conference*, La Jolla, 1984, R. Bryant and B.W. Unger (eds.), p. 122.

Vaucher, J. (1984). 'Process-oriented Simulation in Standard PASCAL'. In *Proceedings of an SCS Conference*, La Jolla, 1984, R. Bryant and B.W. Unger (eds.), pp. 37–43.

Vaucher, J.G. and Duval, P. (1975). 'A Comparison of Simulation Event List Algorithms', *CACM* **18**(4), pp. 223–230.

Washam, W.B. and Pritsker, A.A.B. (1976). *Introduction to GASPPI.* Pritsker Ass., LaFayette.

Weinreb, D. and Moon, D.A. (1981). *LISP Machine LISP Manual* (4th edn.). Cambridge, MA: MIT.

Welsh, J. (1979). 'PASCAL PLUS, another Language for Modular Multiprogramming', *SOFTWARE Practice and Experience* **9**, pp. 947–957.

Wendt, S. (1978). 'BORIS. Block Oriented Interactive Simulation System'. Internal Report, Computer Science Department, University of Kaiserslautern, GFR.

Wickelgren, W.A. (1974). *How to Solve Problems.* San Francisco: W.H. Freeman.

Wilson, L.B. (1972). 'An Analysis of the Stable Marriage Assignment Algorithm', *BIT* **12**, pp. 569–575.

Wilson, J.R. and Pritsker, A.A.B. (1978). 'Evaluation of Starting Policies in Simulation Experiments', *SIMULATION* (September), pp. 79–89.

Winograd, T. (1983). 'Programming Issues for the 1980s', Panel Discussion. At SIGPLAN 1983: *Symposium on Programming Language Issues in Software Systems.* Reprinted in *SIGPLAN Notices* **19**(8), pp. 51–61.

Winston, P.H. (1984). *Artificial Intelligence* (2nd edn.). Reading, MA: Addison-Wesley.

Wirth, N. (1971). 'Program Development by Stepwise Refinement', *Communications of the ACM* **14**(4), pp. 221–227.

Wittgenstein, L. (1953). *Philosophical Investigations.* Oxford: Blackwell.

Wittgenstein, L. (1956). *Remarks on the Foundations of Mathematics.* Oxford: Blackwell.

Wittgenstein, L. (1961). *Tractatus Logico-Philosophicus.* London: Routledge.

Wyman, F.P. (1970). *Simulation Modelling – A Guide to Using SIMSCRIPT.* London/New York/Toronto: John Wiley.

Zaremba, S.K. (ed.) (1972). *Applications of Number Theory to Numerical Analysis.* New York/London: Academic Press.

Zeigler, B.P. (1976). *Theory of Modelling and Simulation.* London/New York/Toronto: John Wiley.

Zeigler, B.P. (1984). *Multifacetted Modelling and Discrete Event Simulation.* New York: Academic Press.

Zeigler, B.P., Elzas, M.S., Klir, G.J. and Oren, T.I. (eds.) (1979). *Methodology in Systems Modelling and Simulation.* Amsterdam/New York: North-Holland.

Index

A. Concepts

B. Sofware Systems

C. Programs